Titu Andreescu • Cristinel Mortici • Marian Tetiva

Mathematical Bridges

 Birkhäuser

Titu Andreescu
University of Texas at Dallas Natural
  Sciences and Mathematics
Richardson, Texas, USA

Cristinel Mortici
Valahia University of Targoviste
Targoviste, Romania

Marian Tetiva
Gheorghe Rosca Codreanu National College
Barlad, Romania

ISBN 978-0-8176-4394-2   ISBN 978-0-8176-4629-5   (eBook)
DOI 10.1007/978-0-8176-4629-5

Library of Congress Control Number: 2017932379

Mathematics Subject Classification (2010): 00A07, 08A50, 15-XX, 26-XX, 97H60, 97I20, 97I30

Printed on acid-free paper

This book is published under the trade name Birkhäuser, www.birkhauser-science.com
The registered company is Springer Science+Business Media LLC
The registered company address is: 233 Spring Street, New York, NY 10013, U.S.A.

Preface

Many breakthroughs in research and, more generally, solutions to problems come as the result of someone making *connections*. These connections are sometimes quite subtle, and at first blush, they may not appear to be plausible candidates for part of the solution to a difficult problem. In this book, we think of these connections as *bridges*. A bridge enables the possibility of a solution to a problem that may have a very elementary statement but whose solution may involve more complicated realms that may not be directly indicated by the problem statement. Bridges extend and build on existing ideas and provide new knowledge and strategies for the solver. The ideal audience for this book consists of ambitious students who are seeking useful tools and strategies for solving difficult problems (many of olympiad caliber), primarily in the areas of real analysis and linear algebra.

The opening chapter (aptly called "Chapter 1") explores the metaphor of bridges by presenting a myriad of problems that span a diverse set of mathematical fields. In subsequent chapters, it is left to the *reader* to decide what constitutes a bridge. Indeed, different people may well have different opinions of whether something is a (useful) bridge or not. Each chapter is composed of three parts: the theoretical discussion, proposed problems, and solutions to the proposed problems. In each chapter, the theoretical discussion sets the stage for at least one bridge by introducing and motivating the themes of that chapter—often with a review of some definitions and proofs of classical results. The remainder of the theoretical part of each chapter (and indeed the majority) is devoted to examining illustrative examples—that is, several problems are presented, each followed by at least one solution. It is assumed that the reader is intimately familiar with real analysis and linear algebra, including their theoretical developments. There is also a chapter that assumes a detailed knowledge of abstract algebra, specifically, group theory. However, for the not so familiar with higher mathematics reader, we recommend a few books in the bibliography that will surely help, like [5–9, 11, 12].

Bridges can be found everywhere—and not just in mathematics. One such final bridge is from us to our friends who carefully read the manuscript and made extremely valuable comments that helped us a lot throughout the making of the book. It is, of course, a bridge of acknowledgments and thanks; so, last but not least, we must say that we are deeply grateful to Gabriel Dospinescu and Chris Jewell for all their help along the way to the final form of our work.

In closing, as you read this book, we invite you to discover some of these bridges and embrace their power in solving challenging problems.

Richardson, TX, USA   Titu Andreescu
Targoviste, Romania   Cristinel Mortici
Barlad, Romania   Marian Tetiva

Contents

Chapter 1
Mathematical (and Other) Bridges

Many people who read this book will probably be familiar with the following result (very folkloric, if we may say so).

Problem 1. The midpoints of the bases of a trapezoid, the point at which its lateral sides meet, and the point of intersection of its diagonals are four collinear points.

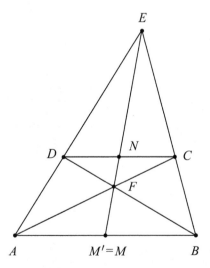

Solution. Indeed, let $ABCD$ be a trapezoid with $AB \parallel CD$; let M and N be the midpoints of the line segments AB and CD, respectively; and let $\{E\} = AD \cap BC$ and $\{F\} = AC \cap BD$. We intend to prove that M, N, E, and F are four collinear points.

Denote by M' the intersection of EF with AB. By Ceva's theorem, we have

$$\frac{AM'}{M'B} \cdot \frac{BC}{CE} \cdot \frac{ED}{DA} = 1.$$

© Springer Science+Business Media LLC 2017
T. Andreescu et al., *Mathematical Bridges*, DOI 10.1007/978-0-8176-4629-5_1

Also, Thales' interception theorem says that

$$\frac{DE}{DA} = \frac{CE}{CB} \Leftrightarrow \frac{CB}{CE} \cdot \frac{DE}{DA} = 1,$$

and by putting together the above two equations, we get $M'A = M'B$, that is, M' is actually the midpoint of AB; therefore, $M = M'$ belongs to EF, which is (part of) what we intended to prove. $\square$

The reader will definitely find a similar way (or will be able to use the already proved fact about the collinearity of M, E, and F) to show that N, E, and F are also collinear.

One can also prove that a converse of this theorem is valid, that is, for instance, if M, E, and F are collinear, then AB and CD are parallel (just proceed analogously, but going in the opposite direction). Or try to prove that if the midpoints of two opposite sides of a trapezoid and the intersection point of its diagonals are three collinear points, then the quadrilateral is actually a trapezoid (the sides whose midpoints we are talking about are the parallel sides); this could be more challenging to prove.

As we said, this is a well-known theorem in elementary Euclidean geometry, so why bother to mention it here? Well, this is because we find in it a very good example of a problem that needs a *(mathematical) bridge*. Namely, you noticed that the problem statement is very easy to understand even for a person who only has a very humble background in geometry—but that person wouldn't be able to *solve* the problem. You could be familiar with basic notions as collinearity and parallelism, you could know such things as properties of angles determined by two parallel lines and a transversal, but any attempt to solve the problem with such tools will fail. One needs much more in order to achieve such a goal, namely, one needs a *new theory*— we are talking about the theory of similarity. In other words, if you want to solve this problem, you have to raise your knowledge to new facts that are not mentioned in its statement. You need to throw a *bridge* from the narrow realm where you are stuck to a larger extent.

The following problem illustrates the same situation.

Problem 2. Determine all monotone functions $f : \mathbb{N}^* \to \mathbb{R}$ such that

$$f(xy) = f(x) + f(y) \text{ for all } x, y \in \mathbb{N}^*$$

($\mathbb{N}^*$ denotes the set of positive integers, while $\mathbb{R}$ denotes the reals).

Solution. All that one can get from the given relation satisfied by f is $f(1) = 0$ and the obvious generalization $f(x_1 \cdots x_k) = f(x_1) + \cdots + f(x_k)$ for all positive integers $x_1, \ldots, x_k$ (an easy and canonical induction leads to this formula) with its corollary $f(x^k) = kf(x)$ for all positive integers x and k. But nothing else can be done if you don't step into a higher domain (mathematical analysis, in this case) and if you don't come up with an idea. The idea is possible in that superior domain, being somehow natural if you want to use limits.

Let n be a positive integer (arbitrary, but fixed for the moment), and let us consider, for any positive integer k, the unique nonnegative integer n_k such that $2^{n_k} \leq n^k < 2^{n_k+1}$. Rewriting these inequalities in the form

$$\frac{\ln n}{\ln 2} - \frac{1}{k} < \frac{n_k}{k} \leq \frac{\ln n}{\ln 2}$$

one sees immediately that $\lim_{k \to \infty} n_k/k = \ln n/\ln 2$.

Because if f is increasing, $-f$ is decreasing and satisfies the same functional equation (and conversely), we can assume, without loss of generality, that f is increasing. Then from the inequalities satisfied by the numbers n_k and by applying the noticed property of f, we obtain

$$f(2^{n_k}) \leq f(n^k) \leq f(2^{n_k+1}) \Leftrightarrow \frac{n_k}{k}f(2) \leq f(n) \leq \left(\frac{n_k}{k} + \frac{1}{k}\right)f(2).$$

Now we can let k go to infinity, yielding

$$f(n) = f(2)\frac{\ln n}{\ln 2}$$

for any positive integer n. So, all solutions are given by a formula of type $f(n) = a \ln n$, for a fixed real constant a. If f is strictly increasing (or strictly decreasing), we get $f(2) > f(1) = 0$ (respectively, $f(2) < f(1) = 0$); thus $f(2) \neq 0$, and, with $b = 2^{1/f(2)}$, the formula becomes $f(n) = \log_b n$ (with greater, respectively lesser than 1 base b of the logarithm according to whether f is strictly increasing, or strictly decreasing). The null function ($f(n) = 0$ for all n) can be considered among the solutions, if we do not ask only for strictly monotonic functions. By the way, if we drop the monotonicity condition, we can find numerous examples of functions that only satisfy the first condition. For instance, define $f(n) = a_1 + a_2 + \cdots + a_k$ for $n = p_1^{a_1} \cdots p_k^{a_k}$, with $p_1, \ldots, p_k$ distinct primes and $a_1, \ldots, a_k$ positive integers and, of course, $f(1) = 0$, and we have a function with property $f(mn) = f(m) + f(n)$ for all positive integers m and n. The interested reader can verify for himself this condition and the fact that f is not of the form $f(n) = \log_b n$, for some positive $b \neq 1$ (or, equivalently, that this function is not monotone). $\square$

Again, one sees that in order to solve such a problem, one needs to build a bridge between the very elementary statement of the problem and the much more involved realm of mathematical analysis, where the problem can be solved.

However, there is more about this problem for the authors of this book. Namely, it also demonstrates another kind of bridge—a bridge over the troubled water of time, a bridge connecting moments of our lives. As youngsters are preoccupied by mathematics, we had (behind the Iron Curtain, during the Cold War) very few sources of information and very few periodicals to work with. There were, say, in the 80s of the former century, *Gazeta Matematică* and *Revista matematică din Timişoara*—only two mathematical magazines. The first one was a monthly magazine founded long ago, in 1895, by a few enthusiastic mathematicians and engineers among which Gheorghe Țițeica is most widely known. The second magazine used to appear twice a year and was much younger than its sister, but also had a national spreading due again to some enthusiastic editors. Anyway, this is all we had, and with some effort, we could also get access to Russian magazines such as *Kvant* or *Matematika v Şkole*, or the Bulgarian *Matematika*. Two of the authors of this book were at the time acquainted with problem 2 through *Revista Matematică din Timişoara*. They were high school students at that time and thoroughly followed up the problem column of this magazine, especially a "selected problems" column where they first met this problem (and couldn't solve it). The third author was the editor of that column—guess who is who! Anyway, for all three of us, a large amount of the problems in this book represent as many (nostalgic) bridges between past and present. Problem 2 is one of them, and we have many more, from which a few examples are presented below.

Problem 3. Are there continuous functions $f : \mathbb{R} \to \mathbb{R}$ such that

$$f(f(x)) + f(x) + x = 0 \text{ for every real } x?$$

Solution. No, there is no such function. The first observation is that if a function with the stated properties existed, then it would be strictly monotone. This is because such a function must be injective (the reader will immediately check that $f(x_1) = f(x_2)$ implies $f(f(x_1)) = f(f(x_2))$; therefore, by the given equation, $x_1 = x_2$). Now, injectivity and continuity together imply strict monotonicity; so if such a function existed, it would be either strictly increasing or strictly decreasing.

However, if f is strictly increasing, then $f \circ f$ and $f \circ f + f + 1_{\mathbb{R}}$ are also strictly increasing, which is impossible, because $f \circ f + f + 1_{\mathbb{R}}$ must equal the identically 0 function (by $1_{\mathbb{R}}$ we mean the identity function of the reals defined by $1_{\mathbb{R}}(x) = x$ for every real x). On the other hand, by replacing x with $f(x)$ in the given equation, we get $f(f(f(x))) + f(f(x)) + f(x) = 0$ for all x, and subtracting the original equation from this one yields $f(f(f(x))) = x$ for all x or $f \circ f \circ f = 1_{\mathbb{R}}$. This equality is a contradiction when f (and $f \circ f \circ f$ also) is strictly decreasing and the solution ends here. □

By the way, note that if $a_n, a_{n-1}, \ldots, a_0$ are real numbers such that the equation $a_n x^n + a_{n-1} x^{n-1} + \cdots + a_0 = 0$ has no real solutions, then there exists no continuous function $f : \mathbb{R} \to \mathbb{R}$ such that $a_n f^{[n]} + a_{n-1} f^{[n-1]} + \cdots + a_0 f^{[0]} = 0$. Here $f^{[n]}$ is the nth iterate of f (with $f^{[0]} = 1_{\mathbb{R}}$), and 0 represents the identically 0 function. This was a (pretty challenging at the time) problem that we had on a test in the mentioned above eighties, on a preparation camp. A bridge, isn't it?

Problem 4. Prove that among any 79 consecutive positive integers, there exists at least one such that the sum of its digits is divisible by 13. Find the smallest 78 consecutive positive integers such that none of them has its sum of digits divisible by 13.

Solution. By $S(N)$ we will denote the sum of digits of the natural number N. We can always find among 79 consecutive natural numbers 40 of the form $100k + \overline{a0}$, $100k + \overline{a1}, \ldots, 100k + \overline{(a + 3)9}$, with k a natural number and $a \leq 6$ a digit. Among the sums of digits of these numbers, there are $S(k)+a$, $S(k)+a+1, \ldots, S(k)+a+12$, that is, there are 13 consecutive natural numbers, one of which has to be divisible by 13.

Now, for the second part, we have to choose the desired numbers in such a way that no forty of them starting with a multiple of 10 are in a segment of natural numbers of the form $\{100k, 100k + 1, \ldots, 100k + 99\}$. This can only happen if the numbers are of the form $100a - 39, 100a - 38, \ldots, 100a + 38$, for some natural number a. Actually we will consider numbers of the form $10^b - 39, 10^b - 38, \ldots, 10^b + 38$, with $b \geq 2$, because it will be important how many nines there are before the last two digits. The sums of digits of the numbers $10^b, 10^b + 1, \ldots, 10^b + 38$ will cover all possibilities from 1 to 12. The sums of digits of the numbers $10^b - 39, 10^b - 38, \ldots, 10^b - 1$ will range from $9(b-2) + 7$ to $9(b-2) + 18$, and it is necessary that they cover exactly the same remainders modulo 13 (from 1 to 12). For this to happen, we need to have $9(b - 2) + 7 \equiv 1 \pmod{13}$, which gives $b \equiv 10 \pmod{13}$. So, the smallest possible 78 such numbers are those obtained for $b = 10$, thus the (78 consecutive) numbers from 9999999961 to 10000000038. $\square$

This is a problem that we know from the good old RMT.

Problem 5 (Erdős-Ginzburg-Ziv theorem). Prove that among any $2n - 1$ integers, one can find n with their sum divisible by n.

Solution. This is an important theorem, and it opened many new approaches in combinatorics, number theory, and group theory (and other branches of mathematics) in the middle of the twentieth century (it has been proven in 1961). However, we first met it in *Kvant*, with no name attached, and it was also *Kvant* that informed us about the original proof. Seemingly the problem looks like that (very known) one which states that from any n integers, one can choose a few with their sum divisible by n. The solution goes like this. If the numbers are $a_1, \ldots, a_n$, consider the n numbers $a_1, a_1 + a_2, \ldots, a_1 + a_2 + \cdots + a_n$. If there is any of them divisible with n, the solution ends; otherwise, they are n numbers leaving, when divided by n, only $n-1$ possible remainders (the nonzero ones); therefore, by the pigeonhole principle, there are two of them, say $a_1 + \cdots + a_i$ and $a_1 + \cdots + a_j$, with, say, $i < j$ that are congruent modulo n. Then their difference $a_{i+1} + \cdots + a_j$ is, of course, divisible by n (and is a sum of a few of the initial numbers). We put here this solution (otherwise, we are sure that it is well-known by our readers) only to see that there is no way to use its idea for solving problem 5 (which is a much deeper theorem). Indeed, the above solution allows no control on the number of elements in the sum that results to be divisible by n; hence, it is of no use for problem 5. The proof that we present now (actually the original proof of the three mathematicians) is very ingenious and, of course, builds a bridge.

The first useful observation is that the property from the theorem is multiplicative, that is, if we name it $P(n)$, we can prove that $P(a)$ and $P(b)$ together imply $P(ab)$. This permits an important reduction of the problem to the case of prime n (and it is used in all the proofs that we know). We leave this as an (easy and nice) exercise for the reader. So, further, we only want to prove (and it suffices, too) that from any $2p - 1$ integers one can always choose p with their sum divisible by p, where p is a positive prime.

The bridge we throw is towards the following:

Theorem. *Let A and B be subsets of $\mathbb{Z}_p$, with p prime, and let*

$$A + B = \{a + b \mid a \in A,\ b \in B\}.$$

Then we have $|A + B| \geq \min\{p, |A| + |B| - 1\}$. (By $|X|$, we mean the number of elements of the set X.)

We skip the proof of this (important) theorem named after Cauchy and Davenport (the second rediscovered it a century after the first one; each of them needed it in his research on other great mathematical results), but we insist on the following:

Corollary. *Let $A_1, \ldots, A_s$ be 2-element subsets of $\mathbb{Z}_p$. Then*

$$|A_1 + \cdots + A_s| \geq \min\{p, s + 1\}.$$

In particular, if $A_1, \ldots, A_{p-1}$ are subsets with two elements of $\mathbb{Z}_p$, then

$$A_1 + \cdots + A_{p-1} = \mathbb{Z}_p;$$

that is, every element from $\mathbb{Z}_p$ can be realized as a sum of elements from $A_1, \ldots, A_{p-1}$ (one in each set).

This corollary is all one needs to prove Erdős-Ginzburg-Ziv's theorem, and it can be demonstrated by a simple induction over s. The base case $s = 1$ being evident, let's assume that the result holds for s two-element subsets of $\mathbb{Z}_p$ and prove it for $s + 1$ such subsets $A_1, \ldots, A_{s+1}$. If $s + 1 \geq p$, we have nothing to prove; hence, we may assume that the opposite inequality holds. In this case, by the induction hypothesis, there are at least $s + 1$ distinct elements $x_1, \ldots, x_{s+1}$ in $A_1 + \cdots + A_s$. Let $A_{s+1} = \{y, z\}$; then the set $A_1 + \cdots + A_s + A_{s+1}$ surely contains $x_1 + y, \ldots, x_{s+1} + y$ and $x_1 + z, \ldots, x_{s+1} + z$. But the sets $\{x_1 + y, \ldots, x_{s+1} + y\}$ and $\{x_1 + z, \ldots, x_{s+1} + z\}$ cannot be equal, because in that case, we would have

$$(x_1 + y) + \cdots + (x_{s+1} + y) = (x_1 + z) + \cdots + (x_{s+1} + z),$$

which means $(s + 1)y = (s + 1)z$. As $1 \leq s + 1 \leq p - 1$, this implies $y = z$ in $\mathbb{Z}_p$, which is impossible (because y and z are the two distinct elements of A_{s+1}). Consequently, among the elements $x_1 + y, \ldots, x_{s+1} + y$ and $x_1 + z, \ldots, x_{s+1} + z$ of $A_1 + \cdots + A_{s+1}$, there are at least $s + 2$ mutually distinct elements, finishing the proof.

Now for the proof of Erdős-Ginzburg-Ziv theorem, consider $a_1 \leq a_2 \leq \ldots \leq a_{2p-1}$ to be the remainders of the given $2p - 1$ integers when divided by p, in increasing order. If, for example, $a_1 = a_p$, then $a_1 = a_2 = \ldots = a_p$, and the sum of the p numbers that leave the remainders $a_1, \ldots, a_p$ is certainly divisible by p; similarly, the problem is solved when any equality $a_j = a_{j+p-1}$ holds (for any $1 \leq j \leq p$). Thus we can assume further that (for every $1 \leq j \leq p$) a_j and a_{j+p-1} are distinct. Now we can consider the two-element subsets of $\mathbb{Z}_p$ defined by $A_j = \{a_j, a_{j+p-1}\}$, $1 \leq j \leq p - 1$. (We do not use a special notation for the residue class modulo p of the number x, which is also denoted by x.)

According to the above corollary of the Cauchy-Davenport theorem, $A_1 + \cdots + A_{p-1}$ has at least p elements; therefore, it covers all $\mathbb{Z}_p$. Consequently, there exist $i_1, \ldots, i_{p-1}$ such that i_j is either j or $j + p - 1$ for any $j \in \{1, \ldots, p - 1\}$ and $a_{i_1} + \cdots + a_{i_{p-1}} = -a_{2p-1}$ in $\mathbb{Z}_p$. This means that the sum $a_{i_1} + \cdots + a_{i_{p-1}} + a_{2p-1}$ (where, clearly, all indices are different) is divisible by p, that is, the sum of the corresponding initial numbers is divisible by p, finishing the proof. $\square$

One can observe that the same argument applies to prove the stronger assertion that among any $2p - 1$ given integers, there exist p with their sum giving any remainder we want when divided by p. Also, note that the numbers $0, \ldots, 0, 1, \ldots, 1$ ($n - 1$ zeros and $n - 1$ ones) are $2n - 2$ integers among which one cannot find any n with their sum divisible by n (this time n needs not be a prime). Thus, the number $2n - 1$ from the statement of the theorem is minimal with respect to n and the stated property.

There are now many proofs of this celebrated theorem, each and every one bringing its amount of beauty and cleverness. For instance, one of them uses the congruence

$$\sum_{1 \leq i_1 < \cdots < i_p \leq 2p-1} (x_{i_1} + \cdots + x_{i_p})^{p-1} \equiv 0 \pmod{p}$$

(the sum is over all possible choices of a subset of p elements of the set $\{1, \ldots, 2p - 1\}$; in other words, it contains all sums of p numbers among the $2p - 1$ given integers, which we denoted by $x_1, \ldots, x_{2p-1}$). Knowing this congruence and Fermat's Theorem, one gets $N \equiv 0 \pmod{p}$, where N means the number of those sums of p of the given $2p - 1$ integers that are not divisible by p. However, if all the possible sums weren't divisible by p, we would have $N = \dbinom{2p-1}{p-1} \equiv 1$ (mod p), a contradiction—hence there must exist at least one sum of p numbers that is divisible by p.

This proof is somehow simpler than the previous one, but it relies on the above congruence, which, in turn, can be obtained from the general identity

$$\sum_{S \subseteq \{1, \ldots, m\}} (-1)^{m-|S|} \left(\sum_{i \in S} x_i \right)^k = 0,$$

valid for all elements $x_1, \ldots, x_m$ of a commutative ring and for any $1 \leq k \leq m - 1$. For $k = m$, we need to replace the 0 from the right hand side with $m! x_1 \ldots x_m$, and one can find results for the corresponding sum obtained by letting $k = m + 1$, $k = m + 2$, and so on, but this is not interesting for us here. Let us only remark how another bridge (a connection between this identity and the Erdős-Ginzburg-Ziv theorem) appeared, seemingly out of the blue. The reader can prove the identity for himself (or herself) and use it then for every group of p of the given $2p - 1$ integers, with exponent $p - 1$, and then add all the yielded equalities; then try to figure out (it is not hard at all) how these manipulations lead to the desired congruence and, finally, to the second (very compact) proof of the Erdős-Ginzburg-Ziv's theorem. However, we needed a bridge. What this book tries to say is that there are bridges everywhere (in mathematics and in the real life). At least nostalgic bridges, if none other are evident.

Let us see now a few more problems whose solutions we'll provide after the reader has already tried (a bit or more) to solve independently. As the whole book, the collection is eclectic and very subjective—and it is based on the good old sources from our youth, such as *Gazeta Matematică* (GM), *Revista matematică din Timişoara* (RMT), *Kvant*, the Romanian olympiad or TSTs, and so on. Most of the problems are folklore (and their solutions, too), but they first came to us from these sources. When the problems have proposers we mention them; otherwise, as they can be found in many books and magazines, we avoid any references—every reader, we are sure, knows where to find them.

Proposed Problems

1. (Mihai Bălună, RMT) Find all positive integers n such that any permutation of the digits of n (in base ten) produces a perfect square.
2. Let $a_1, \ldots, a_n$ be real numbers situated on a circumference and having zero sum. Prove that there exists an index i such that the n sums a_i, $a_i + a_{i+1}$, ..., $a_i + a_{i+1} + \cdots + a_{i+n-1}$ are all nonnegative. Here, all indices are considered modulo n.
3. Prove that there exist integers a, b, and c, not all zero and with absolute values less than one million, such that $|a + b\sqrt{2} + c\sqrt{3}| < 10^{-11}$.
4. Prove that, for any positive integer k, there exist k consecutive natural numbers such that each of them is not square-free.
5. Find the largest possible side of an equilateral triangle with vertices within a unit square. (The vertices can be inside the square or on its boundary.)
6. Let A and B be square matrices of the same order such that $AB - BA = A$. Prove that $A^m B - BA^m = mA^m$ for all $m \in \mathbb{N}^*$ and that A is nilpotent.
7. (Dorel Miheţ, RMT) Prove that from the set $\{1^k, 2^k, 3^k, \ldots\}$ of the powers with exponent $k \in \mathbb{N}^*$ of the positive integers, one cannot extract an infinite arithmetic progression.

8. Let $1, 4, 8, 9, 16, 27, 32, \ldots$ be the sequence of the powers of natural numbers with exponent at least 2. Prove that there are arbitrarily long (nonconstant) arithmetic progressions with terms from this sequence, but one cannot find such a progression that is infinite.

9. (Vasile Postolică, RMT) Let $(a_n)_{n \geq 1}$ be a convergent increasing sequence. Prove that the sequence with general term

$$(a_{n+1} - a_n)(a_{n+1} - a_{n-1}) \ldots (a_{n+1} - a_1)$$

is convergent, and find its limit.

What can we say if we only know that $(a_n)_{n \geq 1}$ is increasing?

10. Let f be a continuous real function defined on $[0, \infty)$ such that $\lim_{n \to \infty} f(nt) = 0$ for every t in a given open interval (p, q) $(0 < p < q)$. Prove that $\lim_{x \to \infty} f(x) = 0$.

11. (Mihai Onucu Drimbe, GM) Find all continuous functions $f : \mathbb{R} \to \mathbb{R}$ such that

$$f(x + y + z) + f(x) + f(y) + f(z) = f(x + y) + f(x + z) + f(y + z)$$

for all $x, y, z \in \mathbb{R}$.

12. (Dorel Miheţ, RMT) Let $f : [a, b] \to [a, b]$ (where $a < b$ are real numbers) be a differentiable function for which $f(a) = b$ and $f(b) = a$. Prove that there exist $c_1, c_2 \in (a, b)$ such that $f'(c_1)f'(c_2) = 1$.

13. Evaluate

$$\int_0^{\pi/2} \frac{1}{1 + (\tan x)^{\sqrt{2}}} dx.$$

14. Show that

$$\lim_{n \to \infty} \int_a^b \left(1 + \frac{x}{n}\right)^n e^{-x} dx = b - a$$

for all real numbers a and b.

Solutions

1. Only the one-digit squares (that is, 1, 4, and 9) have (evidently) this property. Suppose a number with at least two digits has the property. It is well known that a number with at least two digits and for which all digits are equal cannot be a square; therefore, there must be at least two distinct digits, say a and b, with $a < b$. Then if $\overline{\ldots ab} = k^2$ and $\overline{\ldots ba} = l^2$, we clearly have $k < l$, hence $l \geq k + 1$, and

$$2k + 1 = (k + 1)^2 - k^2 \leq l^2 - k^2 = 9(b - a) \leq 81.$$

So $k \leq 40$ and a direct (boring, but simple) inspection show that no perfect square (with at least two digits) until $40^2 = 1600$ has the required property.

2. **Solution I.** Among all sums of the form $a_j + a_{j+1} + \cdots + a_{j+k-1}$ (for different integers $j \in \{1, \ldots, n\}$ and $k \geq 1$), there must be a minimal one. Because any cyclic permutation of the numbers affects neither the hypothesis nor the conclusion of the problem, we can assume that this minimal sum is $a_1 + \cdots + a_{i-1}$. We claim that in this case, $a_i + a_{i+1} + \cdots + a_{i+k-1} \geq 0$ for all k. Indeed, as long as $i + k - 1 \leq n$, this means that $a_1 + \cdots + a_{i+k-1} \geq a_1 + \cdots + a_{i-1}$. When $i + k - 1$ exceeds n, we use the hypothesis $a_1 + \cdots + a_n = 0$ to see that $a_i + a_{i+1} + \cdots + a_{i+k-1} \geq 0$ is equivalent to $a_{n+1} + \cdots + a_{i+k-1} \geq a_1 + \cdots + a_{i-1}$.

Solution II. We proceed by contradiction and suppose that for every $i \in \{1, \ldots, n\}$, there exists some $j \geq 1$ such that $a_i + \cdots + a_{i+j-1} < 0$. So we find a sequence $0 = i_0 < i_1 < \cdots < i_n$ with property $A_s = a_{i_s+1} + \cdots + a_{i_{s+1}} < 0$ for all $s \in \{0, 1, \ldots, n\}$. By the pigeonhole principle, there are j and k, with $j < k$, such that $i_j \equiv i_k \pmod{n}$, and in that case, $A_j + \cdots + A_{k-1} < 0$ represents a contradiction, because $A_j + \cdots + A_{k-1}$ is a multiple of $a_1 + \cdots + a_n$; therefore it is, in fact, 0.

3. There exist 10^{18} numbers of the form $x + y\sqrt{2} + z\sqrt{3}$, with x, y, and z integers from the set $\{0, 1, \ldots, 10^6 - 1\}$. Any two such numbers are distinct (we discuss this a little bit later), and any such number is at most equal to $(10^6 - 1)(1 + \sqrt{2} + \sqrt{3}) < 5 \cdot 10^6$. Divide the interval $[0, 5 \cdot 10^6)$ into $10^{18} - 1$ disjoint subintervals with equal lengths $5 \cdot 10^6 / (10^{18} - 1)$, and observe that there must be two numbers in the same interval; thus, there exist two of these numbers having the absolute value of their difference less than $5 \cdot 10^6 / (10^{18} - 1) < 10^{-11}$. If the numbers are $x_1 + y_1\sqrt{2} + z_1\sqrt{3}$ and $x_2 + y_2\sqrt{2} + z_2\sqrt{3}$, we get $|a + b\sqrt{2} + c\sqrt{3}| < 10^{-11}$ for $a = x_1 - x_2$, $b = y_1 - y_2$, and $c = z_1 - z_2$, which are integers (at least one of them being nonzero) with absolute values less than one million.

There are two more issues about this problem that we want (and have) to discuss. One of them is necessary to complete the proof, namely, we still need to show that two numbers of the form $x + y\sqrt{2} + z\sqrt{3}$ can only be equal whenever the corresponding coefficients x, y, and z are equal. More specifically, if $x_1 + y_1\sqrt{2} + z_1\sqrt{3} = x_2 + y_2\sqrt{2} + z_2\sqrt{3}$, and $x_1, y_1, z_1, x_2, y_2, z_2$ are integers, then $x_1 = x_2$, $y_1 = y_2$, and $z_1 = z_2$. Equivalently, if $x + y\sqrt{2} + z\sqrt{3} = 0$, with integers x, y, and z, then $x = y = z = 0$. (We need this fact at the beginning of the above proof, when we number the numbers of the form $x + y\sqrt{2} + z\sqrt{3}$, with $0 \leq x, y, z \leq 10^{18} - 1$: they have to be mutually distinct when they differ by at least a component. We also need it in the last step of the proof, in order to show that at least one of the obtained a, b, and c is nonzero.)

Actually, one can prove a more general statement, namely, that if x, y, z, t, are rational numbers and $x + y\sqrt{2} + z\sqrt{3} + t\sqrt{6} = 0$, then $x = y = z = t = 0$. Indeed, we have $z + t\sqrt{2} = 0$ if and only if $z = t = 0$ (this is well-known and actually is another way to state the irrationality of $\sqrt{2}$). So, if we have $z + t\sqrt{2} = 0$, the original equation yields $x + y\sqrt{2} = 0$, too, and the desired conclusion easily follows. Otherwise, we can rearrange the equation as

$$\sqrt{3} = -\frac{x + y\sqrt{2}}{z + t\sqrt{2}} = u + v\sqrt{2},$$

where u and v are rational numbers. But this implies $3 = u^2 + 2v^2 + 2uv\sqrt{2}$, hence, by invoking the irrationality of $\sqrt{2}$ again, we get $2uv = 0$ and $u^2 + 2v^2 = 3$, which, as the reader can check, immediately leads to contradiction.

Actually a much more general statement is valid, namely, that $a_1 \sqrt[n]{b_1} + \cdots + a_m \sqrt[n]{b_m} = 0$, with $a_1, \ldots, a_m, b_1, \ldots, b_m \in \mathbb{Q}$ ($b_1, \ldots, b_m$ being nonzero) implies $a_1 = \cdots = a_m = 0$ whenever there do not exist $i \neq j$ among the numbers $1, \ldots, m$ such that b_i / b_j is an nth power of a rational number. It is the so-called *linear independence of radicals*, a rather folkloric result. When $m = 2$, one can give a similar proof to the one above; the reader can try to see that, after a few reductions, it is enough to prove the following: if $p_1, \ldots, p_N$ are positive primes and a_I are rational numbers indexed after the subsets I of $\{1, \ldots, N\}$ such that

$$\left(\sum_{I \subseteq \{1, \ldots, N\}} a_I \sqrt{\prod_{i \in I} p_i} \right)^2 \in \mathbb{Q},$$

then (at least) $2^N - 1$ of the coefficients a_I are null. (The product corresponding to the empty set is 1.) We arrived pretty far away from the initial point, didn't we?

The second issue doesn't belong to this proof—it is more like a reminder. What we want to say is that this proof looks very similar to the proof of Dirichlet's approximation theorem. This theorem says that, given the real numbers $a_1, \ldots, a_k$ and given $\epsilon > 0$, there exist integers n and $m_1, \ldots, m_k$ such that $|na_i - m_i| < \epsilon$ for all $1 \leq i \leq k$. Indeed, let us consider some positive integer N satisfying $N > 1/\epsilon$ and look at the intervals

$$I_1 = \left[0, \frac{1}{N} \right), \ I_2 = \left[\frac{1}{N}, \frac{2}{N} \right), \ldots, \ I_N = \left[\frac{N-1}{N}, 1 \right)$$

that are partitioning $[0, 1)$. Consider also the k-tuples $(b_1^{(j)}, \ldots, b_k^{(j)})$ defined by $b_i^{(j)} = s$ if and only if $\{ja_i\} \in I_s$ for $1 \leq i \leq k$ and $1 \leq j \leq N^k + 1$. Each $b_i^{(j)}$ can only take the N values from the set $\{1, \ldots, N\}$; hence, the $N^k + 1$ k-tuples can have at most N^k values. By the pigeonhole principle, there are two of them, say $(b_1^{(j)}, \ldots, b_k^{(j)})$ and $(b_1^{(l)}, \ldots, b_k^{(l)})$ that are identical. This means that $\{ja_i\}$ and $\{la_i\}$ belong to the same interval I_s (with s depending on i) for each and every i from 1 to k, further yielding

$$|\{ja_i\} - \{la_i\}| < \frac{1}{N} < \epsilon$$

for all $1 \leq i \leq k$, or

$$|(j - l)a_i - ([ja_i] - [la_i])| < \epsilon$$

for all i. It is thus enough to choose $n = j - l$ and $m_i = [ja_i] - [la_i]$ for $1 \leq i \leq k$.

It is more likely that problem 3 (and many similar ones) was created following the idea of Dirichlet's approximation theorem, but for us, the order was reversed: we first met the problem and only later found out about Dirichlet's theorem. A bridge can be crossed in both directions.

4. This is a very easy problem for someone who knows the Chinese remainder theorem. Namely, one can pick some k distinct primes, say $p_1, \ldots, p_k$, then, by the mentioned result, one can conclude that there exists a positive integer x that solves the system of congruences $x \equiv -j \pmod{p_j^2}$, $1 \leq j \leq k$; the numbers $x + 1, \ldots, x + k$ are not square-free and the problem is solved. One can prove in the same way that there exist k consecutive positive integers that are not mth power-free (each of which is divisible by a power of a prime with exponent at least m), or one can prove that there exist k consecutive positive integers each of which is not representable as the sum of two squares (this one is a bit more elaborate, but the reader will find her/his way in order to solve it; the only necessary result is the one stating that a positive integer cannot be represented as the sum of two squares whenever there is a prime congruent to 3 modulo 4 such that its exponent in the factorization of that integer is odd).

The original proof (from RMT, many years ago) makes no use of Chinese remainder theorem, but rather of a rudiment of it. Namely, we will only use the fact that if a and b are relatively prime, then the congruence $ax \equiv c \pmod{b}$ has solutions (or, we can say, ax generates a complete system of residues modulo b whenever x does the same). Although a bit more complicated, we consider this proof to be instructive; hence, we present it here.

We proceed by induction. The base case is clear, so we assume that we dispose of k consecutive natural numbers $n + 1, \ldots, n + k$, each of which is not square-free, and let p_j^2 be some prime square that divides the jth number, for $1 \leq j \leq k$. Let also p_{k+1} be a new prime, different from all p_j, $1 \leq j \leq k$. Then all numbers $xp_1^2 \cdots p_k^2 + n + j$, $1 \leq j \leq k$, are not square-free, for every integer x, and we can choose x in such a way that $xp_1^2 \cdots p_k^2 + n \equiv -k - 1 \pmod{p_{k+1}^2}$. One sees that $xp_1^2 \cdots p_k^2 + n + j$, $1 \leq j \leq k + 1$, are $k + 1$ non-square-free numbers, as we intended to show.

Can this argument be adapted to obtain a proof of the Chinese remainder theorem? You probably already noted that the answer is definitely yes. Do we have another bridge? Yes, we do.

5. The answer is $\sqrt{6} - \sqrt{2}$. Let l be the length of the side of an arbitrary equilateral triangle ABC whose vertices are within the closed surface bounded by a unit square. There is always a vertex of the triangle such that the triangle is situated in the right angle determined by the parallels through that vertex to a pair

of perpendicular sides of the square. Say A is this vertex, and note that the projections of the sides AB and AC of the triangle on two sides sharing a common vertex of the square have lengths $l\cos\alpha$ and $l\cos(\pi/6 - \alpha)$, where α is the angle that one of them forms with the side on which it is projected. Also, $0 \le \alpha \le \pi/6$. The projections being included in the corresponding unit sides of the initial square, we obtain

$$l\cos\alpha \le 1 \quad \text{and} \quad l\cos\left(\frac{\pi}{6} - \alpha\right) \le 1.$$

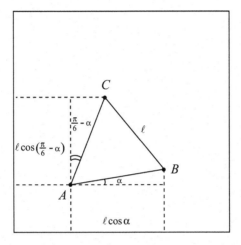

However, one of α and $\pi/6 - \alpha$ is at most $\pi/12$, thus either $\cos\alpha \ge \cos(\pi/12) = (\sqrt{6} + \sqrt{2})/4$, or $\cos(\pi/6 - \alpha) \ge \cos(\pi/12) = (\sqrt{6} + \sqrt{2})/4$. In both cases, by using one of the above inequalities, we conclude that

$$\frac{\sqrt{6} + \sqrt{2}}{4}l \le 1 \Leftrightarrow l \le \sqrt{6} - \sqrt{2}.$$

The existence of an equilateral triangle inscribed in the unit square with precisely this length of its side is easy to prove. The triangle has a vertex in a vertex of the square, the other two vertices on two sides of the square, and its sides emerging from that vertex form angles of measure $\pi/12$ with the sides of the square.

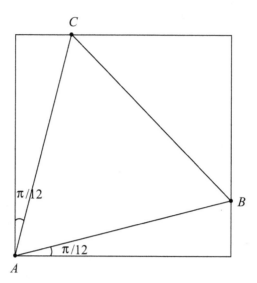

By the way, one can state the similar problem in which only the word "square" is replaced by "cube." If $ABCDA'B'C'D'$ is a unit cube (with two opposite faces $ABCD$ and $A'B'C'D'$ and AA', BB', CC', DD' parallel edges), one can immediately see that triangle ACD' (for instance) is equilateral with $\sqrt{2}$ as length of its side; also, one can readily conclude that there is no bigger equilateral triangle inscribed in a unit cube. As intuitive as this result might be (more intuitive than the planar case, isn't it?), one needs a demonstration for it. The reader is invited to see that if S is the area of a triangle and S_1, S_2, and S_3 are the areas of the projections of the triangle on the faces of a rectangular trihedron (that is, on three planes perpendicular to each other), then $S^2 = S_1^2 + S_2^2 + S_3^2$ (an extension of the Pythagorean theorem, yet some kind of a bridge). In our case, if S is the area of an equilateral triangle situated within a unit a cube and we project the triangle on the planes of three faces of the cube that share a common vertex, we have $S^2 = S_1^2 + S_2^2 + S_3^2$ (S_1, S_2, and S_3 being the areas of the projections), and, moreover, each $S_k \leq 1/2$ (as being the area of a triangle within a unit square). Thus, we obtain $S^2 \leq 3/4$, and, consequently, the length of the side of the equilateral triangle is at most $\sqrt{2}$, finishing the proof. Isn't this a mathematical bridge (from plane to solid geometry)?

6. We prove the first part by induction on m. For $m = 1$, the equality is the given one; thus, let us suppose that it is true for m and prove it for $m + 1$. To do this, it suffices to multiply to the right with A^m the relation $AB - BA = A$ and to multiply to the left with A the induction hypothesis $A^mB - BA^m = mA^m$, and then add side by side the two equalities thus obtained.

The second part relies on the first. From the well-known fact that $XY - YX$ always has zero trace (for X and Y square matrices of the same dimension) and $mA^m = A^mB - BA^m$, we conclude that the trace of A^m is 0 for all positive integers m. If $a_1, \ldots, a_n$ are the eigenvalues of A, this means that $a_1^m + \cdots + a_n^m = 0$ for all

m, and a canonical application of Newton's formulae shows that the symmetric sums of $a_1, \ldots, a_n$ are all 0, therefore that $a_1 = \ldots = a_n = 0$. Now an $n \times n$ matrix with null eigenvalues has characteristic polynomial X^n; therefore, by the Hamilton-Cayley theorem, $A^n = O_n$, that is, A is nilpotent.

We can also give an alternative proof of the fact that A is nilpotent, using the equality $A^m B - B A^m = m A^m$. Namely, if one defines the norm $\|X\|$ of a matrix $X = (x_{ij})_{1 \le i,j \le n}$ by

$$\|X\| = \max_{1 \le i,j \le n} |x_{ij}|$$

(with $|x|$ being the usual norm of the complex number x), then one immediately sees that $\|X + Y\| \le \|X\| + \|Y\|$, $\|\alpha X\| = |\alpha| \|X\|$, and $\|XY\| \le n\|X\|\|Y\|$ for any $n \times n$ complex matrices X and Y and any complex number α. Thus we have

$$m\|A^m\| = \|A^m B - B A^m\| \le \|A^m B\| + \|B A^m\| \le 2n\|A^m\|\|B\|$$

for every positive integer m. By choosing a large enough m, we see that this implies $\|A^m\| = 0$ for that m, therefore $A^m = O_n$, finishing this variant of the proof. (Of course, $A^n = O_n$ can be also deduced if one only knows that $A^m = O_n$ for some m, even though the original problem didn't actually ask for that.)

A third approach for solving this problem is presented in Chapter 3.

7. If $a < b < c$ are positive and $a^k + c^k = 2b^k$, then, by Jensen's inequality for the convex function $x \mapsto x^k$, we have

$$b^k = \frac{a^k + c^k}{2} > \left(\frac{a+c}{2}\right)^k \Rightarrow b > \frac{a+c}{2},$$

hence $b - a > c - b$. (Jensen's inequality can be avoided; we can write

$$c^k - b^k = b^k - a^k$$

in the form

$$(c - b)(c^{k-1} + c^{k-2}b + \cdots + b^{k-1}) = (b - a)(b^{k-1} + b^{k-2}a + \cdots + a^{k-1}),$$

then use $a < b < c$.)

Suppose now that $n_1^k < n_2^k < \cdots$ is an infinite arithmetic progression with terms from the sequence of powers with exponent k. According to the above observation, we obtain $n_2 - n_1 > n_3 - n_2 > \cdots$, which would be an infinite strictly decreasing sequence of positive integers—an impossibility.

There is a second approach (that will be useful for the next problem—which we cannot solve by other means). Namely, if $n_1^k, n_2^k, \ldots$ is an infinite arithmetic progression, then there are a and b such that $n_j^k = aj + b$ for all $j \ge 1$. Then

$$\sum_{j=1}^{\infty} \frac{1}{aj+b} = \sum_{j=1}^{\infty} \frac{1}{n_j^k} < \sum_{n=1}^{\infty} \frac{1}{n^k} < \infty$$

because the k-series (with $k > 1$) converges. On the other hand, the generalized harmonic series $\sum_{j=1}^{\infty} \frac{1}{aj+b}$ diverges to ∞, as it is again well-known (and one gets easily, by comparing with the harmonic series). Thus, we have a contradiction and the problem is solved.

8. We can start with $1^2, 5^2, 7^2$ as such a three-term progression. Or, if we want to avoid the term 1, we can choose $28^2, 42^2, 14^3$—a progression with common difference $5 \cdot 14^2 = 980$. If we have the arithmetic progression $n_1^{k_1}, \ldots, n_s^{k_s}$ with common difference d, then, for $k = [k_1, \ldots, k_s]$ (the least common multiple of $k_1, \ldots, k_s$) and $n = n_s^{k_s} + d$, the numbers

$$n^k n_1^{k_1}, \ldots, n^k n_s^{k_s}, n^{k+1}$$

are $s + 1$ powers with exponent at least 2 forming an arithmetic progression with common difference $n^k d$. Thus, inductively, we get such progressions with as many terms as we want.

However, an infinite such progression does not exist, due to the same reason that we used in the second solution of the previous problem. Indeed, if we had $n_j^{k_j} = aj + b$ for all $j \geq 1$, then, on one hand, we would have

$$\sum_{j \geq 1} \frac{1}{n_j^{k_j}} = \sum_{j \geq 1} \frac{1}{aj+b} = \infty,$$

and, on the other hand,

$$\sum_{j \geq 1} \frac{1}{n_j^{k_j}} < 1 + \sum_{n \geq 2} \sum_{k \geq 2} \frac{1}{n^k} = 1 + \sum_{n \geq 2} \frac{1}{n(n-1)} = 2.$$

9. We have

$$0 < (a_{n+1} - a_n)(a_{n+1} - a_{n-1}) \cdots (a_{n+1} - a_1) \leq \left(a_{n+1} - \frac{a_1 + \cdots + a_n}{n}\right)^n$$

and the conclusion

$$\lim_{n \to \infty} (a_{n+1} - a_n)(a_{n+1} - a_{n-1}) \cdots (a_{n+1} - a_1) = 0$$

follows by the squeeze principle. In order to infer this, one has to know (apart from the arithmetic mean-geometric mean inequality) that if $\lim_{n \to \infty} a_n = a$, then

$\lim_{n\to\infty} (a_1 + \cdots + a_n)/n = a$, too (an immediate consequence of the Stolz-Cesàro theorem) and, of course, to know that $0^\infty = 0$ (that is, to know that if $\lim_{n\to\infty} x_n = 0$ and $\lim_{n\to\infty} y_n = \infty$, then $\lim_{n\to\infty} x_n^{y_n} = 0$).

If $(a_n)_{n\geq 1}$ is only increasing it can also have limit infinity. For $a_n = n$, we obtain

$$b_n = (a_{n+1} - a_n)(a_{n+1} - a_{n-1}) \cdots (a_{n+1} - a_1) = n!,$$

which has the limit ∞. For $a_n = n/2$ for even n and $a_n = (n+1)/2$ for odd n, $(b_n)_{n\geq 1}$ is divergent, having a subsequence with limit 0, and another with limit ∞. Finally, for $a_n = 1 + \dfrac{1}{2} + \cdots + \dfrac{1}{n}$ (the harmonic series), we have

$$0 < b_n = \frac{1}{n+1}\left(\frac{1}{n+1} + \frac{1}{n}\right)\cdots\left(\frac{1}{n+1} + \cdots + \frac{1}{2}\right)$$

$$< \frac{1}{n+1}\frac{2}{n}\cdots\frac{n}{2} = \frac{1}{n+1},$$

hence $\lim_{n\to\infty} b_n = 0$. So, anything can happen if we drop the convergence condition for the initial sequence. (This part was not in the original problem. But, when you deal with mathematics for so long, you learn to ask—and this is not bad at all. In this case, we found the last limit, which we did not meet before.)

10. This statement is sometimes called Croft's lemma. We prove it by contradiction, namely, if we assume that f does not have limit 0 at infinity, there exist a positive number ϵ and a sequence (x_n) of real numbers, with limit infinity such that $|f(x_n)| > \epsilon$ for all n. Because f is continuous, one can find intervals I_n such that $x_n \in I_n$ and $|f(x)| > \epsilon$ for all $x \in I_n$ and all n. Clearly, the intervals I_n can be chosen as small as we want, and in particular, we can assume that their lengths are all smaller than a fixed positive number α. Further, we use the following:

Lemma. *Let $0 < c < d$ and let $\alpha > 0$ be given. Then, for every sufficiently large $a > 0$, any interval (a, b) of length $b - a < \alpha$ can be covered by a "multiple" of (c, d). (A "multiple" of (c, d) is $n(c, d) = \{nx \mid x \in (c, d)\}$.)*

Proof of the lemma. Indeed, let us choose $a > c(d+\alpha)/(d-c)$ and $b < a+\alpha$. Then

$$\frac{a}{c} - \frac{a}{d} > 1 + \frac{\alpha}{d}$$

is a rephrasing of the inequality satisfied by a; therefore, we have

$$\frac{a}{c} - \frac{b}{d} > \frac{a}{c} - \frac{a+\alpha}{d} > 1,$$

yielding the existence of an integer n between b/d and a/c; because $b/d < n < a/c$, we have $nc < a < b < nd$; hence $(a, b) \subset n(c, d)$.

Now back to our problem. Denote $I_n = (a_n, b_n)$, where, of course, $\lim_{n\to\infty} a_n = \lim_{n\to\infty} b_n = \infty$. Start with an interval $J_1 = [c_1, d_1] \subset (p, q)$ (with $c_1 < d_1$) and positive integers m_1 and n_1 (that exist according to the lemma) such that $n_1 J_1$ includes I_{m_1}. Then find a closed interval $J_2 = [c_2, d_2]$ (with $c_2 < d_2$) such that $n_1 J_2$ is a subset of I_{m_1}. This is possible: all we have to do is to choose c_2 and d_2 in such a way that

$$\frac{a_{m_1}}{n_1} < c_2 < d_2 < \frac{b_{m_1}}{n_1} \Leftrightarrow a_{m_1} < n_1 c_2 < n_1 d_2 < b_{m_1}.$$

Note that, since $I_{m_1} \subset n_1 J_1$ means

$$n_1 c_1 < a_{m_1} < b_{m_1} < n_1 d_1,$$

$c_1 < c_2 < d_2 < d_1$ and, consequently, $J_2 \subset J_1$ follows. Then (again, by the lemma) we can find m_2 and n_2 such that $n_2 J_2$ includes I_{m_2}, and we can pick n_2 and m_2 as large as we want; hence, we choose $m_2 > m_1$ and $n_2 > n_1$. Then we define a nondegenerate compact interval $J_3 = [c_3, d_3]$ such that $n_2 J_3 \subset I_{m_2} \subset n_2 J_2$; we conclude that $J_3 \subset J_2$.

In general, we can define two increasing sequences of positive integers (m_k) and (n_k) and a sequence of compact nested intervals (J_k) satisfying

$$n_k J_{k+1} \subset I_{m_k} \subset n_k J_k$$

for all k. Now, the nested intervals J_k must have (at least) a common point t_0 that belongs to (p, q), also, because J_1 was chosen inside (p, q). Since $n_k t_0 \in n_k J_{k+1} \subset I_{m_k}$, we conclude that $|f(n_k t_0)| > \epsilon$ for all k, which is a contradiction to the hypothesis that $\lim_{n\to\infty} f(n t_0) = 0$—and finishes our proof.

11. **Solution I.** The solutions are second degree polynomial functions of the form $f(x) = Ax^2 + Bx$ (taking 0 value for $x = 0$). For $x = mt$, $y = t$, $z = t$, with $m \in \mathbb{Z}$ and $t \in \mathbb{R}$, we get the recurrence relation

$$f((m+2)t) - 2f((m+1)t) + f(mt) = f(2t) - 2f(t)$$

satisfied by the sequence $(f(mt))$, for every real t. Solving this linear recurrence, we obtain

$$f(mt) = \frac{f(2t) - 2f(t)}{2} m^2 + \frac{4f(t) - f(2t)}{2} m$$

for all reals t and all integers m. In particular, $f(m) = Am^2 + Bm$ holds for all $m \in \mathbb{Z}$, with

$$A = \frac{f(2) - 2f(1)}{2} \quad \text{and} \quad B = \frac{4f(1) - f(2)}{2}.$$

Next we have, for a nonzero integer n,

$$f(1) = f\left(n \cdot \frac{1}{n}\right) = \frac{f(2/n) - 2f(1/n)}{2}n^2 + \frac{4f(1/n) - f(2/n)}{2}n$$

and

$$f(2) = f\left(2n \cdot \frac{1}{n}\right) = \frac{f(2/n) - 2f(1/n)}{2}(2n)^2 + \frac{4f(1/n) - f(2/n)}{2}(2n);$$

some tedious algebra shows that solving for $f(1/n)$ and $f(2/n)$ yields

$$f\left(\frac{1}{n}\right) = A\left(\frac{1}{n}\right)^2 + B\left(\frac{1}{n}\right) \quad \text{and} \quad f\left(\frac{2}{n}\right) = A\left(\frac{2}{n}\right)^2 + B\left(\frac{2}{n}\right),$$

with, of course, the same A and B as before. Now replacing these in the above relation for $f(mt)$, with $t = 1/n$, gives

$$f\left(\frac{m}{n}\right) = A\left(\frac{m}{n}\right)^2 + B\left(\frac{m}{n}\right)$$

for all integers m and $n \neq 0$. Or we can say that $f(r) = Ar^2 + Br$ for every rational number r. For finalizing, there is now a standard procedure (based on the continuity of f). For an arbitrary real x, there exists a sequence $(r_n)_{n \geq 1}$ of rational numbers such that $\lim_{n \to \infty} r_n = x$. As we have $f(r_n) = Ar_n^2 + Br_n$ for all n and f is continuous, we can pass to the limit for $n \to \infty$ and get $f(x) = \lim_{n \to \infty} f(r_n) = \lim_{n \to \infty} (Ar_n^2 + Br_n) = Ax^2 + Bx$.

Solution II. Let, for any real x and y, $f_x(y) = f(x + y) - f(x) - f(y)$; of course, $f_x(y) = f_y(x)$ for all x and y. We see that the functional equation can be written in the form $f_y(x + z) = f_y(x) + f_y(z)$, for all $x, y, z \in \mathbb{R}$. Because f_y is continuous and satisfies Cauchy's functional equation, it must be of the form $f_y(x) = k_y x$ for all x, with a fixed real constant k_y. Of course, the constant depends on y, and we rather prefer to use the functional notation $k_y = g(y)$. Thus $f_y(x) = g(y)x$ for all $x, y \in \mathbb{R}$. From the initial equation, we get (for $x = y = z = 0$) $f(0) = 0$, therefore $0 = -f(0) = f_0(x) = g(0)x$ for all x implies $g(0) = 0$, too. For arbitrary nonzero x and y, the equality $g(y)x = g(x)y$ can be also expressed as $g(y)/y = g(x)/x$, which means that $x \mapsto g(x)/x$ is a constant function; thus, there exists $k \in \mathbb{R}$ such that $g(x) = kx$ for all $x \neq 0$. However, this formula works for $x = 0$, too, as long as we know $g(0) = 0$. Thus, we obtained $f(x + y) - f(x) - f(y) = g(y)x = g(x)y = kxy$ for all real numbers x and y.

Now consider the (also continuous) function h defined by $h(x) = f(x) - kx^2/2$ for all x, and note that it also satisfies Cauchy's functional equation: $h(x + y) = h(x) + h(y)$ for all $x, y \in \mathbb{R}$, therefore there exists $B \in \mathbb{R}$ such that $h(x) = Bx$ for all x. With $A = k/2$, we thus conclude that $f(x) = Ax^2 + Bx$ for all real x, as in the first solution.

Of course, any of the solutions must end with the verification of the fact that the functions given by such a formula are, indeed, solutions of the problem, which is not at all complicated. A moment of attention shows that the verification relies on two identities, namely,

$$(x + y + z) + x + y + z = (x + y) + (x + z) + (y + z)$$

and

$$(x + y + z)^2 + x^2 + y^2 + z^2 = (x + y)^2 + (x + z)^2 + (y + z)^2$$

that hold for all $x, y, z \in \mathbb{R}$. Knowing these identities is, of course, helpful for solving the problem (at least for guessing the solutions)—also, they could lead us to some generalizations of it. We are sure that the reader already recognized some particular (and simple) cases of the identity that appeared in the solution of problem 5 from the text (the Erdős-Ginzburg-Ziv theorem). Actually, we are sure that the reader will be able to extend the above and solve the following more general problem: prove that the only continuous real functions f satisfying

$$\sum_{S \subseteq \{1,\dots,m\}} (-1)^{m-|S|} f\left(\sum_{i \in S} x_i\right) = 0$$

for all $x_1, \dots, x_m \in \mathbb{R}$ are the polynomials of degree at most $m - 1$ taking value 0 for $x = 0$. Of course, this generalizes (besides our problem) Cauchy's functional equation for continuous functions, but actually, everything is based on it (so that, in the end, we don't get much of a generalization)—the result follows inductively, in the vein of the second solution. However, this statement is a nice converse of the mentioned identity (that can be useful in one proof of the Erdős-Ginzburg-Ziv theorem).

As said before, in mathematics (and not only in mathematics) there are bridges everywhere.

12. This is a clever application of the mean value theorem, combined with the intermediate value theorem for continuous functions. The second allows us to prove that f has a fixed point (just apply the theorem to the continuous function g defined by $g(x) = f(x) - x$; since $g(a) \geq 0$ and $g(b) \leq 0$, there surely exists $c \in [a, b]$ with $g(c) = 0$)—this is actually a particular case of Brouwer's fixed point theorem. So, there is $c \in [a, b]$ such that $f(c) = c$. The conditions $f(a) = b \neq a$ and $f(b) = a \neq b$ show that, in fact, c is (strictly) between a

and b. Now apply the mean value theorem to f on the intervals $[a, c]$ and $[c, b]$; accordingly, there exist $c_1 \in (a, c)$ and $c_2 \in (c, b)$ such that

$$f'(c_1) = \frac{f(c) - f(a)}{c - a} = \frac{c - b}{c - a}$$

and

$$f'(c_2) = \frac{f(b) - f(c)}{b - c} = \frac{a - c}{b - c}.$$

Of course, $f'(c_1)f'(c_2) = 1$ follows.

13. With the change of variable $t = \pi/2 - x$, we obtain

$$\int_0^{\pi/2} \frac{1}{1 + (\tan x)^{\sqrt{2}}} dx = -\int_{\pi/2}^0 \frac{1}{1 + (\tan(\pi/2 - t))^{\sqrt{2}}} dt$$

$$= \int_0^{\pi/2} \frac{1}{1 + 1/(\tan t)^{\sqrt{2}}} dt = \int_0^{\pi/2} \frac{1}{1 + 1/(\tan x)^{\sqrt{2}}} dx$$

$$= \int_0^{\pi/2} \frac{(\tan x)^{\sqrt{2}}}{1 + (\tan x)^{\sqrt{2}}} dx = \frac{\pi}{2} - \int_0^{\pi/2} \frac{1}{1 + (\tan x)^{\sqrt{2}}} dx,$$

whence

$$\int_0^{\pi/2} \frac{1}{1 + (\tan x)^{\sqrt{2}}} dx = \frac{\pi}{4}.$$

Of course, there is no special significance of the exponent $\sqrt{2}$ in this problem, and, of course, one has to note that the function under the integral is always defined and continuous in the entire interval $[0, \pi/2]$—although it seems not to be defined at $\pi/2$ (or at 0, if the exponent in place of $\sqrt{2}$ is negative).

The trick that we learned from it is that you can make the change of variable $t = a + b - x$ in an integral $\int_a^b f(x)dx$ and thus infer the equality

$$\int_a^b f(x)dx = \int_a^b f(a + b - x)dx,$$

which is often useful in the evaluation of definite integrals—when other approaches fail. For example, one can calculate with this trick

$$\int_0^{\pi/4} \ln(1 + \tan x)dx \ (= (\pi/8) \ln 2),$$

or

$$\int_{-1}^{1} x \ln(1 + e^x)dx \ (= 1/3),$$

or one can find the useful result that the integral of an odd integrable function on an interval symmetric with respect to the origin is 0.

14. First one sees that the function f defined by

$$f(x) = \left(1 + \frac{x}{n}\right)^n e^{-x}$$

has derivative

$$f'(x) = -\frac{x}{n}\left(1 + \frac{x}{n}\right)^n e^{-x} \le 0$$

on $[0, \infty)$. Thus, in the case $0 \le a \le b$, we have, by the monotonicity of the Riemann integral,

$$(b-a)\left(1 + \frac{b}{n}\right)^n e^{-b} \le \int_a^b \left(1 + \frac{x}{n}\right)^n e^{-x}dx \le (b-a)\left(1 + \frac{a}{n}\right)^n e^{-a}.$$

The conclusion follows in this first case by using the squeeze theorem and the well-known fact that $\lim_{n\to\infty} (1 + t/n)^n = e^t$ for every real number t.

Consider now that $a \le b \le 0$. We have (by changing the variable with $x = -t$)

$$\int_a^b \left(1 + \frac{x}{n}\right)^n e^{-x}dx = -\int_{-a}^{-b} \left(1 - \frac{t}{n}\right)^n e^t dt = \int_{-b}^{-a} \left(1 - \frac{x}{n}\right)^n e^x dx$$

and one can see that the function g defined by

$$g(x) = \left(1 - \frac{x}{n}\right)^n e^x$$

is again decreasing (although not on the whole interval $[0, \infty)$) because it has derivative

$$g'(x) = -\frac{x}{n}\left(1 - \frac{x}{n}\right)^n e^x$$

which is ≤ 0 on $[-b, -a]$ for $n > -a$. So, for such n, we have similar inequalities to those above:

$$(b-a)\left(1 - \frac{a}{n}\right)^n e^a \le \int_a^b \left(1 + \frac{x}{n}\right)^n e^{-x}dx \le (b-a)\left(1 - \frac{b}{n}\right)^n e^b,$$

yielding the conclusion in this case, too.

Finally, when $a \leq 0 \leq b$, we can split the integral as follows:

$$\int_a^b \left(1 + \frac{x}{n}\right)^n e^{-x} dx = \int_a^0 \left(1 + \frac{x}{n}\right)^n e^{-x} dx + \int_0^b \left(1 + \frac{x}{n}\right)^n e^{-x} dx,$$

and we get the result by applying the previous, already proved, cases; accordingly, the integral has limit $0 - a + b - 0 = b - a$, finishing the proof. Of course, the result remains true for $a > b$, too.

The last few problems are from the category of nostalgic bridges—they could mean nothing to other people, although they are deeply deposited in our minds and souls. However, the reader will recall his own problems of this kind, and he or she will definitely agree with us when we say that nostalgic bridges appear at every step we take in this world, during our (more or less mathematical) lives.

Chapter 2
Cardinality

We say that two nonempty sets A, B are *equivalent* or *of the same cardinality* or *of the same power* if there is a bijection from A to B. We write this as $A \sim B$.

If $f : A \to B$ is a bijection, then we also denote $A \overset{f}{\sim} B$. Note that "$\sim$" is an equivalence relation. Indeed, we have

$$\begin{cases} \text{(reflexivity) } A \overset{1_A}{\sim} A. \\ \text{(symmetry) if } A \overset{f}{\sim} B, \text{ then } B \overset{f^{-1}}{\sim} A. \\ \text{(transitivity) if } A \overset{f}{\sim} B \text{ and } B \overset{g}{\sim} C, \text{ then } A \overset{g \circ f}{\sim} C. \end{cases}$$

The equivalence class $\widehat{A} = \{B \mid A \sim B\}$ is called *the cardinality of* A, denoted by $|A|$ or $\operatorname{card} A$.

A nonempty set A is called *finite* if $A \sim \{1, 2, \ldots, n\}$, for some positive integer n. In this case, A has n elements and we put $|A| = n$.

For finite sets A, B with $|A| = |B|$ and function $f : A \to B$ we have:

$$f \text{ injective } \Leftrightarrow f \text{ bijective } \Leftrightarrow f \text{ surjective.}$$

As a nice application, we give the following:

Problem. Let p and q be primes, $p \neq q$. Then for all integers $0 \leq r_1 \leq p - 1$, $0 \leq r_2 \leq q - 1$, there exists an integer n which gives the remainders r_1, r_2 when divided by p and q, respectively.

Solution. Denote by

$$\mathbb{Z}_p = \left\{ \overline{0}, \overline{1}, \ldots, \overline{p-1} \right\} , \ \mathbb{Z}_q = \left\{ \widetilde{0}, \widetilde{1}, \ldots, \widetilde{q-1} \right\} , \ \mathbb{Z}_{pq} = \left\{ \widehat{0}, \widehat{1}, \ldots, \widehat{pq-1} \right\}$$

the remainder (or residue) class sets relative to p, q,, respectively pq. Remember that, for any positive integer m, we can define (on the set $\mathbb{Z}$ of the integers) the

© Springer Science+Business Media LLC 2017
T. Andreescu et al., *Mathematical Bridges*, DOI 10.1007/978-0-8176-4629-5_2

relation of congruence modulo m by $a \equiv b \bmod m$ if and only if $a - b$ is divisible by m (or, equivalently, if a and b give equal remainders when divided by m). This is an equivalence relation on $\mathbb{Z}$, and the remainder (or residue, or congruence) class modulo m of the integer x (that is, its equivalence class with respect to the congruence relation) is readily seen to be the set $\widehat{x} = \{\ldots, x - 2n, x - n, x, x + n, x + 2n, \ldots\}$. The set $\mathbb{Z}/m\mathbb{Z} = \mathbb{Z}_m$ of all the residue classes modulo m is then a ring with respect to addition and multiplication defined by $\widehat{a} + \widehat{b} = \widehat{a + b}$ and $\widehat{a} \cdot \widehat{b} = \widehat{a \cdot b}$. The reader is invited to verify that these operations are well defined (they do not depend on choosing the representatives of the remainder classes) and that they indeed provide a ring structure for the set $\mathbb{Z}_m$. Also note that $\mathbb{Z}_m = \{\widehat{0}, \widehat{1}, \ldots, \widehat{m-1}\}$.

Now we go on further with the solution of the problem and define the function $\varphi : \mathbb{Z}_{pq} \to \mathbb{Z}_p \times \mathbb{Z}_q$, by the law $\varphi(\widehat{r}) = (\overline{r}, \widetilde{r})$, $\widehat{r} \in \mathbb{Z}_{pq}$. We have

$$\left| \mathbb{Z}_{pq} \right| = \left| \mathbb{Z}_p \times \mathbb{Z}_q \right| = pq.$$

The surjectivity of φ follows if φ is injective. Thus, we have the implications

$$\varphi(\widehat{r}) = \varphi(\widehat{r'}) \Rightarrow (\overline{r}, \widetilde{r}) = (\overline{r'}, \widetilde{r'})$$

$$\Rightarrow \begin{cases} \overline{r} = \overline{r'} \\ \widetilde{r} = \widetilde{r'} \end{cases} \Rightarrow \begin{cases} p \mid r - r' \\ q \mid r - r' \end{cases} \Rightarrow pq \mid r - r',$$

so $\widehat{r} = \widehat{r'}$. This means that φ is injective and consequently surjective. There is $n \in \{0, 1, \ldots, pq - 1\}$ such that

$$\varphi(\widehat{n}) = (\overline{r}_1, \widetilde{r}_2) \Leftrightarrow (\overline{n}, \widetilde{n}) = (\overline{r}_1, \widetilde{r}_2) \Rightarrow \overline{n} = \overline{r}_1 \text{ and } \widetilde{n} = \widetilde{r}_2,$$

thus $p \mid n - r_1$ and $q \mid n - r_2$. Observe that this argument can be easily extended in order to obtain a proof of the very useful Chinese remainder theorem: if $a_1, a_2, \ldots, a_n$ are pairwise relatively prime integers, then for any integers $b_1, b_2, \ldots, b_n$, the system $x \equiv b_1 \pmod{a_1}, \ldots, x \equiv b_n \pmod{a_n}$ has a unique solution modulo $a_1 a_2 \ldots a_n$. $\square$

If A, B are finite and there is an injective map $f : A \to B$, then we put $|A| \le |B|$. If moreover there is an injective map $g : B \to A$, then $|A| = |B|$. This result, the Cantor-Bernstein theorem, is difficult when A and B are infinite. Here is a proof.

Theorem (Cantor-Bernstein). *If A, B are nonempty sets and there are injections $f : A \to B$, $g : B \to A$, then $|A| = |B|$, i.e., there exists a bijection $\phi : A \to B$.*

Proof. We say that $b \in B$ is an *ancestor* of $a \in A$ if

$$\underbrace{(g \circ f \circ g \circ \ldots \circ f \circ g)}_{2k+1 \text{ times}}(b) = a,$$

for some k. Similarly, $a \in A$ is an *ancestor* of $b \in B$ if

$$\underbrace{(f \circ g \circ f \circ \ldots \circ g \circ f)(a) = b,}_{2k+1 \text{ times}}$$

for some k. Moreover, $a' \in A$ is an *ancestor* of $a \in A$ if $f(a') \in B$ is an ancestor of a and $b' \in B$ is an *ancestor* of $b \in B$ if $g(b') \in A$ is an ancestor of b.

Denote by M_1, M_2, M_∞ the set of all elements of A which have an odd, even, respectively an infinite number of ancestors. Define analogously N_1, N_2, N_∞ for B.

We prove that the function $\phi : A \to B$, given by

$$\phi(x) = \begin{cases} g^{-1}(x), & x \in M_1 \cup M_\infty \\ f(x), & x \in M_2 \end{cases}$$

is bijective. In this sense, we prove that its inverse is $\psi : B \to A$,

$$\psi(y) = \begin{cases} f^{-1}(y), & y \in N_1 \\ g(y), & y \in N_2 \cup N_\infty \end{cases}.$$

These functions ϕ, ψ are well defined because f, g are injective.

Let $x \in A$. If $x \in M_1 \cup M_\infty$, then

$$\phi(x) = g^{-1}(x) \in N_2 \cup N_\infty,$$

so

$$\psi(\phi(x)) = g(\phi(x)) = g(g^{-1}(x)) = x.$$

If $x \in M_2$, then $\phi(x) = f(x) \in N_1$ and

$$\psi(\phi(x)) = f^{-1}(\phi(x)) = f^{-1}(f(x)) = x.$$

Hence $\psi \circ \phi = \mathbf{1}_B$.

Let $y \in B$. If $y \in N_1$, then

$$\psi(y) = f^{-1}(y) \in M_2$$

and

$$\phi(\psi(y)) = f(\psi(y)) = f(f^{-1}(y)) = y.$$

If $y \in N_2 \cup N_\infty$, then

$$\psi(y) = g(y) \in M_1 \cup M_\infty$$

and

$$\phi(\psi(y)) = g^{-1}(\psi(y)) = g^{-1}(g(y)) = y.$$

Hence $\phi \circ \psi = 1_A$. In conclusion, $\phi^{-1} = \psi$ and consequently, $|A| = |B|$. $\square$

This theorem allows us to define an order relation by the law

$$|A| \leq |B| \text{ if and only if there is } f : A \to B \text{ injective.}$$

As a direct consequence, we have

$$|A| \leq |B| \text{ if and only if there is } g : B \to A \text{ surjective.}$$

Using Zorn's lemma, one can prove that this order relation is actually total. A set A is called *countable* if A is equivalent to the set $\mathbb{N}$ of nonnegative integers. A is called *at most countable* if it is finite or countable.

A set is countable if and only if its elements can be written as a sequence. This does not happen for the set of the reals or any of its (nondegenerate) intervals (see problems 2 and 7 below).

Nevertheless, a countable union of countable sets is also a countable set. Indeed, let $A = \bigcup_{n \geq 1} A_n$, where each A_n is countable. Let $A_n = \{a_{n1}, a_{n2}, \ldots\}$ be an enumeration of A_n, for every natural number $n \geq 1$, and note that

$$a_{11}, a_{12}, a_{21}, a_{13}, a_{22}, a_{31}, \ldots$$

is an enumeration of A (basically, the same argument shows that the set of positive rational numbers is countable, as we will immediately see). Obviously, the result remains true if every A_n is *at most* countable.

For instance, the set $\mathbb{Z}$ of all integers is countable because

$$\mathbb{Z} = \{0, 1, -1, 2, -2, 3, -3, 4, -4, \ldots\}.$$

We can also note that

$$\mathbb{Z} = \bigcup_{n \in \mathbb{N}} \{-n, -n+1, -n+2, \ldots\},$$

which is a countable union of countable sets.

For the set $\mathbb{Q}$ of rationals we have the decomposition

$$\mathbb{Q} = \bigcup_{n \in \mathbb{Z}^*} \left\{ \frac{1}{n}, \frac{2}{n}, \frac{3}{n}, \frac{4}{n}, \ldots \right\},$$

so $\mathbb{Q}$ is countable. In another way, the set of positive rationals can be ordered as follows:

$$
\begin{array}{cccccc}
1/1 & \rightarrow & 1/2 & \quad 1/3 & \rightarrow & 1/4 & \cdots \\
 & \swarrow & & \nearrow & \swarrow & & \nearrow \cdots \\
2/1 & \quad 2/2 & & 2/3 & \quad 2/4 & & \cdots \\
\downarrow & \nearrow & \swarrow & & \nearrow & \swarrow & \cdots \\
3/1 & \quad 3/2 & & 3/3 & \quad 3/4 & & \cdots \\
 & \swarrow & & \nearrow & \swarrow & & \nearrow \cdots \\
4/1 & \quad 4/2 & & 4/3 & \quad 4/4 & & \cdots \\
\downarrow & \nearrow & \swarrow & & \nearrow & \swarrow & \cdots \\
5/1 & \quad 5/2 & & 5/3 & \quad 5/4 & & \cdots
\end{array}
$$

$$\cdots \ \cdots \ \cdots \ \cdots \ \cdots \ \cdots \ \cdots \ \cdots \cdots$$

Each element appears many times in the table, but we consider each of them only for the first time.

The set $\mathbb{N} \times \mathbb{N}$ of pairs of nonnegative integers is countable. Indeed,

$$f : \mathbb{N} \to \mathbb{N} \times \mathbb{N} \ , \quad f(n) = (n, 0)$$

is injective and

$$g : \mathbb{N} \times \mathbb{N} \to \mathbb{N} \ , \quad g(m, n) = 2^m \cdot 3^n$$

is injective. According to the Cantor-Bernstein theorem, $\mathbb{N} \times \mathbb{N} \sim \mathbb{N}$. In addition, note that the map

$$\varphi : \mathbb{N}^* \times \mathbb{N}^* \to \mathbb{N}^* \ , \quad \varphi(m, n) = 2^{m-1} \cdot (2n - 1)$$

is bijective. One can even find a polynomial bijection between $\mathbb{N} \times \mathbb{N}$ and $\mathbb{N}$, which we leave as an interesting exercise for the reader.

Proposed Problems

1. Let A be an infinite set. Prove that for every positive integer n, A has a finite subset with n elements. Deduce that every infinite set has at least one countable subset.
2. Prove that $(0, 1)$ is not countable. Infer that $\mathbb{R}$ and $\mathbb{R} \setminus \mathbb{Q}$ are not countable.
3. Let X, A, B, be pairwise disjoint sets such that A, B are countable. Prove that

$$X \cup A \cup B \sim X \cup A.$$

Deduce that for every countable set B of real numbers, $\mathbb{R} \setminus B \sim \mathbb{R}$.
4. Prove that $\mathbb{N} \times \mathbb{N} \times \mathbb{N}$ is countable and so is $\mathbb{N} \times \mathbb{N} \times \ldots \times \mathbb{N}$.

5. Let $a < b$ be real numbers. Prove that $(0, 1) \sim (a, b) \sim \mathbb{R}$.
6. Let A be a countable set and $a \in A$. Prove that $A \setminus \{a\} \sim A$. Is this result true for every infinite set A?
7. Let $a < b$ be real numbers. Prove that $[a, b] \sim [a, b) \sim (a, b] \sim (a, b)$.
8. Prove that every ε-discrete set of real numbers is at most countable. (A set $A \subset \mathbb{R}$ is called ε-discrete if $|a - b| > \varepsilon$, for any different elements a, b of A).
9. Let S be a set of real numbers with the property that for all real numbers $a < b$, the set $S \cap [a, b]$ is finite, possibly empty. Prove that S is at most countable. Is every set S with the above property an ε-discrete set, for some positive real ε?
10. Let S be an infinite and uncountable set of real numbers. For each real number t, we put

$$S^-(t) = S \cap (-\infty, t], \quad S^+(t) = S \cap [t, \infty).$$

Prove that there exists a real number t_0 for which both sets $S^-(t_0)$ and $S^+(t_0)$ are infinite and uncountable.
11. A set M of positive real numbers has the property that the sum of any finite number of its elements is not greater than 7. Prove that the set M is at most countable.
12. Prove that the set of polynomials with integer coefficients is countable.
13. Prove that the set of algebraic numbers is countable. Deduce that the set of transcendental numbers is not countable. (A real number α is called an algebraic number if there exists a polynomial $P \neq 0$ with integer coefficients such that $P(\alpha) = 0$. Otherwise, α is called transcendental.)
14. Prove that for each set X, we have $|X| < |\mathcal{P}(X)|$. We denote by $\mathcal{P}(X)$ the power set of X (that is, the set of all subsets of X, including the empty set and X). However, the set of all finite subsets of $\mathbb{N}$ is countable (thus $|\mathbb{N}| = |\mathcal{P}(\mathbb{N})|$).
15. Let $p_1, p_2, \ldots, p_k$ be distinct primes. Prove that for all integers $r_1, r_2, \ldots, r_k$ there is an integer n such that $n \equiv r_i \pmod{p_i}$, for all $1 \leq i \leq k$.
16. Prove that there are no functions $f : \mathbb{R} \to \mathbb{R}$ with the property

$$|f(x) - f(y)| \geq 1,$$

for all $x, y \in \mathbb{R}$, $x \neq y$.
17. Prove that there are no functions $f : \mathbb{R} \to \mathbb{R}$ with the property

$$|f(x) - f(y)| \geq \frac{1}{x^2 + y^2},$$

for all $x, y \in \mathbb{R}$, $x \neq y$.
18. Prove that the discontinuity set of a monotone function $f : \mathbb{R} \to \mathbb{R}$ is at most countable.
19. Prove that the set of all permutations of the set of positive integers is uncountable.

20. Let f, g be two real functions such that $f(x) < g(x)$ for all real numbers x. Prove that there exists an uncountable set A such that $f(x) < g(y)$ for all $x, y \in A$.

21. Let $\mathbb{R}$ be the real line with the standard topology. Prove that every uncountable subset of $\mathbb{R}$ has uncountably many limit points.

22. Find a function $f : [0, 1] \to [0, 1]$ such that for each nontrivial interval $I \subseteq [0, 1]$ we have $f(I) = [0, 1]$.

23. Let a and k be positive integers. Prove that for every positive integer d, there exists a positive integer n such that d divides $ka^n + n$.

Solutions

1. First we will prove by induction the following proposition:

$$P(n) : \text{"The set } A \text{ has a finite subset with } n \text{ elements."}$$

The set A is nonempty, so we can find an element $a_1 \in A$. Then $A_1 = \{a_1\}$ is a finite subset of A with one element, thus $P(1)$ is true.

Assume now that $P(k)$ is true, so A has a finite subset with k elements,

$$A_k = \{a_1, a_2, \dots, a_k\} \subset A.$$

The set A is infinite, while A_k is finite, so the set $A \setminus A_k$ is nonempty. If we choose an element $a_{k+1} \in A \setminus A_k$, then the set

$$A_k = \{a_1, a_2, \dots, a_k, a_{k+1}\}$$

is a finite subset of A, with $k + 1$ elements. Hence $P(k + 1)$ is true.

Further, we prove that A has a countable subset. As we proved, for every positive integer n, we can find a finite subset $A_n \subset A$ with n elements. Then the set

$$S = \bigcup_{n \geq 1} A_n \subseteq A$$

is an infinite subset of A. Moreover, S is countable, as a countable union of finite sets.

2. Let us assume by contradiction that $A = (0, 1)$ is countable, say

$$A = \{x_n \mid n \in \mathbb{N}, \ n \geq 1\}.$$

Let us consider the decimal representations of the elements of A,

$$x_1 = 0.a_{11}a_{12}a_{13}\ldots a_{1n}\ldots$$

$$x_2 = 0.a_{21}a_{22}a_{23}\ldots a_{2n}\ldots$$

$$\ldots\ldots\ldots\ldots\ldots\ldots$$

$$x_n = 0.a_{n1}a_{n2}a_{n3}\ldots a_{nn}\ldots$$

$$\ldots\ldots\ldots\ldots\ldots\ldots$$

For each integer $k \geq 1$, we choose a digit denoted b_k so that

$$b_k \neq a_{kk} \;,\;\; b_k \neq 9 \;,\;\; b_k \neq 0.$$

Now let us define the number

$$x = 0.b_1 b_2 b_3 \ldots b_n \ldots \;.$$

Obviously, $x \in (0, 1)$, so there is an integer $m \geq 1$ for which $x = x_m$. But the equality $x = x_m$ implies

$$0.b_1 b_2 b_3 \ldots b_n \ldots = 0.a_{m1} a_{m2} a_{m3} \ldots a_{mm} \ldots \;,$$

which is impossible, because the decimal of rank m are different, $b_m \neq a_{mm}$. Now, let us assume by way of contradiction that $\mathbb{R} = \{x_n \mid n \geq 1\}$ is countable. Obviously, there exists a subsequence $(x_{k_n})_{n\geq 1}$ of $(x_n)_{n\geq 1}$ such that

$$(0, 1) = \{x_{k_n} \mid n \geq 1\}.$$

This means that $(0, 1)$ is countable, which we just showed to be false.

Finally, we use the fact that the union of two countable sets is also a countable set. If $\mathbb{R} \setminus \mathbb{Q}$ is countable, it should follow that $\mathbb{R} = \mathbb{Q} \cup (\mathbb{R} \setminus \mathbb{Q})$ is countable, as union of two countable sets. This contradiction shows that $\mathbb{R} \setminus \mathbb{Q}$ is not countable.

3. The sets A, B are countable so we can assume that

$$A = \{a_1, a_2, \ldots, a_n, \ldots\}, \quad B = \{b_1, b_2, \ldots, b_n, \ldots\}.$$

Let us define the map

$$f : X \cup A \cup B \to X \cup A$$

given by the formula

$$f(x) = \begin{cases} x, & \text{if } x \in X \\ a_{2n}, & \text{if } x \in A, \; x = a_n \\ a_{2n-1}, & \text{if } x \in B, \; x = b_n \end{cases}.$$

First, the map f is well defined: it takes (all) values in $X \cup A$. Hence it is surjective. For injectivity, note that f is injective on each restriction to X, A, and B. Then f is injective on $X \cup A \cup B$, if we take into account that any two of the sets

$$f(X) = X, \quad f(A) = \{a_2, a_4, \ldots, a_{2n}, \ldots\}, \quad f(B) = \{a_1, a_3, \ldots, a_{2n-1}, \ldots\}$$

are disjoint.

For the second part of the problem, let A be a countable subset of $\mathbb{R} \setminus B$. This choice is possible, because the set $\mathbb{R} \setminus B$ is nonempty and infinite. If $X = \mathbb{R} \setminus (A \cup B)$, then

$$\mathbb{R} \setminus B = X \cup A, \quad \mathbb{R} = X \cup A \cup B,$$

with A, B countable, and the conclusion follows, according to the first part of the problem.

4. We begin by proving the implication

$$A \sim B \Rightarrow A \times C \sim B \times C,$$

for all sets A, B, C. Indeed, if $f : A \to B$ is a bijection, then the map

$$\phi : A \times C \to B \times C$$

given by

$$\phi(a, c) = (f(a), c) \ , \quad a \in A, \ c \in C,$$

is also a bijection, so $A \times C \sim B \times C$.

We have already proved that $\mathbb{N} \sim \mathbb{N} \times \mathbb{N}$. According to the above remark,

$$\mathbb{N} \times \mathbb{N} \sim \mathbb{N} \times \mathbb{N} \times \mathbb{N}.$$

Finally, by transitivity,

$$\mathbb{N} \sim \mathbb{N} \times \mathbb{N} \sim \mathbb{N} \times \mathbb{N} \times \mathbb{N},$$

so $\mathbb{N} \sim \mathbb{N} \times \mathbb{N} \times \mathbb{N}$.

In a similar way, the sets $\mathbb{N} \times \mathbb{N} \times \ldots \times \mathbb{N}$ are countable, too. As well, note that

$$f : \mathbb{N} \to \mathbb{N} \times \mathbb{N} \times \mathbb{N} \ , \quad f(n) = (n, 0, 0)$$

is injective and

$$g : \mathbb{N} \times \mathbb{N} \times \mathbb{N} \to \mathbb{N} \ , \quad g(m, n, p) = 2^m \cdot 3^n \cdot 5^p$$

is injective. The conclusion follows by Cantor-Bernstein theorem. This method can also be used to prove that $\mathbb{N} \times \mathbb{N} \times \ldots \times \mathbb{N}$ is countable.

5. One idea is to search a linear function $f(x) = mx + n$ from $(0, 1)$ onto (a, b). In order to determine the values m, n, we impose the condition $f(0) = a$ and $f(1) = b$. It gives

$$\begin{cases} n = a \\ m + n = b \end{cases} \Rightarrow \begin{cases} n = a \\ m = b - a \end{cases}.$$

Consequently, the function $f : (0, 1) \to (a, b)$, given by

$$f(x) = (b - a)x + a$$

is bijective, so $(0, 1)$ is equivalent to every interval (a, b).

For the other part, it is sufficient to prove that $\mathbb{R}$ is equivalent to some open interval. Indeed, we can see that the function $\phi : \mathbb{R} \to \left(-\frac{\pi}{2}, \frac{\pi}{2}\right)$, given by $\phi(x) = \arctan x$ is bijective.

6. Assume that $A = \{a_n \mid n \in \mathbb{N}\}$, so that $a_0 = a$.

Then the bijection $f : A \setminus \{a\} \to A$ given by the formula $f(a_n) = a_{n-1}$, $n \in \mathbb{N}, n \geq 1$, shows us that $A \setminus \{a\} \sim A$.

$$
\begin{array}{ccccccc}
a_1 & a_2 & a_3 & a_4 & a_5 & & \cdots \\
\swarrow & \swarrow & \swarrow & \swarrow & \swarrow & \swarrow & \\
a_0 & a_1 & a_2 & a_3 & a_4 & a_5 & \cdots
\end{array}
$$

The result remains true if A is an arbitrary infinite set.

Indeed, let $B = \{x_n \mid n \in \mathbb{N}\}$ be a countable subset of A. We choose $x_0 = a$, then define the function $\phi : A \setminus \{a\} \to A$ by the formula

$$\phi(x) = \begin{cases} x, & x \in A \setminus B \\ x_{n-1}, & x \in B, \ x = x_n, \ n \geq 1 \end{cases}.$$

In a classical way, we can easily prove that ϕ is bijective. Moreover, we can indicate its inverse $\phi^{-1} : A \to A \setminus \{a\}$ with

$$\phi^{-1}(x) = \begin{cases} x, & x \in A \setminus B \\ x_{n+1}, & x \in B, \ x = x_n, \ n \geq 1 \end{cases}.$$

7. We have already proved that $A \setminus \{x\} \sim A$, for every infinite set A and $x \in A$. In our case,

$$[a, b] \sim [a, b] \setminus \{b\} \Leftrightarrow [a, b] \sim [a, b),$$

$$[a, b] \sim [a, b] \setminus \{a\} \Leftrightarrow [a, b] \sim (a, b]$$

and further

$$(a, b] \sim (a, b] \setminus \{b\} \Leftrightarrow (a, b] \sim (a, b).$$

Finally, from transitivity,

$$[a, b] \sim [a, b) \sim (a, b] \sim (a, b).$$

8. For each element $x \in A$, consider the interval

$$I_x = \left(x - \frac{\varepsilon}{2}, x + \frac{\varepsilon}{2}\right).$$

If $x, y \in A$, $x \neq y$, then $I_x \cap I_y = \emptyset$.
   Indeed, if there is c in $I_x \cap I_y$, then

$$\varepsilon < |x - y| \leq |x - c| + |y - c| < \frac{\varepsilon}{2} + \frac{\varepsilon}{2} = \varepsilon,$$

which is false. Now, for each $x \in A$, we choose a rational number $r_x \in I_x$. As we have proved, $x \neq y \Rightarrow r_x \neq r_y$, which can be expressed that the map $\phi : A \to \mathbb{Q}$ given by the law $\phi(x) = r_x$, for all $x \in A$, is injective. Finally, A is at most countable because $\mathbb{Q}$ is countable.

9. For every integer n, we put $S_n = S \cap [n, n + 1]$. According to the hypothesis, all sets S_n, $n \in \mathbb{Z}$ are finite. Thus the set

$$S = \bigcup_{n \in \mathbb{Z}} S_n$$

is at most countable, as a countable union of finite sets.
   The answer to the question is negative. There exist sets S with the property from the hypothesis, which are not ε-discrete. An example is

$$S = \{\ln n \mid n \in \mathbb{N}^*\}.$$

10. First we prove that there exists r such that the set $S^-(r)$ is infinite and uncountable. If we assume the contrary, then the decomposition

$$S = \bigcup_{n \in \mathbb{Z}} S^-(n),$$

is a countable union of at most countable sets. Hence S is countable, a contradiction. Let

$$\alpha = \inf \{r \mid S^-(r) \text{ infinite and uncountable}\}$$

and similarly, we can define

$$\beta = \sup \left\{ r \mid S^+(r) \text{ infinite and uncountable} \right\},$$

where cases $\alpha = -\infty$ or $\beta = \infty$ are accepted.

We prove that $\alpha \leq \beta$. If $\beta < \alpha$, then let $\beta < t < \alpha$. According to the definition of α, the set $S^-(t)$ is countable, and from the definition of β, the set $S^+(t)$ is countable. Hence S is countable, as union of two countable sets,

$$S = S^-(t) \cup S^+(t).$$

Consequently, $\alpha \leq \beta$. Then for every $\alpha \leq t_0 \leq \beta$, the sets $S^-(t_0)$ and $S^+(t_0)$ are infinite and uncountable, because

$$S^-(t_0) \supseteq S^-(\alpha), \quad S^+(t_0) \supseteq S^+(\beta).$$

11. For each integer $n \geq 1$, define the set

$$A_n = \left\{ x \in M \mid x > \frac{1}{n} \right\}.$$

Easily, $M = \bigcup_{n \geq 1} A_n$. We will prove that every set A_n is finite or empty, so M is countable as a countable union of finite sets. Now we can prove that A_n has at most $7n$ elements. If for some n, the set A_n has at least $7n + 1$ elements, say $x_1, x_2, \ldots x_{7n+1} \in A$, then

$$x_1 > \frac{1}{n}, \quad x_2 > \frac{1}{n}, \quad \ldots, \quad x_{7n+1} > \frac{1}{n}.$$

By adding,

$$x_1 + x_2 + \cdots + x_{7n+1} > \frac{7n + 1}{n} > 7,$$

which is a contradiction.

12. For each polynomial $P \in \mathbb{Z}[X]$,

$$P = a_0 + a_1 X + \cdots + a_n X^n,$$

define and denote by

$$h(P) = n + |a_0| + |a_1| + \cdots + |a_n|$$

the height of P, $h(0) = 0$. Let us put for each nonnegative integer k,

$$\mathcal{P}_k = \{ P \in \mathbb{Z}[X] \mid h(P) = k \}.$$

Each set $\mathcal{P}_k$ is finite, possibly empty, so

$$\mathbb{Z}[X] = \bigcup_{k \in \mathbb{N}} \mathcal{P}_k$$

is countable, as a countable union of finite sets. Indeed, there are only a finite number of polynomials with $h(P) = k$. First, note that if $h(P) = k$, then $\deg P \leq k$ and $|a_0|, |a_1|, \ldots, |a_n| \leq k$. Consequently, P is defined by a finite number of integer coefficients $a_0, a_1, \ldots, a_n$ which are less than or equal to k in absolute value.

13. The set $\mathcal{A}$ of algebraic numbers is the set of real roots of all nonconstant polynomials with integer coefficients. Using this remark, we can write

$$\mathcal{A} = \bigcup_{P \in \mathbb{Z}_1[X]} \{x \in \mathbb{R} \mid P(x) = 0\},$$

where

$$\mathbb{Z}_1[X] = \mathbb{Z}[X] \setminus \{0\}.$$

Consequently, $\mathcal{A}$ is countable as a countable union of finite sets. Indeed, each set from the union has at most k elements, where $k = \deg P$.

14. The function $\phi : X \to \mathcal{P}(X)$ given by $\phi(x) = \{x\}$, for all $x \in X$, is injective, so $|X| \leq |\mathcal{P}(X)|$. Thus we have to prove that there are no bijections from X onto $\mathcal{P}(X)$. If we assume by contradiction that there is a bijection $f : X \to \mathcal{P}(X)$, then define the set $A = \{x \in X \mid x \notin f(x)\}$, $A \in \mathcal{P}(X)$. Because of the surjectivity of f, we have $A = f(x_0)$, for some $x_0 \in X$. Now the question is

$$x_0 \in A \quad \text{or} \quad x_0 \notin A ?$$

If $x_0 \in A$, then $x_0 \notin f(x_0)$, false because $f(x_0) = A$. If $x_0 \notin A$, then $x_0 \in f(x_0)$, false, because $f(x_0) = A$.

The last two implications follow from the definition of the set A. These contradictions solve the problem.

One way to see that the second statement of the problem is true is to consider the function that maps every finite subset $\{n_1, \ldots, n_k\}$ of $\mathbb{N}$ to the natural number $2^{n_1} + \cdots + 2^{n_k}$ (and maps the empty set to 0). This mapping is clearly a bijection (since every positive integer has a unique binary representation), and the conclusion follows.

Or, one can see that this is an enumeration of all finite sets of $\mathbb{N}$:

$$\emptyset, \{0\}, \{1\}, \{0, 1\}, \{2\}, \{0, 2\}, \{1, 2\}, \{0, 1, 2\}, \{3\}, \ldots$$

(we leave to the reader to decipher how the sets are enumerated; we think that he/she will do).

Finally, one can see that the set $\mathcal{P}_f(\mathbb{N})$ of finite parts of $\mathbb{N}$ is the union of the sets P_i, where P_i means the set of finite subsets of $\mathbb{N}$ having the sum of their elements precisely i; since every P_i is finite, $\mathcal{P}_f(\mathbb{N})$ is countable (as a countable union of finite sets). The reader will surely find a few more approaches.

15. Denote $p = p_1 p_2 \ldots p_k$. Let us define the function

$$f : \mathbb{Z}_p \rightarrow \mathbb{Z}_{p_1} \times \mathbb{Z}_{p_2} \times \ldots \times \mathbb{Z}_{p_k}$$

by the formula

$$f(r \bmod p) = (r \bmod p_1, r \bmod p_2, \ldots, r \bmod p_k), \quad r \in \{0, 1, \ldots, p-1\}.$$

The sets $\mathbb{Z}_p$ and $\mathbb{Z}_{p_1} \times \mathbb{Z}_{p_2} \times \ldots \times \mathbb{Z}_{p_k}$ have the same number of elements. In fact, the problem asks to show that f is surjective. Under our hypothesis, it is sufficient to prove that f is injective. In this sense, let $r, s \in \{0, 1, \ldots, p-1\}$ be such that

$$(r \bmod p_1, r \bmod p_2, \ldots, r \bmod p_k) = (s \bmod p_1, s \bmod p_2, \ldots, s \bmod p_k).$$

It follows that

$$r \bmod p_1 = s \bmod p_1, \ldots, r \bmod p_k = s \bmod p_k,$$

or $p_1 | r - s$, $p_2 | r - s$, ..., $p_k | r - s$. Hence $p | r - s$, which is equivalent to $r \bmod p = s \bmod p$. This proves the injectivity of f.

16. For each integer k, there exists at most one element $f(x) \in [k, k+1)$. Therefore to each real number x, we can assign a unique integer $k = k(x)$ such that $f(x) \in [k, k+1)$. Thus, the function

$$\mathbb{R} \ni x \longmapsto k(x) \in \mathbb{Z}$$

is injective. This is impossible, because $\mathbb{R}$ is not countable and $\mathbb{Z}$ is countable.

Another method uses the injectivity of f. Indeed, if $x \neq y$, then $|f(x) - f(y)| \geq 1$, so the equality $f(x) = f(y)$ is not possible. The inequality from the hypothesis says that the image of the function f is a 1-discrete set, so it is at most countable. Now, the map $f : \mathbb{R} \rightarrow \operatorname{Im} f$ is injective, so

$$\operatorname{card} \mathbb{R} \leq \operatorname{card} \operatorname{Im} f.$$

This is impossible, because $\mathbb{R}$ is not countable.

17. Let us assume, by way of contradiction, that such a function f does exist. Define $g : (-1, 1) \rightarrow \mathbb{R}$, given by $g(x) = f(x)$, for all $x \in (-1, 1)$. Then

$$|g(x) - g(y)| \geq \frac{1}{2},$$

for all $x, y \in (-1, 1)$, $x \neq y$. Indeed,

$$|g(x) - g(y)| \geq \frac{1}{x^2 + y^2} \geq \frac{1}{2}.$$

Now define $h : \mathbb{R} \to \mathbb{R}$ by $h = 2 \cdot (g \circ \phi)$, where $\phi : \mathbb{R} \to (-1, 1)$ is

$$\phi(x) = \frac{2}{\pi} \arctan x.$$

Finally, the function $h : \mathbb{R} \to \mathbb{R}$ satisfies

$$|h(x) - h(y)| \geq 1,$$

for all $x, y \in \mathbb{R}$, $x \neq y$, which is impossible, as we have seen in the previous problem.

18. Let D be the discontinuity set of f. It is well known that D contains only discontinuities of the first kind if f is monotone. We mean that for each $x \in D$, there exist finite one-sided limits denoted

$$f_s(x) = \lim_{y \nearrow x} f(y) \quad , \quad f_d(x) = \lim_{y \searrow x} f(y).$$

If f is increasing, then $f_s(x) \leq f_d(x)$, with strict inequality if $x \in D$. Now, for every $x \in D$, we choose a rational number denoted

$$r_x \in (f_s(x), f_d(x))$$

and we define the function

$$D \ni x \overset{f}{\longmapsto} r_x \in \mathbb{Q}.$$

It is injective because of the implication

$$x < y \Rightarrow f_d(x) \leq f_s(y).$$

Finally, D is countable, because $\mathbb{Q}$ is countable.

19. Take any semi-convergent series of real numbers with general term a_n (for instance, $a_n = \frac{(-1)^n}{n}$) and apply Riemann's theorem: for any real number a there exists a permutation π of the set of positive integers such that $a_{\pi(1)} + a_{\pi(2)} + \ldots = a$. This gives an injection from the set of real numbers into the set of permutations of the positive integers. Since the former is uncountable, so is the desired set.

20. Let us consider a rational number $r(x)$ between $f(x)$ and $g(x)$ and look at the sets A_x of those real numbers a such that $r(a) = x$. The union of these sets (taken

over all rational numbers x) is the set of real numbers, which is uncountable. So, at least one of these sets is uncountable, and it clearly satisfies the conditions.

21. Suppose that A is a subset of $\mathbb{R}$ that has a countable set of limit points. The points from A split into two classes: those that are limit points of A (denote by B this set) and those that are not (let C be this second subset of A). But any point c from C must have a neighborhood that does not intersect A (with the exception of c), and this neighborhood may be chosen to be an open interval that contains c and has rational points as extremities. Therefore C is countable (possibly finite).

So, if we assume that the set of limit points of A is at most countable, B is also at most countable; then, since $A = B \cup C$, the countability of A follows, and this proves the problem's claim by contraposition.

22. Let us consider on $[0, 1]$ the relation "$\sim$" defined by

$$a \sim b \Leftrightarrow a - b \in \mathbb{Q}.$$

Clearly, this is an equivalence relation; thus we can find a complete system of representatives of the equivalence classes, that is, a set $A \subseteq [0, 1]$ such that any distinct $a, b \in A$ are not in the relation $\sim$ and for each $t \in [0, 1]$, there is a (unique) $a \in A$ for which $t \sim a$. We have

$$[0, 1] = \bigcup_{a \in A} X_a$$

if we denote by X_a the equivalence class of a. Since

$$X_a = \{y \in [0, 1] \mid y - a \in \mathbb{Q}\} = \{a + t \mid t \in \mathbb{Q} \cap [-a, 1 - a]\},$$

we see that each class X_a is dense in $[0, 1]$ and is a countable set. If A was countable, then $[0, 1]$ would be countable, too, as a countable union of countable sets. Since this is not the case, we infer that A is not countable, hence a bijection $\varphi : [0, 1] \to A$ can be found.

Then we can define the desired function $f : [0, 1] \to [0, 1]$ by setting

$$f(x) = t,$$

for all $x \in X_{\varphi(t)}$ and for suitably chosen $t \in [0, 1]$. Since each $x \in [0, 1]$ belongs to exactly one set X_a and for each $a \in A$ there is a unique $t \in [0, 1]$ such that $a = \varphi(t)$, the function f is well defined.

Now, let $I \subseteq [0, 1]$ be an interval which is not reduced to one point. I contains elements from any set X_a, $a \in A$ (since the classes of equivalence are dense in $[0, 1]$). An arbitrary $t \in [0, 1]$ being given, the intersection of I with $X_{\varphi(t)}$ is nonempty; so we can consider an $x \in I \cap X_{\varphi(t)}$ for which we have $f(x) = t$. Thus, we see that f takes in I any value $t \in [0, 1]$, that is, the desired result.

This is problem 1795, proposed by Jeff Groah in *Mathematics Magazine*, 2/2008. Two more solutions can be found in the same *Magazine*, 2/2009.

23. We prove the slightly more general statement that, for positive integers a, k, d, and N, there exist positive integers n_i, $0 \le i \le d-1$ such that $n_i > N$ and $ka^{n_i} + n_i \equiv i \pmod{d}$ for all $0 \le i \le d-1$. Of course, $n = n_0$ is the solution to our problem.

The proof of the general result is by induction on d. For $d = 1$, we have nothing to prove (and even for $d = 2$, one can easily prove the statement). So, let's assume it is true for all positive integers $d < D$ and deduce it for D. Also, assume that some positive integer N has been fixed.

We can consider the (smallest) period p of a modulo D; that is, p is the smallest positive integer such that $a^{m+p} \equiv a^m$ for all $m > M$, M being a certain nonnegative integer. We then also have $a^{m+lp} \equiv a^m \pmod{D}$ for all nonnegative integers $m > M$ and l. Yet, note that $p < D$ because the sequence of powers of a modulo D either contains 0 (and then $p = 1$), or it doesn't (and then, surely, p is at most $D-1$). Consequently, $d = (D, p)$ is also less than D and we can apply the induction hypothesis to infer that there exist positive integers m_i such that $m_i > \max\{M, N\}$ and $ka^{m_i} + m_i \equiv i \pmod{d}$ for all $0 \le i \le d-1$.

We claim that the numbers $ka^{m_i+sp} + (m_i + sp)$, with $0 \le i \le d-1$, and $0 \le s \le D/d - 1$ are mutually distinct modulo D. Indeed, suppose that

$$ka^{m_i+sp} + (m_i + sp) \equiv ka^{m_j+tp} + (m_j + tp) \pmod{D},$$

for $i, j \in \{0, 1, \ldots, d-1\}$, and $s, t \in \{0, 1, \ldots, D/d - 1\}$. Since p is a period of a modulo D and m_i are (by choice) greater than M, we get $ka^{m_i} + (m_i + sp) \equiv ka^{m_j} + (m_j + tp) \pmod{D}$, and because d is a divisor of D, this congruence is also true modulo d. Again by the choice of the m_i, $ka^{m_i} + (m_i + sp) \equiv ka^{m_j} + (m_j + tp) \pmod{d}$ becomes $i + sp \equiv j + tp \pmod{d}$, and then $i \equiv j \pmod{d}$ (as d is also a divisor of p). But i, j are from the set $\{0, 1, \ldots, d-1\}$, thus $i = j$. Going back to the initial congruence, we see that it becomes $sp \equiv tp \pmod{D}$, yielding $s(p/d) \equiv t(p/d) \pmod{D/d}$. But p/d and D/d are relatively prime, hence we get $s \equiv t \pmod{D/d}$ which, together with $s, t \in \{0, 1, \ldots, D/d-1\}$, implies $s = t$ and the fact that the two original numbers are equal.

Thus we have the $d \cdot (D/d) = D$ numbers $ka^{m_i+sp} + (m_i + sp)$, with $0 \le i \le d-1$, and $0 \le s \le D/d - 1$ that are mutually distinct modulo D; therefore they produce all possible remainders when divided by D (and here is our cardinality argument; remember problem 15, that is, roughly, the Chinese remainder theorem). This means that we can rename by n_h, $0 \le h \le D-1$, the numbers $m_i + sp$, $0 \le i \le d, 0 \le s \le D/d-1$ in such a way that $ka^{n_h} + n_h \equiv h \pmod{D}$ for each $0 \le h \le D - 1$, and this is exactly what we wanted to prove for completing the induction.

This is Problem 11789, proposed by Gregory Galperin and Yury J. Ionin in *The American Mathematical Monthly*. A different solution by Mark Wildon appeared in the same *Monthly* from August to September 2016. Note, however, that our proof is nothing but a rewording of the official solution of the seventh shortlisted number theory problem from the 47th IMO, Slovenia, 2006. (The shortlisted problems can be found on the official site of the IMO.) That problem asks to show that, given a positive integer d, there exists a positive integer n such that $2^n + n$ is divisible by d.

Chapter 3
Polynomial Functions Involving Determinants

Some interesting properties of determinants can be established by defining polynomial functions of the type

$$f(X) = \det(A + XB), \qquad (3.1)$$

where A, B are $n \times n$ matrices with complex entries. Under this hypothesis, f is a polynomial of degree $\leq n$. The coefficient of X^n is equal to $\det B$ and the constant term of f is equal to $\det A$. Indeed, the constant term of the polynomial f is equal to

$$f(0) = \det(A + 0 \cdot B) = \det A.$$

The coefficient of X^n can be calculated with the formula

$$\lim_{x \to \infty} \frac{f(x)}{x^n} = \lim_{x \to \infty} \frac{1}{x^n} \det(A + xB) = \lim_{x \to \infty} \det\left(\frac{1}{x}A + B\right) = \det B.$$

In the case $n = 2$, the polynomial f is of second (or first) degree, so we can use the properties of quadratic functions, respectively linear functions.

A special case of such polynomials is the characteristic polynomial $\chi_A(X) = \det(XI_n - A)$. Its zeros are called the *eigenvalues* of A. Using the theory of linear systems, one shows that if $A \in \mathcal{M}_n(\mathbb{C})$, then $\lambda \in \mathbb{C}$ is an eigenvalue of A if and only if there exists a nonzero vector $v \in \mathbb{C}^n$ such that $Av = \lambda v$. One can give a more abstract interpretation of the above objects, using vector spaces. Namely, if V is a vector space over a field F, then any linear map $f : V \to V$ can be represented by a matrix if the dimension of V is finite, say n. Indeed, once we fix a basis $e_1, e_2, \ldots, e_n$ of V, f is determined by its values on $e_1, \ldots, e_n$, thus by a matrix A such that $f(e_i) = \sum_{j=1}^{n} a_{ji} e_j$ for all j. Because for any other basis $e'_1, e'_2, \ldots, e'_n$ in which one can find an

© Springer Science+Business Media LLC 2017
T. Andreescu et al., *Mathematical Bridges*, DOI 10.1007/978-0-8176-4629-5_3

invertible matrix $P \in \mathcal{M}_n(F)$ such that $e'_i = \displaystyle\sum_{j=1}^{n} p_{ji} e_j$, one can easily see that two matrices A, B represent the same linear map f in two different bases of V if and only if there exists $P \in GL_n(F)$, the set of invertible matrices in $\mathcal{M}_n(F)$, such that $A = PBP^{-1}$. In this case, we say that A and B are *similar*. If f is a linear map represented by a matrix A, we then call the polynomial $\chi_f(X) = \chi_A(X) = \det(XI_n - A)$ the *characteristic polynomial* of f. It is easy to see that χ_f does not depend on A, because two similar matrices have the same determinant.

Before passing to some problems, let us point out some other useful facts. First, observe that if $A, B \in \mathcal{M}_n(\mathbb{C})$ and A is invertible, then AB and BA are similar. Indeed, $AB = A(BA)A^{-1}$. Thus $\chi_{AB} = \chi_{BA}$ in this case. But this result holds without the assumption $A \in GL_n(\mathbb{C})$. Indeed, let $z \in \mathbb{C}$ and observe that there are infinitely many $\varepsilon \in \mathbb{C}$ such that $A + \varepsilon I_n \in GL_n(\mathbb{C})$, because $\det(A + XI_n)$ is a nonzero polynomial. Actually, for the same reason, there is such a sequence ε_k with $\lim_{k \to \infty} \varepsilon_k = 0$. Let $A_k = A + \varepsilon_k I_n$; then the previous argument shows that

$$\det(zI_n - A_k B) = \det(zI_n - BA_k).$$

It is enough to make $k \to \infty$ in the previous relation to deduce that

$$\det(zI_n - AB) = \det(zI_n - BA),$$

and so $\chi_{AB} = \chi_{BA}$. The reader should try to understand this technique very well, since it can be used in many situations. Finally, take $A \in \mathcal{M}_n(\mathbb{C})$ and let $\lambda_1, \lambda_2, \ldots, \lambda_n$ be its eigenvalues, counted with multiplicities. By definition,

$$\det(XI_n - A) = (X - \lambda_1)(X - \lambda_2) \ldots (X - \lambda_n).$$

On the other hand,

$$\det(XI_n - A) = \begin{vmatrix} X - a_{11} & -a_{12} & \ldots & -a_{1n} \\ -a_{21} & X - a_{22} & \ldots & -a_{2n} \\ \ldots & \ldots & \ldots & \ldots \\ -a_{n1} & -a_{n2} & \ldots & X - a_{nn} \end{vmatrix}$$

and a direct expansion shows that the coefficient of X^{n-1} is

$$-(a_{11} + a_{22} + \cdots + a_{nn}) = -\operatorname{tr}(A).$$

Thus the trace of a matrix equals the sum of its eigenvalues, counted with multiplicities. Similarly, the determinant is the product of eigenvalues. Also, if $\varepsilon_1, \ldots, \varepsilon_k$ are the kth roots of unity, we have

$$\det(X^k I_n - A^k) = \det\left(\prod_{j=1}^{k}(XI_n - \varepsilon_j A)\right)$$

$$= \prod_{j=1}^{k} \det(XI_n - \varepsilon_j A) = \prod_{j=1}^{k}\left(\varepsilon_j^n f\left(\frac{X}{\varepsilon_j}\right)\right)$$

$$= \prod_{j=1}^{k}\prod_{i=1}^{n}(X - \lambda_i \varepsilon_j) = \prod_{i=1}^{n}\prod_{j=1}^{k}(X - \lambda_i \varepsilon_j) = \prod_{i=1}^{n}(X^k - \lambda_i^k).$$

Thus $\det(YI_n - A^k) = \prod_{i=1}^{n}(Y - \lambda_i^k)$, which means that the eigenvalues of A^k are the kth powers of the eigenvalues of A. Similarly, one can show that if $g \in \mathbb{C}[X]$, then the eigenvalues of $g(A)$ are $g(\lambda_j)$ for $j = 1, 2, \ldots, n$. The reader is encouraged to prove this.

Problem. Let $A, B \in \mathcal{M}_2(\mathbb{R})$ such that $AB = BA$. Prove that

$$\det(A^2 + B^2) \geq 0.$$

Solution. Let us consider the quadratic function

$$f(x) = \det(A + xB) \ , \quad x \in \mathbb{R}.$$

with real coefficients. More precisely, let $a, b, c \in \mathbb{R}$ be such that

$$f(x) = ax^2 + bx + c.$$

Then

$$\det(A^2 + B^2) = \det(A + iB) \cdot \det(A - iB) = f(i)f(-i)$$
$$= (-a + bi + c)(-a - bi + c) = (c - a)^2 - (bi)^2$$
$$= (c - a)^2 + b^2 \geq 0. \ \square$$

Of course, the previous result holds for $n \times n$ matrices. Indeed, the fact that $f(i)f(-i)$ is a nonnegative real number is not specific for polynomials of degree at most 2, but for any polynomial with real coefficients: all we need is to note that $f(i)f(-i) = |f(i)|^2$.

Recall the following basic property of the quadratic function $f(x) = ax^2 + bx + c$:

$$[x, y \in \mathbb{R}, \ f(x) = f(y)] \Rightarrow \left[x = y \quad \text{or} \quad x + y = -\frac{b}{a}\right].$$

In particular, if for some $x, y \in \mathbb{R}$, $x \neq y$, we have $f(x) = f(y)$, then the abscissa of the extremum point of the corresponding parabola is $\dfrac{x + y}{2}$.

Problem. Let $A, B \in \mathcal{M}_2(\mathbb{R})$. Prove that

$$\det\left(I_2 + \frac{2AB + 3BA}{5}\right) = \det\left(I_2 + \frac{3AB + 2BA}{5}\right).$$

Solution. Let us define the quadratic function (or linear, if $\det(BA - AB) = 0$)

$$f(x) = \det(I_2 + AB + x(BA - AB)).$$

Because AB and BA have the same characteristic polynomial,

$$\det(I_2 + AB) = \det(I_2 + BA),$$

so $f(0) = f(1)$. But f is a quadratic function, thus

$$f(x) = f(1 - x) , \quad \forall x \in \mathbb{R}.$$

In particular,

$$f\left(\frac{2}{5}\right) = f\left(\frac{3}{5}\right),$$

and the conclusion is clear. Note that the same argument shows that

$$\det\left(I_2 + \frac{xAB + yBA}{x + y}\right) = \det\left(I_2 + \frac{yAB + xBA}{x + y}\right),$$

for all real numbers x, y with $x + y \neq 0$. $\square$

The following problem is also based on properties of the polynomial $\det(A + xB)$. However, the arguments are more involved.

Problem. Let n be an odd positive integer and let $A, B \in \mathcal{M}_n(\mathbb{R})$ such that $2AB = (BA)^2 + I_n$. Prove that $\det(I_n - AB) = 0$.

Solution. Consider the polynomial $f(X) = \det(XI_n - AB)$. We know that AB and BA have the same characteristic polynomial, therefore $f(x) = \det(xI_n - BA)$. Using the given relation, we conclude that

$$f\left(\frac{x^2 + 1}{2}\right) = \det\left(\frac{x^2 + 1}{2}I_n - AB\right) = \det\left(\frac{x^2 + 1}{2}I_n - \frac{1}{2}(I_n + (BA)^2)\right)$$

$$= -\frac{f(x)f(-x)}{2^n}.$$

Because $\deg(f)$ is odd, f has at least one real root x. The above relation shows that all terms of the sequence $x_1 = x$, $x_{n+1} = \dfrac{x_n^2 + 1}{2}$ are roots of f. This sequence is increasing and takes only a finite number of distinct values, because f is not 0. Thus there exists n such that $x_n = x_{n+1}$. This implies $x_n = 1$ and so 1 is a root of f, which means that $\det(I_n - AB) = 0$. This finishes the proof. $\square$

The following problem was given in the Romanian Mathematical Olympiad in 1999:

Problem. Let A be a 2×2 matrix with complex entries and let $C(A)$ be the set of matrices commuting with A. Prove that $|\det(A + B)| \geq |\det(B)|$ for all $B \in C(A)$ if and only if $A^2 = O_2$.

Solution. For one implication, let r_1, r_2 be the roots of the polynomial $\det(XI_2 - A)$. By taking the matrices $B = -r_i I_2$ we deduce that $|r_i|^2 \leq 0$ and so $r_i = 0$. Thus $\det(XI_2 - A) = X^2$ and from Hamilton-Cayley's relation, it follows that $A^2 = O_2$. Conversely, suppose that $A^2 = O_2$. We will prove more, namely, that $\det(A + B) = \det(B)$ for all $B \in C(A)$. If $\det(B) \neq 0$, we can write $\det(-B^2) = \det(xA + B) \cdot \det(xA - B)$ for all complex numbers x. This shows that the polynomial $\det(XA + B)$ divides a nonzero constant polynomial, so it is constant. Thus $\det(A + B) = \det(B)$. Now, suppose that $\det(B) = 0$. There exists a sequence x_n that converges to 0 and such that $B + x_n I_2$ is invertible for all n. These matrices also belong to $C(A)$ and so $\det(x_n I_2 + A + B) = \det(x_n I_2 + B)$. It is enough to make $n \to \infty$ in order to deduce that $\det(A + B) = 0 = \det(B)$. This finishes the proof. $\square$

The following problem, from Gazeta Matematică's Contest is a nice blending of linear algebra and analysis.

Problem. Let $A, B \in \mathcal{M}_n(\mathbb{R})$ be matrices with the property that

$$|\det(A + zB)| \leq 1$$

for any complex number z such that $|z| = 1$. Prove that

$$(\det(A))^2 + (\det(B))^2 \leq 1.$$

Solution. Consider the polynomial $f(z) = \det(A + zB)$. As we have seen, f can be written in the form $\det(A) + \cdots + \det(B)z^n$. Now, we claim that if a polynomial

$$g(X) = a_n X^n + \cdots + a_1 X + a_0$$

has the property that $|g(z)| \leq 1$ for all $|z| = 1$, then

$$|a_0|^2 + |a_1|^2 + \cdots + |a_n|^2 \leq 1.$$

Indeed, a small computation, based on the fact that $\int_0^{2\pi} e^{int} dt = 0$ if $n \neq 0$, shows that

$$|a_0|^2 + |a_1|^2 + \cdots + |a_n|^2 = \frac{1}{2\pi} \int_0^{2\pi} |g(e^{it})|^2 dt \leq 1.$$

Using this, we can immediately conclude that $(\det(A))^2 + (\det(B))^2 \leq 1$. $\square$

Here is an outstanding consequence of the fact that a nonzero polynomial has only finitely many zeros:

Problem. Let $A, B \in \mathcal{M}_n(\mathbb{C})$ be matrices such that $AB - BA = A$. Prove that A is nilpotent, that is, there exists $k \geq 1$ such that $A^k = O_n$.

Solution. Let us prove by induction that

$$A^k B - BA^k = kA^k \text{ for } k \geq 1.$$

For $k = 1$, this is clear. Assuming that it is true for k, then $A^{k+1}B - ABA^k = kA^{k+1}$. But $AB = BA + A$, hence $A^{k+1}B - BA^{k+1} = (k+1)A^{k+1}$.

Now, consider the map $\phi : \mathcal{M}_n(\mathbb{C}) \to \mathcal{M}_n(\mathbb{C})$, $\phi(X) = XB - BX$. We obtain $\phi(A^k) = kA^k$ for all k. Now, view ϕ as a linear map on a finite dimensional vector space $\mathcal{M}_n(\mathbb{C})$. If A is not nilpotent, it follows that k is an eigenvalue for ϕ and this is true for all k. This is impossible since it would follow that the characteristic polynomial of ϕ, which is nonzero, has infinitely many zeros, namely, all positive integers k. $\square$

Proposed Problems

1. Let $A, B \in \mathcal{M}_2(\mathbb{R})$ be such that $AB = BA$ and $\det(A^2 + B^2) = 0$. Prove that $\det A = \det B$.
2. Let $A, B \in \mathcal{M}_3(\mathbb{R})$ be such that $AB = BA, \det(A^2 + B^2) = 0$, and $\det(A - B) = 0$. Prove that $\det A = \det B$.
3. Let $U, V \in \mathcal{M}_2(\mathbb{R})$. Prove that

$$\det(U + V) + \det(U - V) = 2 \det U + 2 \det V.$$

4. Let $A, B \in \mathcal{M}_2(\mathbb{R})$ be such that $\det(AB + BA) \leq 0$. Prove that

$$\det(A^2 + B^2) \geq 0.$$

5. Let $A, B \in \mathcal{M}_2(\mathbb{Z})$ be two matrices, at least one of them being singular, i.e., of zero determinant. Prove that if $AB = BA$, then $\det(A^3 + B^3)$ can be written as the sum of two cubes of integers.

6. Let A be a 2×2 matrix with rational entries with the property that

$$\det(A^2 - 2I_2) = 0.$$

Prove that $A^2 = 2I_2$ and $\det A = -2$.

7. Let $A, B \in \mathcal{M}_3(\mathbb{R})$ be such that $\det(AB - BA) = 0$. Prove that

$$\det\left(I_3 + \frac{xAB + yBA}{x + y}\right) = \det\left(I_3 + \frac{yAB + xBA}{x + y}\right),$$

for all real numbers x, y with $x + y \neq 0$.

8. Let $A, B \in \mathcal{M}_2(\mathbb{R})$ be such that $\det(AB - BA) \leq 0$. Prove that

$$\det(I_2 + AB) \leq \det\left(I_2 + \frac{AB + BA}{2}\right).$$

9. Let $A, B, C \in \mathcal{M}_2(\mathbb{R})$ be pairwise commuting matrices so that

$$\det(AB + BC + CA) \leq 0.$$

Prove that $\det(A^2 + B^2 + C^2) \geq 0$.

10. Let B be an $n \times n$ nilpotent matrix with real entries. Prove that

$$\det(I_n + B) = 1.$$

(An $n \times n$ matrix B is called *nilpotent* if $B^k = O_n$, for some positive integer k).

11. Let A be a $n \times n$ matrix with real entries so that $A^{k+1} = O_n$, for a given positive integer k. Prove that

$$\det\left(I_n + \frac{1}{1!}A + \frac{1}{2!}A^2 + \cdots + \frac{1}{k!}A^k\right) = 1.$$

12. Let $A \in \mathcal{M}_n(\mathbb{C})$ and $k \geq 1$ be such that $A^k = O_n$. Prove that there exists $B \in \mathcal{M}_n(\mathbb{C})$ such that $B^2 = I_n - A$.

13. Let A be a 3×3 matrix with integer entries such that the sum of the elements from the main diagonal of the matrix A^2 is zero. Prove that we can find some elements of the matrix A^6 whose arithmetic mean is the square of an integer.

14. Let $A, B \in \mathcal{M}_n(\mathbb{R})$ and suppose that there exists $P \in \mathcal{M}_n(\mathbb{C})$ invertible such that $B = P^{-1}AP$. Prove that there exists $Q \in \mathcal{M}_n(\mathbb{R})$ invertible such that $B = Q^{-1}AQ$.

Solutions

1. Let us consider the polynomial $f(X) = \det(A + XB)$. We know that f can be represented as

$$f(x) = \det B \cdot x^2 + mx + \det A,$$

where $m \in \mathbb{R}$. But $\det(A^2 + B^2) = 0$, so

$$\det(A + iB) \det(A - iB) = 0 \Rightarrow f(i) = f(-i) = 0,$$

hence $f(x) = \alpha(x^2 + 1)$, implying $\det A = \det B \,(= \alpha)$.
2. The polynomial $f(X) = \det(A + XB)$ has degree ≤ 3. From

$$\det(A^2 + B^2) = \det(A + iB) \det(A - iB) = 0,$$

we deduce $f(i) = f(-i) = 0$, so f is divisible by $x^2 + 1$. Hence

$$f(x) = (x^2 + 1)(mx + n),$$

for some real numbers m, n. Further, $f(-1) = \det(A - B) = 0$, so $m = n$ and then $f(x) = m(x^2 + 1)(x + 1)$. Now we can see that $\det A = \det B \,(= m)$.
3. The quadratic function $f(x) = \det(U + xV)$ can be written as

$$f(x) = \det V \cdot x^2 + mx + \det U,$$

for some m. Further,

$$\det(U + V) + \det(U - V) = f(1) + f(-1)$$

$$= (\det V + m + \det U) + (\det V - m + \det U) = 2(\det U + \det V).$$

4. Let us define the quadratic function

$$f(x) = \det(A^2 + B^2 + x(AB + BA)).$$

We have

$$f(1) = \det(A^2 + B^2 + AB + BA) = \det(A + B)^2 \geq 0$$

and

$$f(-1) = \det(A^2 + B^2 - AB - BA) = \det(A - B)^2 \geq 0.$$

Now, note that if $\det(AB + BA) < 0$, then f has a maximum. From $f(-1) \geq 0$ and $f(1) \geq 0$, it follows that $f(x) \geq 0$, for all real numbers $x \in [-1, 1]$. In particular,

$$f(0) \geq 0 \Leftrightarrow \det(A^2 + B^2) \geq 0.$$

In case $\det(AB + BA) = 0$, we also have $f(0) \geq 0$, because f is a linear function (possibly constant).

We also can proceed using the identity

$$\det(U + V) + \det(U - V) = 2 \det U + 2 \det V.$$

By taking $U = A^2 + B^2$, $V = AB + BA$, we obtain

$$\det(A^2 + B^2 + AB + BA) + \det(A^2 + B^2 - AB - BA)$$

$$= 2 \det(A^2 + B^2) + 2 \det(AB + BA)$$

or

$$\det(A + B)^2 + \det(A - B)^2 = 2 \det(A^2 + B^2) + 2 \det(AB + BA).$$

It follows that

$$\det(A^2 + B^2) + \det(AB + BA) \geq 0,$$

which finishes the alternative solution.

5. By symmetry, we may assume that $\det(B) = 0$. Consider the linear function $f(x) = \det(A + xB)$. It can be also written as $a + bx$ for some integers a, b. If z is a root of the equation $z^2 + z + 1 = 0$, then the fact that $AB = BA$ implies the identity $A^3 + B^3 = (A + B)(A + zB)(A + z^2B)$. Thus, $\det(A^3 + B^3) = f(1)f(z)f(z^2) = (a + b)(a + bz)(a + bz^2) = a^3 + b^3$, which shows that $\det(A^3 + B^3)$ is indeed the sum of two cubes.

6. Let us define the quadratic function $f(x) = \det(A - xI_2)$, for all real x. The coefficient of x^2 is equal to $\det(-I_2) = 1$ and

$$0 = \det(A^2 - 2I_2) = \det(A - \sqrt{2}I_2) \det(A + \sqrt{2}I_2),$$

so $f(-\sqrt{2})f(\sqrt{2}) = 0$. Now, because f has rational coefficients, we conclude that $f(\sqrt{2}) = f(-\sqrt{2}) = 0$, so $f(x) = x^2 - 2$. Further,

$$\det(A - xI_2) = x^2 - 2, \tag{3.2}$$

for all real numbers x. For $x = 0$, we obtain $\det A = -2$. The relation (3.2) is the characteristic equation of A, so A verifies it, $A^2 - 2I_2 = 0 \Rightarrow A^2 = 2I_2$.

7. Let us consider the polynomial

$$P(t) = \det(I_3 + tAB + (1-t)BA).$$

We have

$$P(t) = \det(t(AB - BA) + I_3 + BA).$$

The coefficient of t^3 is $\det(AB - BA) = 0$, so $\deg P \le 2$. From the equality $P(0) = P(1)$, it follows that

$$P\left(\frac{1}{2} - \lambda\right) = P\left(\frac{1}{2} + \lambda\right),$$

for all real numbers λ. The conclusion follows by taking

$$\lambda = \frac{1}{2} - \frac{x}{x+y}.$$

8. The axis of symmetry of the parabola $y = f(x)$, where

$$f(x) = \det(x(AB - BA) + I_2 + BA).$$

is $x = \dfrac{1}{2}$, because of $f(0) = f(1)$ and the coefficient of x^2 is $\det(AB - BA) < 0$, so $f(x) < f\left(\dfrac{1}{2}\right)$, for all $x \in \mathbb{R} \setminus \{1/2\}$. In particular, $f(1) < f\left(\dfrac{1}{2}\right)$ which is

$$\det(I_2 + AB) < \det\left(I_2 + \frac{AB + BA}{2}\right).$$

In case $\det(AB - BA) = 0$, f is linear and $f(0) = f(1)$, so f is constant. Consequently,

$$\det(I_2 + AB) = \det\left(I_2 + \frac{AB + BA}{2}\right).$$

9. Consider the polynomial function

$$f(x) = \det(A^2 + B^2 + C^2 + x(AB + BC + CA)).$$

The coefficient of x^2 is equal to $\det(AB + BC + CA) \le 0$, thus the parabola $y = f(x)$ is concave. We will prove that $f(-1) \ge 0$ and $f(2) \ge 0$, and then it will follow that $f(0) \ge 0$. Indeed, we have:

$$f(2) = \det(A^2 + B^2 + C^2 + 2AB + 2BC + 2CA) = \det(A + B + C)^2 \ge 0.$$

On the other side,

$$f(-1) = \det(A^2 + B^2 + C^2 - AB - BC - CA)$$

$$= \frac{1}{4} \det \left[(A - B)^2 + (B - C)^2 + (C - A)^2 \right].$$

Now, if $X = A - B$ and $Y = B - C$, then $C - A = -X - Y$ and further,

$$f(-1) = \frac{1}{4} \det \left[X^2 + Y^2 + (X + Y)^2 \right] = \det(X^2 + XY + Y^2)$$

$$= \det \left[\left(X + \frac{1}{2} Y \right)^2 + \left(\frac{\sqrt{3}}{2} Y \right)^2 \right] \geq 0,$$

by the first problem of the theoretical part (note that X and Y commute).

10. Let us define the polynomial function

$$P(x) = \det(I_n - xB) \ , \quad x \in \mathbb{C}.$$

We will prove that P is constant and so $P(-1) = P(0) = 1$. If $B^k = O_n$, then for all complex numbers x, we have

$$I_n = I_n - (zB)^k = (I_n - zB)(I_n + zB + \cdots + (zB)^{k-1}).$$

By passing to determinants, we deduce that P is a divisor of 1 in $\mathbb{C}[X]$, so it is indeed constant.

11. Consider the polynomial function

$$P(x) = \det \left(I_n + \frac{x}{1!} A + \frac{x^2}{2!} A^2 + \cdots + \frac{x^k}{k!} A^k \right).$$

The degree of P is not greater than kn and P satisfies the relation:

$$P(x)P(y) = P(x + y),$$

for all real numbers x, y. Indeed,

$$P(x)P(y) = \det \left(\sum_{i=0}^{k} \frac{x^i}{i!} A^i \right) \cdot \det \left(\sum_{j=0}^{k} \frac{y^j}{j!} A^j \right)$$

$$= \det \left[\left(\sum_{i=0}^{k} \frac{x^i}{i!} A^i \right) \cdot \left(\sum_{j=0}^{k} \frac{y^j}{j!} A^j \right) \right] = \det \left[\sum_{p=0}^{k} \left(\sum_{i+j=p} \frac{x^i y^j}{i! j!} \right) A^p \right]$$

$$= \det \left(\sum_{p=0}^{k} \frac{(x + y)^p}{p!} A^p \right) = P(x + y).$$

In particular, $P(x)P(-x) = P(0) = 1$, so P is a constant polynomial, because it does not have complex roots. It follows that

$$P(1) = P(0) = 1.$$

We can also proceed in a different way. According to the hypotheses, A is nilpotent. The sum of nilpotent matrices which commute with each other is a nilpotent matrix. In particular, the matrix

$$B = \frac{1}{1!}A + \frac{1}{2!}A^2 + \cdots + \frac{1}{k!}A^k$$

is nilpotent. As we have proved in the previous problem,

$$\det(I_n + B) = \det\left(I_n + \frac{1}{1!}A + \frac{1}{2!}A^2 + \cdots + \frac{1}{k!}A^k\right) = 1.$$

12. We search for B in the form $B = P(A)$ for some $P \in \mathbb{C}[X]$. First, let us prove by induction on k that we can find $P_k \in \mathbb{C}[X]$ such that $X^k | P_k^2(X) + X - 1$. Take $P_1 = 1$ and assume that P_k is found. Let $P_k^2(X) + X - 1 = X^k Q(X)$. If $Q(0) = 0$, take $P_{k+1} = P_k$, so assume that $Q(0) \neq 0$. Let us search for

$$P_{k+1}(X) = P_k(X) + \alpha X^k.$$

Then

$$\begin{aligned} P_{k+1}^2(X) + X - 1 &= P_k^2(X) + X - 1 + 2\alpha X^k P_k(X) + \alpha^2 X^{2k} \\ &= X^k(Q(X) + 2\alpha P_k(X) + \alpha^2 X^k). \end{aligned}$$

It is enough to choose α such that $Q(0) + 2\alpha P_k(0) = 0$, which is possible because $P_k^2(0) = 1 \neq 0$.

Now, let $B = P_k(A)$. We know that

$$P_k^2(X) + X - 1 = Q(x) \cdot X^k$$

for some $Q \in \mathbb{C}[X]$. Thus $B^2 + A - I_n = A^k Q(A) = 0$ and so $B^2 = I_n - A$.

13. Let $f(X) = \det(XI_3 - A) = (X - \lambda_1)(X - \lambda_2)(X - \lambda_3)$ and observe that

$$\det(XI_3 - A^2) = (X - \lambda_1^2)(X - \lambda_2^2)(X - \lambda_3^2).$$

Thus the condition $\operatorname{tr} A^2 = 0$ becomes $\lambda_1^2 + \lambda_2^2 + \lambda_3^2 = 0$, and this implies

$$\lambda_1^6 + \lambda_2^6 + \lambda_3^6 = 3\lambda_1^2\lambda_2^2\lambda_3^2.$$

This can be translated as $\operatorname{tr} A^6 = 3(\det A)^2$ and shows that the arithmetic mean of the elements on the main diagonal of A^6 is a square, namely, $(\det A)^2$.

14. Observe that we can write $P = X + iY$ with $X, Y \in \mathcal{M}_n(\mathbb{R})$. Because $PB = AP$ and $A, B \in \mathcal{M}_n(\mathbb{R})$, we deduce that $AX = XB$ and $AY = YB$. We are not done yet, because we cannot be sure that X or Y is invertible. Nevertheless, the polynomial function $f(x) = \det(xY+X)$ is not identically zero because $f(i) \neq 0$ (P being invertible). Therefore, there exists $x \in \mathbb{R}$ such that $f(x) \neq 0$. Clearly, $Q = X + xY$ answers the question.

Chapter 4
Some Applications of the Hamilton-Cayley Theorem

We saw in a previous chapter that the characteristic polynomial of a matrix and its zeros, the eigenvalues of the matrix, can give precious information. The purpose of this chapter is to present a powerful theorem due to Hamilton and Cayley that gives an even stronger relation between a matrix and its characteristic polynomial: in a certain sense, the matrix is a "root" of its characteristic polynomial. After presenting a proof of this theorem, we investigate some interesting applications.

Theorem (Hamilton-Cayley). *Let K be any field and let $A \in \mathcal{M}_n(K)$ be a matrix. Define*

$$\det(XI_n - A) = X^n + a_{n-1}X^{n-1} + \cdots + a_1X + a_0$$

the characteristic polynomial of A. Then

$$A^n + a_{n-1}A^{n-1} + \cdots + a_1A + a_0I_n = O_n.$$

The proof of this result is not easy and we will prove it here only for the field of complex numbers (which is already nontrivial!). Basically, if A is diagonal, the proof is immediate. So is the case when A is diagonalizable, because by a conjugation, one easily reduces the study to diagonal matrices. In order to solve the general case, we will use a density argument: first of all, we will prove that the set of diagonalizable matrices over $\mathbb{C}$ is dense in $\mathcal{M}_n(\mathbb{C})$. This is not difficult: because the characteristic polynomial of A has complex roots, there exists a complex invertible matrix P such that $P^{-1}AP$ is triangular. But then, it is enough to slightly modify the diagonal elements of this triangular matrix so that they become pairwise distinct. The new matrix is diagonalizable (because it is triangular and has pairwise distinct eigenvalues), and it is immediate that it can approximate the initial matrix arbitrarily well. Thus, the claim is proved. Now, consider A a complex matrix and take A_k a

© Springer Science+Business Media LLC 2017

T. Andreescu et al., *Mathematical Bridges*, DOI 10.1007/978-0-8176-4629-5_4

sequence of diagonalizable matrices that converge to A. Let $X^n + a_{n-1}^{(k)} X^{n-1} + \cdots + a_0^{(k)}$ be their characteristic polynomials. We know that

$$A_k^n + a_{n-1}^{(k)} A_k^{n-1} + \cdots + a_0^{(k)} I_n = O_n, \qquad (*)$$

by the argument in the beginning of the solution. On the other hand, the coefficients $a_j^{(k)}$ are polynomials in the coefficients of the matrix A_k (simply because each coefficient of the characteristic polynomial is a polynomial in the coefficients of the matrix, as can be easily seen from the formula of the determinant). Because A_k converges to A, it follows that each $a_j^{(k)}$ converges to a_k, the coefficient of X^k in the characteristic polynomial of A. But then, by taking the limit in the relation $(*)$, we obtain exactly the desired equation satisfied by A. This finishes the proof. $\square$

Now, observe that one can easily compute a_{n-1} in the above formula: all we have to do is to develop the determinant of $XI_n - A$ and take the coefficient of X^{n-1}. Since

$$\det(XI_n - A) = \sum_{\sigma \in S_n} \epsilon(\sigma) \cdot (X\delta_{1\sigma(1)} - a_{1\sigma(1)}) \ldots (X\delta_{n\sigma(n)} - a_{n\sigma(n)})$$

(here, δ_{ij} equals 1 if $i = j$ and 0 otherwise, while $\epsilon(\sigma) = 1$ if σ is even and -1 otherwise), the only term which has a nonzero coefficient of X^{n-1} is $(X - a_{11})(X - a_{22}) \ldots (X - a_{nn})$, which gives the coefficient $-(a_{11} + a_{22} + \cdots + a_{nn})$. We recognize in the last quantity the opposite of the trace of A. Thus, the Hamilton-Cayley relation becomes

$$A^n - \text{tr}(A) \cdot A^{n-1} + \cdots + (-1)^n \det(A) I_n = O_n.$$

In the very easy case $n = 2$ (for which the proof of the theorem itself is immediate by direct computation), we obtain the useful formula

$$A^2 - \text{tr}(A)A + \det(A)I_2 = O_2.$$

Note already that the Hamilton-Cayley theorem comes handy when computing the powers of a matrix. Indeed, we know that A satisfies an equation of the form

$$A^n + a_{n-1}A^{n-1} + \cdots + a_1 A + a_0 I_n = O_n.$$

By multiplying by A^k, we obtain

$$A^{n+k} + a_{n-1}A^{n+k-1} + \cdots + a_0 A^k = O_n,$$

which means that each entry of $(A^k)_{k \geq 1}$ satisfies a linear recurrence with constant coefficients; thus we can immediately obtain formulas for the coefficients of the powers A^k. The computations are not obvious, but in practice, for 2×2 and 3×3 matrices, this gives a very quick way to compute the powers of a matrix. Let us see in more detail what happens for 2×2 matrices. By the previous observation, to each matrix $A \in \mathcal{M}_2(\mathbb{C})$, we can inductively assign two sequences $(a_n)_{n \geq 1}$, $(b_n)_{n \geq 1}$ of complex numbers for which $A^n = a_n A + b_n I_2$, for all positive integers n. Let us find a recurrence satisfied by these sequences. If we assume that $A^k = a_k A + b_k I_2$, since we also have $A^2 = \alpha A + \beta I_2$, where $\alpha = \operatorname{tr} A$ and $\beta = -\det A$, we can write

$$A^{k+1} = A^k \cdot A = (a_k A + b_k I_2)A = a_k A^2 + b_k A$$
$$= a_k(\alpha A + \beta I_2) + b_k A = (a_k \alpha + b_k)A + a_k \beta I_2,$$

so $A^{k+1} = a_{k+1}A + b_{k+1}I_2$ with

$$\begin{cases} a_{k+1} = a_k \alpha + b_k \\ b_{k+1} = a_k \beta \end{cases}.$$

Here is an application of the above observation:

Problem. Let $A, B \in \mathcal{M}_2(\mathbb{C})$ be two matrices such that $A^n B^n = B^n A^n$ for some positive integer n. Prove that either $AB = BA$ or one of the matrices A^n, B^n has the form aI_2 for some complex number a.

Solution. In the framework of the above results, this should be fairly easy. We saw that we can write $A^n = aA + bI_2$ and $B^n = cB + dI_2$ for some complex numbers a, b, c, d. The condition $A^n B^n = B^n A^n$ becomes, after simple computations, $ac(AB - BA) = O_2$. It is clear that this implies $AB = BA$ or $a = 0$ or $c = 0$. But this is precisely what the problem was asking for. $\square$

The Hamilton-Cayley theorem is not only useful to compute powers of matrices but also to solve equations. Here are some examples.

Problem. a) The solutions of the equation

$$X^2 = -I_2, \quad X \in \mathcal{M}_2(\mathbb{C}) \tag{4.1}$$

are $X = iI_2$, $X = -iI_2$ and any matrix of the form

$$X = \begin{pmatrix} a & b \\ c & -a \end{pmatrix}, \text{ with } a^2 + bc = -1.$$

b) The solutions of the equation $X^2 = I_2$, $X \in \mathcal{M}_2(\mathbb{C})$ are $X = I_2$, $X = -I_2$ and any matrix of the form

$$X = \begin{pmatrix} a & b \\ c & -a \end{pmatrix}, \text{ with } a^2 + bc = 1.$$

Solution. First of all, note that part b) follows immediately by using the next equivalence:

$$X^2 = -I_2 \Leftrightarrow (iX)^2 = I_2.$$

By considering the determinant, $\det X^2 = 1$, we deduce that $\det X = \pm 1$. We therefore have two cases. In the first case, $\det X = 1$, and with the notation $\alpha = \operatorname{tr} X$, we have

$$X^2 - \alpha X + I_2 = O_2.$$

Combining this with equation $X^2 = -I_2$ and taking into account that X is invertible, we obtain $\alpha = 0$. Now, with $\det X = 1$ and $\operatorname{tr} X = 0$, it follows that

$$X = \begin{pmatrix} a & b \\ c & -a \end{pmatrix}, \quad \text{with} \quad a^2 + bc = -1.$$

It is easy to check that all these matrices are solutions. In the second case, $\det X = -1$. With $\alpha = \operatorname{tr} X$, we have

$$X^2 - \alpha X - I_2 = O_2.$$

Therefore, since $X^2 = -I_2$, we have $X = -\dfrac{2}{\alpha} I_2$. By considering the trace, it follows that $\alpha = -\dfrac{4}{\alpha} \Rightarrow \alpha = \pm 2i$. Finally, $X = iI_2$ or $X = -iI_2$. $\square$

With these preparations, we can solve the equation

$$AB + BA = O_2. \tag{4.2}$$

when both matrices A and B are invertible.

Problem. The solutions of the equation $AB + BA = O_2$, with $A, B \in M_2(\mathbb{C})$ invertible, are

$$A = \begin{pmatrix} x & y \\ z & -x \end{pmatrix}, \quad B = A^{-1} \cdot \begin{pmatrix} a & b \\ c & -a \end{pmatrix},$$

where x, y, z, a, b, c are any complex numbers with $x^2 + yz \neq 0$, $a^2 + bc \neq 0$, and $2ax + bz + cy = 0$.

Solution. We can assume, without loss of generality, that $\det(AB) = 1$, by multiplying eventually by a scalar.

We have

$$\operatorname{tr}(AB + BA) = 0 \Rightarrow \operatorname{tr}(AB) + \operatorname{tr}(BA) = 0,$$

and from tr $(AB) = \text{tr}(BA)$, we obtain $\text{tr}(AB) = 0$. We also have $\det AB = 1$ and from the Hamilton-Cayley relation, we obtain

$$(AB)^2 = -I_2 \tag{4.3}$$

Now we can use the previous problem to solve the above equation. If X denotes any solution of the equation (4.1), then $AB = X \Rightarrow B = A^{-1}X$. With $B = A^{-1}X$, the equation (4.2) becomes

$$A \cdot A^{-1}X + A^{-1}X \cdot A = O_2.$$

By multiplying by A to the left, we derive

$$AX + XA = O_2. \tag{4.4}$$

The cases $X = iI_2$ and $X = -iI_2$ are not acceptable, because A is invertible. Consequently, if

$$A = \begin{pmatrix} x & y \\ z & t \end{pmatrix}, \quad X = \begin{pmatrix} a & b \\ c & -a \end{pmatrix}$$

with $x, y, z, t, a, b, c \in \mathbb{C}$, $a^2 + bc = -1$, $xt - yz = 1$, then the condition (4.4) becomes

$$\begin{pmatrix} x & y \\ z & t \end{pmatrix} \begin{pmatrix} a & b \\ c & -a \end{pmatrix} + \begin{pmatrix} a & b \\ c & -a \end{pmatrix} \begin{pmatrix} x & y \\ z & t \end{pmatrix} = O_2.$$

In terms of linear systems, we obtain

$$\begin{cases} 2ax + bz + cy = 0 \\ b(t + x) = 0 \\ c(t + x) = 0 \\ -2at + bz + cy = 0. \end{cases}$$

If $t + x \neq 0$, then $b = c = 0$. From the first and the last equations of the system, we deduce $a = 0$. This is impossible, because $a^2 + bc = -1$.

Thus, $t + x = 0$. In this case, the first and the last equations of the system are equivalent. It follows that

$$A = \begin{pmatrix} x & y \\ z & -x \end{pmatrix}, \quad X = \begin{pmatrix} a & b \\ c & -a \end{pmatrix},$$

for any $x, y, z, a, b, c \in \mathbb{C}$, satisfying

$$a^2 + bc = -1, \quad x^2 + yz = -1, \quad 2ax + bz + cy = 0.$$

The general solution of the given equation is $(\zeta A, \mu B)$, where $\zeta, \mu \in \mathbb{C}^*$. This includes also the case (iA, B), when $\det(iAB) = -\det AB$. $\square$

The following problem has a very short solution using the Hamilton-Cayley theorem, but it is far from obvious.

Problem. Let $A, B, C, D \in \mathcal{M}_n(\mathbb{C})$ such that AC is invertible and $A^k B = C^k D$ for all $k \geq 1$. Prove that $B = D$.

Solution. Let

$$f(x) = \det(xI_n - A) = x^n + \cdots + (-1)^n \det A$$

and

$$g(x) = \det(xI_n - C) = x^n + \cdots + (-1)^n \det C$$

be the respective characteristic polynomials of A and C, and let

$$h = \frac{1}{\det(AC)} fg.$$

Then h has the form $h(x) = b_n x^{2n} + \cdots + b_1 x + 1$, and by the Hamilton-Cayley theorem, we know that

$$b_n A^{2n} + \cdots + b_1 A + I_n = O_n$$

and

$$b_n C^{2n} + \cdots + b_1 C + I_n = O_n.$$

It is enough to multiply the first relation by B and the second one by D and to take the difference. $\square$

Here is a beautiful problem proposed by Laurenţiu Panaitopol for the Romanian National Mathematical Olympiad in 1994.

Problem. For a given $n > 2$, find all 2×2 matrices with real entries such that

$$X^n + X^{n-2} = \begin{pmatrix} 1 & -1 \\ -1 & 1 \end{pmatrix}$$

Solution. By taking the determinant of both sides, we observe that we must have $\det X = 0$ or $\det(X + iI_2) = 0$. Assume for a moment that $\det(X + iI_2) = 0$. Thus the characteristic polynomial of X, which has real coefficients, has $-i$ as a root. Thus, it also has i as a root and these are its roots. That is, the characteristic polynomial equals $X^2 + 1$, and by the Hamilton-Cayley theorem, we deduce that $X^2 + I_2 = O_2$, which is clearly impossible. Thus $\det X = 0$ and if $t = \operatorname{tr} X$, we can write (again using Hamilton-Cayley)

$$X^n + X^{n-2} = (t^{n-1} + t^{n-3})X.$$

Therefore, if $f(x) = x^n + x^{n-2} - 2$, we must have $f(t) = 0$. There are now two cases: either n is even, and in this case, -1 and 1 are the only zeros of f (since f is increasing on the set of positive numbers and decreasing on the set of negative numbers), and those two values $t = 1$ and $t = -1$ give two solutions by the above relation. The second case is when n is odd, and in this case, the function f is increasing and has only one zero, at $t = 1$. This gives the corresponding solution X for odd n. $\square$

We continue with a very useful criterion for nilpotent matrices and some applications of it.

Problem. Let $A \in \mathcal{M}_n(\mathbb{C})$. Prove that A is nilpotent if and only if $\operatorname{tr} A^k = 0$ for all $k \geq 1$.

Solution. If A is nilpotent, then any eigenvalue of A is zero. Indeed, if λ is an eigenvalue of A, we know that λ^k is an eigenvalue of A^k, and if $A^k = O_n$, this implies $\lambda^k = 0$ and so $\lambda = 0$. On the other hand, we saw that $\operatorname{tr} A^k$ is the sum of the kth powers of the eigenvalues of A and so it is 0.

Now, suppose that $\operatorname{tr} A^k = 0$ for $k \geq 1$ and let $\lambda_1, \lambda_2, \ldots, \lambda_n$ be the eigenvalues of A. Thus $\lambda_1^k + \lambda_2^k + \cdots + \lambda_n^k = 0$ for all $k \geq 1$. Because $\dfrac{1}{1-z} = 1 + z + z^2 + \ldots$ if $|z| < 1$, we deduce that if z is small enough,

$$\sum_{i=1}^n \frac{1}{1 - z\lambda_i} = \sum_{i=1}^n \sum_{j=0}^{\infty} z^j \lambda_i^j = \sum_{j=0}^{\infty} z^j \sum_{i=1}^n \lambda_i^j = n.$$

Thus the rational function $\displaystyle\sum_{i=1}^n \frac{1}{1 - z\lambda_i} - n$ is identically zero, because it has infinitely many zeros. If all λ_i are nonzero, this rational fraction has limit $-n$ when $|z| \to \infty$, which is absurd. Thus, we may assume that $\lambda_1 = 0$. But then $\lambda_2^k + \cdots + \lambda_n^k = 0$ for $k \geq 1$, and the previous argument shows that one of $\lambda_2, \ldots, \lambda_n$ is 0. Continuing in this way, we deduce that all λ_i are 0. Hamilton-Cayley's theorem shows then that $A^n = O_n$ and so A is nilpotent. $\square$

We present now two beautiful applications of this criterion.

Problem. Let $A \in \mathcal{M}_n(\mathbb{C})$. Prove that A is nilpotent if and only if there is a sequence A_k of matrices similar to A and having limit O_n.

Solution. Suppose that A_k are similar to A and tend to O_n. Now take $k \geq 1$. Then A_j^k is similar to A^k, thus

$$\operatorname{tr} A^k = \lim_{j \to \infty} \operatorname{tr} A_j^k = \operatorname{tr}\left(\lim_{j \to \infty} A_j^k\right) = 0.$$

By the previous problem, A is nilpotent.

Now, suppose that A is nilpotent. By performing a similarity transformation, we may assume that A is upper triangular, with 0 on the main diagonal (because A is nilpotent). It is easy to see that by performing conjugations with matrices of the form

$$\begin{pmatrix} c & & & \\ & c^2 & & 0 \\ & & \ddots & \\ & 0 & & c^n \end{pmatrix}$$

with $c \to \infty$, we obtain a sequence of matrices that converges to O_n. $\square$

Problem. Let $S \subset M_n(\mathbb{C})$ be a set of nilpotent matrices such that whenever $A, B \in S$, we also have $AB \in S$. Prove that for all $A_1, A_2, \ldots, A_k \in S$, we have

$$(A_1 + A_2 + \cdots + A_k)^n = O_n.$$

Solution. Fix a positive integer r and let $X = A_1 + A_2 + \cdots + A_k$. Observe that X^r is a sum of products of the matrices $A_1, A_2, \ldots, A_k$ thus a sum of elements of S, which are nilpotent. By the criterion, it follows that $\operatorname{tr} X^r = 0$ and the same criterion implies that X is nilpotent. It is enough to apply Hamilton-Cayley to deduce that $X^n = O_n$. $\square$

Proposed Problems

1. For a second order matrix $A = (a_{ij})_{1 \leq i,j \leq 2}$, one can prove the Hamilton-Cayley theorem by computing directly

$$A^2 - (a_{11} + a_{22})A + (a_{11}a_{22} - a_{12}a_{21})I_2$$

and obtaining the result O_2. Find a proof of the Hamilton-Cayley theorem by direct computation in the case of third-order matrices.

2. Let A be a second-order complex matrix such that $A^2 = A$, $A \neq O_2$, and $A \neq I_2$. For positive integers n, solve in $M_2(\mathbb{C})$ the equation $X^n = A$.

3. Let $A \in M_2(\mathbb{C})$ be such that $A^n = O_2$, for some integer $n \geq 2$. Prove that $A^2 = O_2$.

4. Let $A, B \in \mathcal{M}_2(\mathbb{C})$ be two matrices, not of the form λI_2, with $\lambda \in \mathbb{C}$. If $AB = BA$, prove that $B = \alpha A + \beta I_2$, for some complex numbers α, β.

5. Let $A, B \in \mathcal{M}_2(\mathbb{C})$ be two matrices. Assume that for some positive integers m, n, we have $A^m B^n = B^n A^m$, and the matrices A^m and B^n are not of the form λI_2, $\lambda \in \mathbb{C}$. Prove that $AB = BA$.

6. Let $\mathcal{M} = \{A \in \mathcal{M}_2(\mathbb{C}) \mid \det(zI_2 - A) = 0 \Rightarrow |z| < 1\}$ and let $A, B \in \mathcal{M}$ with $AB = BA$. Prove that $AB \in \mathcal{M}$.

7. Let $n \geq 2$ and let $A_0, A_1, \ldots, A_n \in \mathcal{M}_2(\mathbb{R})$ be nonzero matrices such that $A_0 \neq aI_2$ for all $a \in \mathbb{R}$ and $A_0 A_k = A_k A_0$ for all k. Prove that

$$\det\left(\sum_{k=1}^{n} A_k^2\right) \geq 0.$$

8. Let $A, B \in \mathcal{M}_2(\mathbb{C})$ be non-invertible matrices and let $n \geq 2$ such that

$$(AB)^n + (BA)^n = O_2.$$

Prove that $(AB)^2 = O_2$ and $(BA)^2 = O_2$.

9. Let $A \in \mathcal{M}_3(\mathbb{C})$ be with $\operatorname{tr} A = 0$. Prove that $\operatorname{tr}(A^3) = 3 \det A$.

10. Prove that for all $X, Y, Z \in \mathcal{M}_2(\mathbb{R})$,

$$ZXYXY + ZYXYX + XYYXZ + YXXYZ$$

$$= XYXYZ + YXYXZ + ZXYYX + ZYXXY.$$

11. Is there a bijective function $f : \mathcal{M}_2(\mathbb{C}) \to \mathcal{M}_3(\mathbb{C})$ such that $f(XY) = f(X)f(Y)$ for all $X, Y \in \mathcal{M}_2(\mathbb{C})$?

12. Let $A, B \in \mathcal{M}_2(\mathbb{C})$ be such that $\operatorname{tr}(AB) = 0$. Prove that $(AB)^2 = (BA)^2$.

13. Let $A, B \in \mathcal{M}_2(\mathbb{C})$ be such that $(AB)^2 = A^2 B^2$. Prove that $(AB - BA)^2 = O_2$.

14. Let $A, B \in \mathcal{M}_n(\mathbb{C})$, with $A + B$ non-invertible, such that $AB + BA = O_n$. Prove that $\det(A^3 - B^3) = 0$.

15. Let $A, B \in \mathcal{M}_n(\mathbb{R})$ such that $3AB = 2BA + I_n$. Prove that $I_n - AB$ is nilpotent.

16. Let $A, B, C \in \mathcal{M}_n(\mathbb{R})$, where n is not divisible by 3, such that

$$A^2 + B^2 + C^2 = AB + BC + CA.$$

Prove that

$$\det\left[(AB - BA) + (BC - CB) + (CA - AC)\right] = 0.$$

17. Let N be the $n \times n$ matrix with all its elements equal to $1/n$, and let $A = (a_{ij})_{1 \leq i,j \leq n} \in \mathcal{M}_n(\mathbb{R})$ be such that for some positive integer k, $A^k = N$. Prove that

$$\sum_{1 \leq i,j \leq n} a_{ij}^2 \geq 1.$$

18. Let $A \in \mathcal{M}_n(\mathbb{C})$, and let adj($A$) denote the adjugate (the transpose of the cofactor matrix) of A. Show that if, for all $k \in \mathbb{N}$, $k \geq 1$, we have

$$\det((\text{adj}(A))^k + I_n) = 1,$$

then $(\text{adj}(A))^2 = O_n$.

Solutions

1. Let $A = (a_{ij})_{1 \leq i,j \leq 3}$ be a 3×3 matrix; we intend to prove that

$$A^3 - s_1 A^2 + s_2 A - s_3 I_3 = O_3,$$

where $s_1 = a_{11} + a_{22} + a_{33}$,

$$s_2 = \begin{vmatrix} a_{22} & a_{23} \\ a_{32} & a_{33} \end{vmatrix} + \begin{vmatrix} a_{11} & a_{13} \\ a_{31} & a_{33} \end{vmatrix} + \begin{vmatrix} a_{11} & a_{12} \\ a_{21} & a_{22} \end{vmatrix},$$

and $s_3 = \det A$. The obvious problem is the calculation of A^3, which seems to be (at least) very unpleasant.

However, we can avoid this laborious calculation! Namely, assume first that A is nonsingular (hence invertible). In this case, the equality that we need to prove is equivalent to

$$A^2 - s_1 A + s_2 I_3 - s_3 A^{-1} = O_3.$$

Taking into account that

$$s_3 A^{-1} = \det(A) A^{-1} = \text{adj}(A) = \begin{pmatrix} d_{11} & -d_{21} & d_{31} \\ -d_{12} & d_{22} & -d_{32} \\ d_{13} & -d_{23} & d_{33} \end{pmatrix},$$

where d_{ij} is the cofactor of a_{ij}, for all $i,j \in \{1, 2, 3\}$ (thus $s_2 = d_{11} + d_{22} + d_{33}$), checking the equation by direct computation becomes a reasonable (and feasible) task. For instance, for the entry in the second row and first column, we need to verify that

$$(a_{21}a_{11} + a_{22}a_{21} + a_{23}a_{31}) - (a_{11} + a_{22} + a_{33})a_{21} + d_{12} = 0,$$

which is true (remember that $d_{12} = a_{21}a_{33} - a_{23}a_{31}$). In a similar manner, we verify that each entry of $A^2 - s_1 A + s_2 I_3 - s_3 A^{-1}$ equals 0.

In order to finish the proof (which is, for the moment, done only for nonsingular matrices), we use a standard argument (that we have already seen before). Namely, suppose that A is a singular matrix, and consider, for each number x, the matrix $A_x = A - xI_3$; A_x is nonsingular for every x in a (small enough) neighborhood of the origin (because $\det(A - xI_3) = 0$ for at most three distinct values of x). Thus the Hamilton-Cayley theorem works for every such A_x (with x in that neighborhood). Letting x tend to 0, we get the validity of the theorem for matrix A as well.

A proof of the Hamilton-Cayley theorem that generalizes the proof for second-order matrices to arbitrary $n \times n$ matrices can be found in the article *"A Computational Proof of the Cayley-Hamilton Theorem"* by Constantin-Nicolae Beli, in *Gazeta Matematică, seria A*, 3-4/2014.

2. The case $n = 1$ being clear, assume that $n \geq 2$. Clearly, A is not invertible, because otherwise, $A = I_2$. Thus $(\det X)^n = \det A = 0$ and so $\det X = 0$. Thus, if $\alpha = \text{tr} X$, by Hamilton-Cayley, we have $X^2 = \alpha X$ and by induction, $X^n = \alpha^{n-1} X$. Clearly, $\alpha \neq 0$ because otherwise, $A = O_2$. Thus $X = \alpha^{1-n} A$, that is, there exists $z \in \mathbb{C}^*$ with $X = zA$. The condition becomes $z^n A^n = A$. But $A^2 = A$ implies $A^n = A$ and since $A \neq 0$, we must have $z^n = 1$. Thus the solutions are $X_k = e^{\frac{2i\pi k}{n}} A$ with $0 \leq k \leq n - 1$.

3. By considering the determinant, we obtain $\det A^n = 0$, so $\det A = 0$. With $\alpha = \text{tr} A$, the Hamilton-Cayley relation becomes $A^2 = \alpha A$, and inductively, $A^n = \alpha^{n-1} \cdot A$, for all integers $n \geq 2$. Now, $A^n = O_2 \Rightarrow \alpha^{n-1} \cdot A = O_2$, so $\alpha = 0$ or $A = O_2$. In both cases, using also $A^2 = \alpha A$, we obtain $A^2 = O_2$.

4. If we denote $A = \begin{pmatrix} a & b \\ c & d \end{pmatrix}$, $B = \begin{pmatrix} x & y \\ z & t \end{pmatrix}$, then from $AB = BA$, we derive

$$\frac{y}{b} = \frac{z}{c} = \frac{x - t}{a - d} = \alpha.$$

(with the convention that if a denominator is zero, then the corresponding numerator is also zero). At least one of the fractions is not $\frac{0}{0}$, because b, c cannot be both zero. Thus α is well defined. It follows that

$$y = \alpha b, \quad z = \alpha c, \quad t = \alpha d + \beta, \quad x = \alpha a + \beta,$$

with $\beta = t - \alpha d$. Finally,

$$B = \begin{pmatrix} x & y \\ z & t \end{pmatrix} = \begin{pmatrix} \alpha a + \beta & \alpha b \\ \alpha c & \alpha d + \beta \end{pmatrix} = \alpha A + \beta I_2.$$

5. As we stated, we can define sequences $(a_n)_{n \geq 1}, (b_n)_{n \geq 1}, (c_n)_{n \geq 1}, (d_n)_{n \geq 1}$ of complex numbers such that

$$A^k = a_k A + b_k I_2, \quad B^k = c_k B + d_k I_2,$$

for all positive integers k. From the hypothesis, $a_m \neq 0$ and $c_n \neq 0$. Then

$$A^m B^n = B^n A^m$$

$$\Rightarrow (a_m A + b_m I_2)(c_n B + d_n I_2) = (c_n B + d_n I_2)(a_m A + b_m I_2)$$

$$\Rightarrow a_m c_n (AB - BA) = O_2.$$

Hence $AB = BA$, because $a_m c_n \neq 0$.

6. First, assume that $A = \alpha I_2$ with $|\alpha| < 1$. Let y_1, y_2 be the roots of the equation $\det(B - xI_2) = 0$; then the roots of $\det(AB - xI_2) = 0$ are αy_1 and αy_2. Clearly $|\alpha y_1| = |\alpha| \cdot |y_1| < 1$ and $|\alpha y_2| < 1$.

Secondly, assume that $A \neq \alpha I_2$ for all $\alpha \in \mathbb{C}$. We saw in the previous problem 4 that there are $a, b \in \mathbb{C}$ with $B = aA + bI_2$.

Let x_1, x_2 be the roots of $\det(A - xI_2) = 0$; then the roots of $\det(B - xI_2) = 0$ are $y_1 = ax_1 + b$, $y_2 = ax_2 + b$ and the roots of $\det(AB - xI_2) = 0$ are $z_1 = x_1 y_1$, $z_2 = x_2 y_2$. Thus again $|z_i| = |x_i| \cdot |y_i| < 1$.

7. By problem 4, there are $\alpha_k, \beta_k \in \mathbb{R}$ with $A_k = \alpha_k A_0 + \beta_k I_2$. If all $\alpha_k = 0$, then

$$\det \left(\sum_{k=1}^{n} A_k^2 \right) = \sum_{k=1}^{n} \beta_k^2 \geq 0.$$

Otherwise, let $\alpha = \sum_{k=1}^{n} \alpha_k^2$. Then

$$\sum_{k=1}^{n} A_k^2 = \alpha A_0^2 + 2 \left(\sum_{k=1}^{n} \alpha_k \beta_k \right) A_0 + \left(\sum_{k=1}^{n} \beta_k^2 \right) I_2 = f(A_0)$$

where

$$f(x) = \alpha x^2 + 2 \left(\sum_{k=1}^{n} \alpha_k \beta_k \right) x + \sum_{k=1}^{n} \beta_k^2.$$

By the Cauchy-Schwarz inequality, the discriminant of f is negative or zero. Thus we can write $f(z) = \alpha(z - z_0)(z - \bar{z}_0)$, and so

$$\det \left(\sum_{k=1}^{n} A_k^2 \right) = \det f(A_0) = \alpha^2 |\det(A_0 - z_0 I_2)|^2 \geq 0.$$

8. With $\det AB = \det BA = 0$, and setting $\alpha = \operatorname{tr} AB = \operatorname{tr} BA$, we have, by Hamilton-Cayley,

$$(AB)^2 = \alpha AB, \quad (BA)^2 = \alpha BA. \tag{4.1}$$

By induction,

$$(AB)^n = \alpha^{n-1}AB, \quad (BA)^n = \alpha^{n-1}BA,$$

for all positive integers n. The given equation $(AB)^n + (BA)^n = O_2$ yields

$$\alpha^{n-1}(AB + BA) = O_2.$$

If $\alpha = 0$, then the conclusion follows directly from (4.1). So suppose that $AB + BA = O_2$. Then

$$2\operatorname{tr}(AB) = \operatorname{tr}(AB + BA) = 0,$$

and again, $\alpha = 0$. In any case, $(AB)^2 = O_2$ and similarly for BA.

9. By the Hamilton-Cayley relation, we obtain the existence of $\alpha \in \mathbb{C}$ with

$$A^3 - \operatorname{tr}A \cdot A^2 + \alpha \cdot A - \det A \cdot I_3 = O_3.$$

By considering the trace, we deduce that

$$\operatorname{tr}(A^3) - \operatorname{tr}A \cdot \operatorname{tr}(A^2) + \alpha \cdot \operatorname{tr}A - \det A \cdot \operatorname{tr}I_3 = 0$$

and with $\operatorname{tr}A = 0$, $\operatorname{tr}I_3 = 3$, the conclusion follows.

Alternatively, let a, b, c be the eigenvalues of A. Then $a + b + c = 0$ is given and we have to prove that $a^3 + b^3 + c^3 = 3abc$, which is immediate.

10. Observe that the identity can be written in a more appropriate form

$$Z(XY - YX)^2 = (XY - YX)^2 Z.$$

This is clear by the Hamilton-Cayley theorem, because $\operatorname{tr}(XY - YX) = 0$, so $(XY - YX)^2 = \alpha I_2$ for some α.

11. The answer is negative. Suppose that f is such a function and consider

$$A = \begin{pmatrix} 0 & 1 & 0 \\ 0 & 0 & 1 \\ 0 & 0 & 0 \end{pmatrix}.$$

One can easily check that $A^3 = O_3$ and $A^2 \neq O_3$. Let $B \in \mathcal{M}_2(\mathbb{C})$, $B = f^{-1}(A)$. Then $(f(B))^3 = O_3$, thus $f(B^3) = O_3$ and so $B^3 = O_2$, because $f(O_2) = O_3$ (indeed, let $X_0 = f^{-1}(O_3)$; then $f(X_0) = O_3$, and so $f(2X_0) = f(2I_2)f(X_0) = O_3 = f(X_0)$, thus $2X_0 = X_0$ and $X_0 = O_2$). But previous exercise 3 implies that $B^2 = O_2$, thus $O_3 = f(O_2) = f(B^2) = (f(B))^2 = A^2$, a contradiction.

12. By the Hamilton-Cayley relation for the matrices AB and BA, we have

$$(AB)^2 - \text{tr}(AB) \cdot AB + \det(AB) \cdot I_2 = O_2$$

and

$$(BA)^2 - \text{tr}(BA) \cdot BA + \det(BA) \cdot I_2 = O_2.$$

By a well-known property of the trace, we have $\text{tr}(AB) = \text{tr}(BA) = 0$, so

$$(AB)^2 = -\det(AB) \cdot I_2, \quad (BA)^2 = -\det(BA) \cdot I_2.$$

But $\det(AB) = \det(BA)$, hence $(AB)^2 = (BA)^2$.

13. Using the Hamilton-Cayley relation for the matrix $AB - BA$, we have

$$(AB - BA)^2 - \text{tr}(AB - BA) \cdot (AB - BA) + \det(AB - BA) \cdot I_2 = O_2$$

or

$$(AB - BA)^2 = \lambda I_2, \tag{4.1}$$

with $\lambda = -\det(AB - BA)$ because $\text{tr}(AB - BA) = 0$. From (4.1), it follows that

$$\text{tr}(AB - BA)^2 = 2\lambda. \tag{4.2}$$

Now the conclusion of the problem follows from (4.1) if we can show that $\lambda = 0$ or equivalently, according to (4.2), $\text{tr}(AB - BA)^2 = 0$. In this sense, note that

$$(AB - BA)^2 = (AB)^2 - ABBA - BAAB + (BA)^2. \tag{4.3}$$

By the hypothesis, $\text{tr}\,(AB)^2 = \text{tr}(A^2B^2)$. Using $\text{tr}(UV) = \text{tr}(VU)$, it follows that $\text{tr}(AB)^2 = \text{tr}(BA)^2$ and

$$\text{tr}(ABB \cdot A) = \text{tr}(B \cdot AAB) = \text{tr}(A^2B^2).$$

Finally, from (4.3),

$$\text{tr}(AB - BA)^2 = \text{tr}\,(AB)^2 - \text{tr}(A^2B^2) - \text{tr}(A^2B^2) + \text{tr}(AB)^2$$
$$= 2\,\text{tr}\left[(AB)^2 - A^2B^2\right] = 0.$$

14. Let $x \in \mathbb{C}^n$, $x \neq 0$ be such that

$$(A + B)x = O_n.$$

It follows that $Ax = -Bx$, so, using yet the identity $AB = -BA$, we have

$$A^3x = -A^2Bx = ABAx = -BA^2x = B^3x.$$

Hence $(A^3 - B^3)x = O_n$ for some nonzero vector $x \in \mathbb{C}^n$, so $\det(A^3 - B^3) = 0$. A direct proof is provided by the equality

$$A^3 - B^3 = (A + B)(A^2 - AB - B^2).$$

(Prove it by using $AB + BA = O_n$!) From $\det(A + B) = 0$, the conclusion follows.

15. Let $f(x) = \det(xI_n - AB)$ be the characteristic polynomial of AB (and BA, too). Then

$$f(x) = \det\left(xI_n - \frac{2BA + I_n}{3}\right) = \frac{2^n}{3^n}\det\left(\frac{3x-1}{2}I_n - BA\right)$$

$$= \frac{2^n}{3^n}\det\left(\frac{3x-1}{2}I_n - AB\right) = \frac{2^n}{3^n}f\left(\frac{3x-1}{2}\right).$$

Thus, if z is a root of f, so are the terms of the sequence

$$x_1 = z, \ x_{n+1} = \frac{3x_n - 1}{2}.$$

Because f has a finite number of roots, the previous sequence has finitely many terms, which easily implies $x_1 = 1$. Thus all roots of f are equal to 1 and so $f(x) = (x - 1)^n$. By Hamilton-Cayley, $(AB - I_n)^n = O_n$.

16. Let us consider the complex number

$$\varepsilon = \cos\frac{2\pi}{3} + i\sin\frac{2\pi}{3}$$

and define the matrix

$$X = A + \varepsilon B + \varepsilon^2 C.$$

We have

$$\overline{X} = A + \overline{\varepsilon}B + \overline{\varepsilon}^2C = A + \varepsilon^2 B + \varepsilon C.$$

Then the number

$$\det(X \cdot \overline{X}) = \det X \cdot \det \overline{X} = \det X \cdot \overline{\det X} = |\det X|^2 \geq 0$$

is real and nonnegative. On the other hand, by hypothesis and by $\varepsilon^2 + \varepsilon + 1 = 0$, we get

$$X \cdot \overline{X} = (A + \varepsilon B + \varepsilon^2 C)(A + \varepsilon^2 B + \varepsilon C)$$

$$= (\varepsilon^2 + 1)(AB + BC + CA) + \varepsilon(BA + CB + AC)$$

$$= \varepsilon\left[(AB - BA) + (BC - CB) + (CA - AC)\right].$$

Now, because $\det(X \cdot \overline{X}) \in \mathbb{R}$, we infer that

$$\varepsilon^n \det\left[(AB - BA) + (BC - CB) + (CA - AC)\right] \in \mathbb{R},$$

so $\det\left[(AB - BA) + (BC - CB) + (CA - AC)\right] = 0$.

17. **Solution I.** First note that the equality $A^k = N$ implies $AN = NA$ and this shows that the sums of entries in each row and column of A are all equal. Then, if a is the common value of these sums, an easy induction shows that A^s (for positive integer s) has the same property as A, that is, all sums of entries in each row and column are equal to the same number, which is a^s. In particular, since $A^k = N$, $a^k = 1$ follows; thus, being real, a can be either 1 or -1.

Next, one immediately sees that $\displaystyle\sum_{1 \leq i,j \leq n} a_{ij}^2$ represents the sum of entries from the principal diagonal of AA^t, A^t being the transpose of A; that is, $\displaystyle\sum_{1 \leq i,j \leq n} a_{ij}^2$ is the trace of AA^t, which also equals the sum of the eigenvalues of AA^t. But, clearly, AA^t is a symmetric real matrix, implying that all its eigenvalues are nonnegative real numbers.

Finally, for the vector u with all entries 1, we have $Au = u$ and $A^t u = u$, or $Au = -u$ and $A^t u = -u$ (remember that the sums of entries in each row and in each column of A are either all equal to 1 or are all equal to -1). In both cases, $AA^t u = u$ follows, and this shows that 1 is one of the eigenvalues of AA^t (with corresponding eigenvector u).

Let us summarize: AA^t has all eigenvalues nonnegative, and one of them is 1. Of course, this being the case, their sum (i.e., the trace of AA^t) is at least 1, and thus we can finish our proof:

$$\sum_{1 \leq i,j \leq n} a_{ij}^2 = \mathrm{tr}(AA^t) \geq 1.$$

This is problem 341 proposed by Lucian Țurea in *Gazeta Matematică, seria A*, 3-4/2011, and here follows his solution.

Solution II. For any real $n \times n$ matrix $X = (x_{ij})_{1 \le i,j \le n}$, define

$$m(X) = \sum_{1 \le i,j \le n} x_{ij}^2;$$

thus $m(X)$ is the sum of the squares of all entries of X. One can prove that $m(XY) \le m(X)m(Y)$ for all $X, Y \in \mathcal{M}_n(\mathbb{R})$, with the obvious consequence $m(X^s) \le (m(X))^s$ for every real matrix X and every positive integer s. In our case,

$$1 = m(N) = m(A^k) \le (m(A))^k$$

follows, that is, $m(A) \ge 1$—which is precisely what we wanted to prove.

For the sake of completeness, let us prove that $m(XY) \le m(X)m(Y)$ for all real $n \times n$ matrices $X = (x_{ij})_{1 \le i,j \le n}$ and $Y = (y_{ij})_{1 \le i,j \le n}$. Let $XY = (z_{ij})_{1 \le i,j \le n}$. Thus

$$z_{ij}^2 = \left(\sum_{k=1}^n x_{ik} y_{kj} \right)^2 \le \left(\sum_{k=1}^n x_{ik}^2 \right) \left(\sum_{k=1}^n y_{kj}^2 \right),$$

by the Cauchy-Schwarz inequality. Summing upon j, we have

$$\sum_{j=1}^n z_{ij}^2 \le \left(\sum_{k=1}^n x_{ik}^2 \right) \left(\sum_{j=1}^n \sum_{k=1}^n y_{kj}^2 \right) = \left(\sum_{k=1}^n x_{ik}^2 \right) m(Y),$$

and, if we sum up again, this time over i, we get exactly the desired result $m(XY) \le m(X)m(Y)$.

Of course, this solution shows that in place of N, we can consider any real matrix such that the sum of the squares of its entries is (at least) 1, and the statement of the problem remains true. On the other hand, from the first solution, we see that for any real matrix A such that AA^t has an eigenvalue equal to 1, the sum of the squares of the entries of A is at least 1. We used the equality $A^k = N$ only to infer that AA^t has such an eigenvalue. Thus each solution provides a kind of generalization of the original problem statement.

18. Let $P(x) = \det(xI_n - A) = x^n + c_{n-1}x^{n-1} + \cdots + c_0$ be the characteristic polynomial of the matrix A, and let $\lambda_1, \ldots, \lambda_n$ be the eigenvalues of A (i.e., the zeros of P). The following three facts are well-known and we will use them soon (prove the first and the second!).

1) The eigenvalues of adj(A) are $\lambda_2 \cdots \lambda_n, \ldots, \lambda_1 \cdots \lambda_{n-1}$ (that is, the products of $n - 1$ of the eigenvalues of A).
2) adj$(A) = (-1)^{n-1}(A^{n-1} + c_{n-1}A^{n-2} + \cdots + c_1 I_n)$.
3) Hamilton-Cayley theorem: $A^n + c_{n-1}A^{n-1} + \cdots + c_1 A + c_0 I_n = O_n$.

Now we proceed to solve the problem. Let $\mu_1 = \lambda_2 \cdots \lambda_n, \ldots, \mu_n = \lambda_1 \cdots \lambda_{n-1}$ be the eigenvalues of the adjugate of A. For any positive integer k, the matrix $(\mathrm{adj}(A))^k + I_n$ has the eigenvalues $\mu_1^k + 1, \ldots, \mu_n^k + 1$ (also well-known), and the determinant of any such matrix is the product of its eigenvalues. Thus the condition given in the hypothesis can be written in the form

$$(\mu_1^k + 1) \cdots (\mu_n^k + 1) = 1$$

for all natural numbers $k \geq 1$. After expanding and using Newton's formulas, one sees that this happens if and only if $\mu_1 = \cdots = \mu_n = 0$. This means that

$$c_1 = (-1)^{n-1}(\mu_1 + \cdots + \mu_n) = 0,$$

and at least one of $\lambda_1, \ldots, \lambda_n$ is zero, hence

$$c_0 = (-1)^n \det(A) = (-1)^n \lambda_1 \cdots \lambda_n = 0.$$

Now, the Hamilton-Cayley theorem yields

$$A^n + c_{n-1}A^{n-1} + \cdots + c_2 A^2 = O_n$$

and from the second fact mentioned in the beginning, we get

$$(\mathrm{adj}(A))^2 = (A^{n-1} + c_{n-1}A^{n-2} + \cdots + c_2 A)^2 =$$

$$= (A^n + c_{n-1}A^{n-1} + \cdots + c_2 A^2)(A^{n-2} + c_{n-1}A^{n-3} + \cdots + c_2 I_n) = O_n,$$

because the first factor is O_n. The solution ends here.

Note that the core of the problem is the fact that if the adjugate of a matrix is nilpotent (which we obtained when we found that all its eigenvalues are 0), then actually the square of the adjugate is the zero matrix. This can be also proved as follows. Because $\mathrm{adj}(A)$ is nilpotent, there exists a positive integer s such that $(\mathrm{adj}(A))^s = O_n$, and the determinant of $\mathrm{adj}(A)$ is zero, which means that A is singular too (because $A\mathrm{adj}(A) = \det(A)I_n$). If the rank of A is less than $n - 1$, then $\mathrm{adj}(A)$ is the zero matrix and there is nothing else to prove. Otherwise, Sylvester's inequality shows that $\mathrm{adj}(A)$ has rank at most 1, and, in fact, the rank is 1, because $\mathrm{adj}(A)$ is not the zero matrix—look for problem 3 in the Chapter 5. Thus (see problem 4 in the same chapter), there exist $B \in \mathcal{M}_{n,1}(\mathbb{C})$ and $C \in \mathcal{M}_{1,n}(\mathbb{C})$ such that $\mathrm{adj}(A) = BC$. In this case, we have $CB = t \in \mathbb{C}$.

The equality $(\mathrm{adj}(A))^s = O_n$ can now be written as $(BC)^s = O_n$, or $B(CB)^{s-1}C = O_n$, or $t^{s-1}\mathrm{adj}(A) = O_n$, and it implies $t = 0$ (as the adjugate is not the zero matrix, and only when $s \geq 2$—actually when $s \geq 3$—there is really something to prove). Thus $CB = 0$, and $(\mathrm{adj}(A))^2 = (BC)^2 = B(CB)C = O_n$.

This is problem 360, proposed by Marius Cavachi and Cezar Lupu, in *Gazeta Matematică, seria A*, 1-2/2012 (and solved in 1-2/2013 by the second method).

Chapter 5
A Decomposition Theorem Related to the Rank of a Matrix

Here, for sake of simplicity, we often assume that the matrices we are dealing with are square matrices. Indeed, an arbitrary matrix can be transformed into a square matrix by attaching zero rows (columns), without changing its rank. Let us consider for the beginning the following operations on a square matrix, which does not change its rank:

1. permutation of two rows (columns).
2. multiplication of a row (column) with a nonzero real number.
3. addition of a row (column) multiplied by a real number to another row (column).

We will call these operations *elementary operations*. We set the following problem: are these elementary operations of algebraic type? For example, we ask if the permutation of the rows (columns) i and j of an arbitrary matrix A is in fact the result of multiplication to the left (right) of the matrix A with a special matrix denoted U_{ij}. If such a matrix U_{ij} exists, then it should have the same effect on the identity I_n. Hence the matrix $U_{ij}I_n$ is obtained from the identity matrix by permutating the rows i and j. But $U_{ij}I_n = U_{ij}$, so

© Springer Science+Business Media LLC 2017
T. Andreescu et al., *Mathematical Bridges*, DOI 10.1007/978-0-8176-4629-5_5

$$U_{ij} = \begin{pmatrix} 1 & & & & & & & & \\ & \ddots & & & & & & & \\ & & 1 & & & & & & \\ & & & 0 \dots \dots \dots 1 \dots \dots \dots & & & & \dots i \\ & & & \vdots & 1 & & \vdots & & \vdots \\ & & & \vdots & & \ddots & \vdots & & \vdots \\ & & & \vdots & & & 1 \; \vdots & & \vdots \\ & & & 1 \dots \dots \dots 0 \dots \dots \dots & & & & \dots j \\ & & & & 1 & & & & \\ & & & & & & \ddots & & \\ & & & & & & & 1 \end{pmatrix}$$

Thus U_{ij} is the matrix obtained from the identity matrix by interchanging its ith and jth rows (all the missing entries in the above expression of U_{ij} are equal to 0). Now it can be easily seen that the matrix $U_{ij}A$, respectively AU_{ij} is the matrix A with the rows, respectively the columns i and j permuted. The matrix U_{ij} is invertible, because $U_{ij}^2 = I_n$. Moreover, $\det U_{ij} = -1$, because the permutation of two rows (columns) changes the sign of the determinant. In an analogous way, we now search for a matrix $V_i(\alpha)$ for which the multiplication with an arbitrary matrix A leads to the multiplication of the ith row (column) of A by the nonzero real α. In particular, $V_i(\alpha)I_n$ will be the identity matrix having the ith row multiplied by α. But $V_i(\alpha)I_n = V_i(\alpha)$, so $V_i(\alpha)$ must be the matrix obtained by performing one single change to the identity matrix; namely, its element 1 from the ith row and ith column is replaced by α:

$$V_i(\alpha) = \begin{pmatrix} 1 & & & & & & \\ & \ddots & & & & & \\ & & 1 & & & & \\ & & & \alpha \dots \dots \dots & & & \dots i \\ & & & 1 & & & \\ & & & & & \ddots & \\ & & & & & & 1 \end{pmatrix}$$

Now it can be easily seen that the matrix $V_i(\alpha)A$, respectively $AV_i(\alpha)$ is the matrix A having the ith row, respectively the ith column multiplied by α. Obviously, $\det V_i(\alpha) = \alpha \neq 0$, so the matrix $V_i(\alpha)$ is invertible. Similarly, let us remark that if, for $i \neq j$, we add the jth row of the identity matrix multiplied by λ to the ith row, we obtain the matrix (again, all missing entries are zeros):

$$W_{ij}(\lambda) = \begin{pmatrix} 1 & & & & & & \\ & \ddots & & & & & \\ & & 1 \dots \lambda \dots \dots & & & \dots i \\ & & & 1 & \vdots & & \vdots \\ & & & & 1 \dots \dots & & \dots j \\ & & & & & \ddots & \\ & & & & & & 1 \end{pmatrix} \begin{matrix} \\ \\ \dots i \\ \vdots \\ \dots j \\ \\ \end{matrix}$$

Now, we can easily see that the matrix $W_{ij}(\lambda)A$, respectively $AW_{ij}(\lambda)$ is obtained from the matrix A by adding the jth row multiplied by λ to the ith row, respectively by adding the ith column multiplied by λ to the jth column.

All the matrices U_{ij}, $V_i(\alpha)$, $W_{ij}(\lambda)$, $\alpha \in \mathbb{R}^*$, $\lambda \in \mathbb{R}$ are invertible, and we will call them *elementary matrices*. Now we can give the following basic result:

Theorem 1. *Each matrix $A \in \mathcal{M}_n(\mathbb{C})$ can be represented in the form*

$$A = PQR,$$

where $P, R \in \mathcal{M}_n(\mathbb{C})$ are invertible and $Q = \begin{pmatrix} I_r & 0 \\ 0 & 0 \end{pmatrix} \in \mathcal{M}_n(\mathbb{C})$, with $r = \mathrm{rank}(A)$.

To prove this, let us first note that every matrix A can be transformed into a matrix Q by applying the elementary operations 1–3. If, for example, $a_{11} \neq 0$, then multiply the first column by a_{11}^{-1} to obtain 1 on the position $(1, 1)$. Then add the first line multiplied by $-a_{i1}$ to the ith row, $i \geq 2$ to obtain zeros in the other places of the first column. Similarly, we can obtain zeros in the other places of the first row. Finally, a matrix Q is obtained, and in algebraic formulation, we can write

$$S_1 \dots S_p A T_1 \dots T_q = Q,$$

where S_i, T_j, $1 \leq i \leq p$, $1 \leq j \leq q$ are elementary matrices. Hence

$$A = (S_1 \dots S_p)^{-1} Q (T_1 \dots T_q)^{-1}$$

and we can take

$$P = (S_1 \dots S_p)^{-1}, \quad R = (T_1 \dots T_q)^{-1}. \quad \square$$

The rank is invariant under elementary operations, so

$$\mathrm{rank}(A) = \mathrm{rank}(Q) = r.$$

We can see that for every matrix X, the matrix QX, respectively XQ is the matrix X having all elements of the last $n - r$ rows, respectively the last $n - r$ columns equal to zero. Theorem 1 is equivalent to the following:

Proposition. *Let $A, B \in \mathcal{M}_n(\mathbb{C})$. Then* $\text{rank}(A) = \text{rank}(B)$ *if and only if there exist invertible matrices $X, Y \in \mathcal{M}_n(\mathbb{C})$ such that $A = XBY$.*

If $\text{rank}(A) = r$, then $\text{rank}(A) = \text{rank}(Q)$, and according to theorem 1, there exist X, Y invertible such that $A = XQY$. By multiplication by an invertible matrix X, the rank remains unchanged. Indeed, this follows from the proposition and from the relations

$$XB = XBI_n, \quad BX = I_nBX. \quad \square$$

As a direct consequence, we give a proof of the famous Sylvester's inequality:

Proposition. *If $A, B \in \mathcal{M}_n(\mathbb{C})$, then*

$$\text{rank}(AB) \geq \text{rank}(A) + \text{rank}(B) - n.$$

Let $r_1 = \text{rank}(A)$, $r_2 = \text{rank}(B)$ and consider the decompositions

$$A = P_1Q_1R_1, \quad B = P_2Q_2R_2,$$

with P_i, R_i invertible, $\text{rank}(Q_i) = r_i$, $i = 1, 2$. Then

$$AB = P_1 (Q_1R_1P_2Q_2) R_2,$$

so

$$\text{rank}(AB) = \text{rank} (Q_1R_1P_2Q_2).$$

The matrix $Q_1R_1P_2Q_2$ is obtained from the (invertible) matrix R_1P_2 by replacing the last $n - r_1$ rows and last $n - r_2$ columns with zeros. Consequently,

$$\text{rank}(AB) \geq n - (n - r_1) - (n - r_2) = r_1 + r_2 - n. \quad \square$$

We continue with some applications of the above results. The first one was a problem proposed in 2002 at the Romanian National Mathematical Olympiad:

Problem. Let $A \in \mathcal{M}_n(\mathbb{C})$ be a matrix such that $\text{rank}(A) = \text{rank}(A^2) = r$ for some $1 \leq r \leq n - 1$. Prove that $\text{rank}(A^k) = r$ for all positive integers k.

Solution. Let us consider the decomposition $A = PQR$, as in theorem 1. Then $A^2 = P(QRPQ)R$, and we can easily see that $QRPQ = \begin{pmatrix} D & 0 \\ 0 & 0 \end{pmatrix}$ for some $D \in \mathcal{M}_r(\mathbb{C})$. The hypothesis implies that D is invertible. Now, observe that

$A^2 = PXQR$, where X is the invertible matrix with diagonal blocks D and I_{n-r}. Thus

$$A^3 = PXQRPQR = PXXQR = PX^2QR.$$

An immediate induction shows that $A^n = PX^{n-1}QR$, and thus $\text{rank}(A^k) = r$, because X, P, and R are invertible. $\square$

The next example is a very classical result, due to Moore-Penrose about the generalized inverse of a matrix. It appeared, however, as a problem in the Romanian National Mathematical Olympiad in 2005:

Problem. Prove that for any matrix $A \in \mathcal{M}_n(\mathbb{C})$, there exists a matrix B such that $ABA = A$ and $BAB = B$. Prove that if A is not invertible, then such a matrix is not uniquely determined.

Solution. Let us again write A in the form $A = PQR$, as in theorem 1. Let us search for B in the form $B = R^{-1}XP^{-1}$. This form is natural, since by writing the conditions $ABA = A$ and $BAB = B$, the matrix RBP, denoted by X, plays a central role. With this substitution, the conditions that should be satisfied by B are simply $QXQ = Q$ and $XQX = X$. This shows that we can already assume that $A = Q$! Now, let us search for X as a block matrix: $\begin{pmatrix} U & V \\ W & T \end{pmatrix}$.

The above conditions on X become $U = I_r$ and $T = WV$. Thus, we have surely at least one solution (just take $X = Q$), and moreover, if $r < n$, there are infinitely many choices for the matrices V, W, thus infinitely many such X. Therefore, if A is not invertible, it has infinitely many generalized inverses. $\square$

Problem. Let $A \in \mathcal{M}_n(\mathbb{C})$ be a matrix such that

$$\det(A + X) = \det(A) + \det(X)$$

for all $X \in \mathcal{M}_n(\mathbb{C})$. If $n \geq 2$, prove that $A = O_n$.

Solution. Take $X = A$ to deduce that $(2^n - 2)\det(A) = 0$; thus, $\det(A) = 0$. Therefore, if $A \neq O_n$, we can write $A = PQR$, where $P, R \in GL_n(\mathbb{C})$ and $Q = \begin{pmatrix} I_r & 0 \\ 0 & 0 \end{pmatrix}$, $r = \text{rank}(A) \in \{1, 2, \ldots, n-1\}$. Let $X = PSR$ where $S = I_n - Q$. Then $\det(X) = 0$ because $I_n - Q$ is diagonal and has at least one zero on the main diagonal. However, $\det(A+X) = \det(P) \cdot \det(R) \neq 0$, a contradiction. Thus $A = O_n$. $\square$

Problem. Let $A, B \in \mathcal{M}_n(\mathbb{C})$. Prove that the map $f : \mathcal{M}_n(\mathbb{C}) \to \mathcal{M}_n(\mathbb{C})$, $f(X) = AX - XB$ is bijective if and only if A and B have no common eigenvalues.

Solution. Because $\mathcal{M}_n(\mathbb{C})$ is a finite-dimensional $\mathbb{C}$-vector space, f is bijective if and only if it is injective.

Suppose that A, B have no common eigenvalues and let $X \neq O_n$ be such that $f(X) = O_n$, that is, $AX = XB$. Let $X = PQR$ with $P, R \in GL_n(\mathbb{C})$ and $Q = \begin{pmatrix} I_r & 0 \\ 0 & 0 \end{pmatrix}$, $1 \leq r \leq n$.

Write now $P^{-1}AP$ and RBR^{-1} as block matrices

$$P^{-1}AP = \begin{pmatrix} A_1 & A_2 \\ A_3 & A_4 \end{pmatrix}, \quad RBR^{-1} = \begin{pmatrix} B_1 & B_2 \\ B_3 & B_4 \end{pmatrix}.$$

The condition $AX = XB$ becomes $P^{-1}APQ = QRBR^{-1}$. This in turn implies $A_3 = 0, B_2 = 0, A_1 = B_1$. Thus

$$P^{-1}AP = \begin{pmatrix} A_1 & A_2 \\ 0 & A_4 \end{pmatrix}, \quad RBR^{-1} = \begin{pmatrix} A_1 & 0 \\ B_3 & B_4 \end{pmatrix}.$$

By the multiplicative property of determinants in block-triangular matrices and the fact that similar matrices have the same characteristic polynomial, it follows that

$$\chi_A = \chi_{A_1} \chi_{A_4}, \quad \chi_B = \chi_{A_1} \chi_{B_4},$$

where χ_Z is the characteristic polynomial of Z. But since $A_1 \in \mathcal{M}_r(\mathbb{C})$, χ_{A_1} has at least one zero, which will be a common eigenvalue of A and B, a contradiction. This shows one implication.

Assume now that A and B have a common eigenvalue λ and observe that one can make A and B triangular with λ on the first place of the main diagonal by suitable similarities, That is, there are $P, R \in GL_n(\mathbb{C})$ such that

$$P^{-1}AP = \begin{pmatrix} \lambda & x_{12} & \dots & x_{1n} \\ 0 & x_{22} & \dots & x_{2n} \\ \vdots & \vdots & \ddots & \vdots \\ 0 & \dots & \dots & x_{nn} \end{pmatrix}, \quad RBR^{-1} = \begin{pmatrix} \lambda & 0 & \dots & 0 \\ y_{21} & y_{22} & \dots & 0 \\ \vdots & \vdots & \ddots & \vdots \\ y_{n1} & y_{n2} & \dots & y_{nn} \end{pmatrix}.$$

But then we easily see that $f(X) = 0$, with

$$X = P \begin{pmatrix} 1 & & & \\ & 0 & & 0 \\ & & \ddots & \\ 0 & & & 0 \end{pmatrix} R \neq 0,$$

thus f is not injective in this case. $\square$

We end this theoretical part with a classical result.

Problem. Let $A \in \mathcal{M}_n(\mathbb{C})$. Find, as a function of A, the smallest integer k such that A can be written as a sum of k matrices of rank 1.

Solution. As you can easily guess, $k = \text{rank}(A)$. Indeed, if A has rank equal to r, then $A = PQR$ for some $P, R \in GL_n(\mathbb{C})$ and $Q = \begin{pmatrix} I_r & 0 \\ 0 & 0 \end{pmatrix}$. Thus

$$A = A_1 + A_2 + \cdots + A_r,$$

where

$$A_i = P \begin{pmatrix} 0 & & & & & & \\ & \ddots & & & & & \\ & & 1 & & & 0 & \\ & & & \ddots & & & \\ & & & & 0 & & \\ & & & & & \ddots & \\ & & 0 & & & & 0 \end{pmatrix} R,$$

the only 1 being on position (i, i). Clearly, $\text{rank}(A_i) = 1$ because

$$\begin{pmatrix} 0 & & & & 0 \\ & \ddots & & & \\ & & 1 & & \\ & & & \ddots & \\ 0 & & & & 0 \end{pmatrix}$$

has rank 1 and $P, R \in GL_n(\mathbb{C})$. Thus $k \leq r$.

On the other hand, $\text{rank}(X + Y) \leq \text{rank}(X) + \text{rank}(Y)$ thus if $A = A_1 + \cdots + A_k$ with $\text{rank}(A_i) = 1$, we have

$$r = \text{rank}(A) = \text{rank}\left(\sum_{i=1}^{k} A_i\right) \leq \sum_{i=1}^{k} \text{rank}(A_i) = k.$$

It follows that, actually, $k = \text{rank}(A)$. $\square$

Proposed Problems

1. Consider the matrices

$$
A = \begin{pmatrix} a & b \\ c & d \end{pmatrix}, \quad
B = \begin{pmatrix} x & y \\ z & t \end{pmatrix}, \quad
C = \begin{pmatrix}
ax & bx & az & bz \\
ay & by & at & bt \\
cx & dx & cz & dz \\
cy & dy & ct & dt
\end{pmatrix}
$$

with complex entries. Prove that if $\operatorname{rank} A = \operatorname{rank} B = 2$, then $\operatorname{rank} C = 4$.

2. Let $A = (a_{ij})_{1 \le i,j \le n}$, $B = (b_{ij})_{1 \le i,j \le n}$ be so that

$$
a_{ij} = 2^{i-j} \cdot b_{ij}
$$

for all integers $1 \le i,j \le n$. Prove that $\operatorname{rank} A = \operatorname{rank} B$.

3. Let $A \in \mathcal{M}_n(\mathbb{C})$ be singular. Prove that the rank of the adjugate matrix $\operatorname{adj}(A)$ (the transpose of the cofactor matrix, sometimes called the adjoint) of A equals 0 or 1.

4. Let $A \in \mathcal{M}_n(\mathbb{C})$ be with $\operatorname{rank}(A) = r$, $1 \le r \le n - 1$. Prove that there exist $B \in \mathcal{M}_{n,r}(\mathbb{C})$, $C \in \mathcal{M}_{r,n}(\mathbb{C})$ with

$$
\operatorname{rank}(B) = \operatorname{rank}(C) = r,
$$

such that $A = BC$. Deduce that every matrix A of rank r satisfies a polynomial equation of degree $r + 1$.

5. Let $A \in \mathcal{M}_n(\mathbb{C})$, $A = (a_{ij})_{1 \le i,j \le n}$ be with $\operatorname{rank} A = 1$. Prove that there exist complex numbers $x_1, x_2, \ldots, x_n, y_1, y_2, \ldots, y_n$ such that $a_{ij} = x_i y_j$ for all integers $1 \le i,j \le n$.

6. Let $A, B \in \mathcal{M}_n(\mathbb{C})$ be two matrices such that $AB = A + B$. Prove that $\operatorname{rank}(A) = \operatorname{rank}(B)$.

7. Let A be a complex $n \times n$ matrix such that $A^2 = A^*$, where A^* is the conjugate transpose of A. Prove that the matrices A and $A + A^*$ have equal ranks.

8. Let $A \in \mathcal{M}_n(\mathbb{C})$ be a matrix with $\operatorname{rank}(A) = 1$. Prove that

$$
\det(I_n + A) = 1 + \operatorname{tr}(A).
$$

Moreover,

$$
\det(\lambda I_n + A) = \lambda^n + \lambda^{n-1} \cdot \operatorname{tr}(A),
$$

for all complex numbers λ.

9. Let $A \in \mathcal{M}_n(\mathbb{Z})$, $n \ge 3$, with $\det A = 1$ and let $B \in \mathcal{M}_n(\mathbb{Z})$ have all its entries equal to 1. If $\det(I + AB) = 1$, prove that the sum of all entries of the matrix A is equal to 0.

10. Let $A \in \mathcal{M}_n(\mathbb{R})$, $A = (a_{ij})$, and let X be the square matrix of order n with all entries equal to a real number x. Prove that

$$\det(A + X) \cdot \det(A - X) \leq (\det A)^2.$$

11. Let $m \leq n$ be positive integers, and let $A \in \mathcal{M}_{m,n}(\mathbb{R})$ and $B \in \mathcal{M}_{n,m}(\mathbb{R})$ be matrices such that rank A = rank B = m. Show that there exists an invertible matrix $C \in \mathcal{M}_n(\mathbb{R})$ such that $ACB = I_m$, where I_m denotes the $m \times m$ identity matrix.

12. Let A and B be complex $n \times n$ matrices having the same rank and such that $BAB = A$. Prove that $ABA = B$ (i.e., B is a pseudoinverse of A).

13. Let A and B be complex $n \times n$ matrices of the same rank. Show that if $A^2B = A$, then $B^2A = B$.

14. Prove that if A is an $n \times n$ complex matrix with zero trace, then there exist $n \times n$ complex matrices X and Y such that $A = XY - YX$.

Solutions

1. We have to prove that if A, B are invertible, then C is invertible. If

$$(ad - bc)(xt - yz) \neq 0,$$

then

$$\begin{vmatrix} ax & bx & az & bz \\ ay & by & at & bt \\ cx & dx & cz & dz \\ cy & dy & ct & dt \end{vmatrix} = - \begin{vmatrix} ax & az & bx & bz \\ ay & at & by & bt \\ cx & cz & dx & dz \\ cy & ct & dy & dt \end{vmatrix}$$

$$= - \begin{vmatrix} a & 0 & b & 0 \\ 0 & a & 0 & b \\ c & 0 & d & 0 \\ 0 & c & 0 & d \end{vmatrix} \cdot \begin{vmatrix} x & z & 0 & 0 \\ y & t & 0 & 0 \\ 0 & 0 & x & z \\ 0 & 0 & y & t \end{vmatrix} = \begin{vmatrix} a & b & 0 & 0 \\ 0 & 0 & a & b \\ c & d & 0 & 0 \\ 0 & 0 & c & d \end{vmatrix} \cdot \begin{vmatrix} x & z & 0 & 0 \\ y & t & 0 & 0 \\ 0 & 0 & x & z \\ 0 & 0 & y & t \end{vmatrix}$$

$$= - \begin{vmatrix} a & b & 0 & 0 \\ c & d & 0 & 0 \\ 0 & 0 & a & b \\ 0 & 0 & c & d \end{vmatrix} \cdot \begin{vmatrix} x & z & 0 & 0 \\ y & t & 0 & 0 \\ 0 & 0 & x & z \\ 0 & 0 & y & t \end{vmatrix} = -(ad - bc)^2(xt - yz)^2 \neq 0.$$

2. We will show here only that $\det A = \det B$, because in the same way, it can be proved that every minor of A is equal to the corresponding minor of B. We have, using the definition of the determinant,

$$\det A = \sum_{\sigma \in S_n} \varepsilon(\sigma) a_{1\sigma(1)} a_{2\sigma(2)} \cdots a_{n\sigma(n)}$$

$$= \sum_{\sigma \in S_n} \varepsilon(\sigma) (2^{1-\sigma(1)} b_{1\sigma(1)}) (2^{2-\sigma(2)} b_{2\sigma(2)}) \cdots (2^{n-\sigma(n)} b_{n\sigma(n)})$$

$$= \sum_{\sigma \in S_n} \varepsilon(\sigma) \cdot 2^{1+2+\cdots+n-(\sigma(1)+\sigma(2)+\cdots+\sigma(n))} \cdot b_{1\sigma(1)} b_{2\sigma(2)} \cdots b_{n\sigma(n)}$$

$$= \sum_{\sigma \in S_n} \varepsilon(\sigma) b_{1\sigma(1)} b_{2\sigma(2)} \cdots b_{n\sigma(n)} = \det B.$$

Alternatively, note that $B = DAD^{-1}$, where

$$D = \begin{pmatrix} 2^1 & & & \\ & 2^2 & & 0 \\ & & \ddots & \\ & 0 & & 2^n \end{pmatrix}.$$

3. If $\operatorname{rank}(A) \le n - 2$, then $\operatorname{adj}(A) = O_n$, because all minors of order $n - 1$ of the matrix A are equal to zero. Further, let us assume that $\operatorname{rank}(A) = n - 1$. Then

$$\operatorname{rank}(A\operatorname{adj}(A)) \ge \operatorname{rank}(A) + \operatorname{rank}(\operatorname{adj}(A)) - n.$$

Using also $A\operatorname{adj}(A) = O_n$, we derive

$$0 \ge (n - 1) + \operatorname{rank}(\operatorname{adj}(A)) - n$$

so $\operatorname{rank}(\operatorname{adj}(A)) \le 1$.

4. Let $A = PQR$, where P, R are invertible and

$$Q = \begin{pmatrix} I_r & 0 \\ 0 & 0 \end{pmatrix}.$$

Because $Q^2 = Q$, we have $A = (PQ)(QR)$. Observe that $\operatorname{rank}(PQ) = \operatorname{rank}(Q) = r$ and $\operatorname{rank}(QR) = \operatorname{rank}(Q) = r$ because P, Q are invertible. Also, we can write $P = \begin{pmatrix} B & 0 \end{pmatrix}$ and $QR = \begin{pmatrix} C \\ 0 \end{pmatrix}$ for $B \in \mathcal{M}_{n,r}(\mathbb{C})$ and $C \in \mathcal{M}_{r,n}(\mathbb{C})$. Clearly, $BC = A$ and $\operatorname{rank}(B) = \operatorname{rank}(C) = r$.

For the second part, we use the Hamilton-Cayley theorem. For the matrix $\overline{A} = CB \in \mathcal{M}_r(\mathbb{C})$, we can find complex numbers $a_1, \ldots, a_r$ such that

$$\overline{A}^r + a_1 \overline{A}^{r-1} + \cdots + a_r I = 0.$$

By multiplying by B to the left and by C to the right, we obtain

$$B\overline{A}^r C + a_1 B\overline{A}^{r-1} C + \cdots + a_r BC = 0.$$

Now,

$$B\overline{A}^k C = (BC)^{k+1} = A^{k+1}, \quad 1 \le k \le r,$$

so

$$A^{r+1} + a_1 A^r + \cdots + a_r A = 0.$$

5. According to the previous problem, in case $r = 1$, there exist two matrices

$$B \in \mathcal{M}_{n,1}(\mathbb{C}), \quad C \in \mathcal{M}_{1,n}(\mathbb{C})$$

so that $A = BC$. If

$$B = \begin{pmatrix} x_1 \\ x_2 \\ \cdots \\ x_n \end{pmatrix}, \quad C = (y_1 \ y_2 \ \cdots \ y_n),$$

then

$$A = \begin{pmatrix} x_1 \\ x_2 \\ \cdots \\ x_n \end{pmatrix} (y_1 \ y_2 \ \cdots \ y_n) = \begin{pmatrix} x_1 y_1 & x_1 y_2 & \cdots & x_1 y_n \\ x_2 y_1 & x_2 y_2 & \cdots & x_2 y_n \\ \cdots & \cdots & \cdots & \cdots \\ x_n y_1 & x_n y_2 & \cdots & x_n y_n \end{pmatrix}.$$

6. The given equality can be written as

$$(A - I)(B - I) = I,$$

so the matrices $A - I$ and $B - I$ are invertible. Now the conclusion follows from the equality $A = (A - I)B$, taking into account that $A - I$ is invertible.

7. First we prove that $I_n + A$ is invertible. Let x be any (column) vector such that $(I_n + A)x = 0$ (the 0 from the right is the $n \times 1$ zero matrix). Thus $Ax = -x$, and $A^2 x = -Ax = x$ and $A^* Ax = -A^* x = -A^2 x = -x$. Because all eigenvalues of $A^* A$ are nonnegative (indeed, if $A^* Av = \alpha v$, with nonzero vector v, then $(Av)^*(Av) = \alpha v^* v$, and $v^* v > 0$, $(Av)^*(Av) \ge 0$), it follows that $x = 0$, meaning that $I_n + A$ is invertible.

Next, $A + A^* = A + A^2 = (I_n + A)A$, with nonsingular $I_n + A$, and the conclusion follows. This is problem 1793 by Götz Trenkler in Mathematics Magazine 2/2008 (solved in 2/2009 by Eugene A. Herman).

8. Let us consider the matrices B, C of rank 1,

$$
B = \begin{pmatrix} x_1 \\ x_2 \\ \cdots \\ x_n \end{pmatrix} \in \mathcal{M}_{n,1}(\mathbb{C}), \quad C = \begin{pmatrix} y_1 & y_2 & \cdots & y_n \end{pmatrix} \in \mathcal{M}_{1,n}(\mathbb{C})
$$

such that

$$
A = BC = \begin{pmatrix} x_1 \\ x_2 \\ \cdots \\ x_n \end{pmatrix} \begin{pmatrix} y_1 & y_2 & \cdots & y_n \end{pmatrix} = \begin{pmatrix} x_1y_1 & x_1y_2 & \cdots & x_1y_n \\ x_2y_1 & x_2y_2 & \cdots & x_2y_n \\ \cdots & \cdots & \cdots & \cdots \\ x_ny_1 & x_ny_2 & \cdots & x_ny_n \end{pmatrix}.
$$

We also have $A = B'C'$, where

$$
B' = \begin{pmatrix} x_1 & 0 & \cdots & 0 \\ \cdots & \cdots & \cdots & \cdots \\ x_n & 0 & \cdots & 0 \end{pmatrix} \in \mathcal{M}_n(\mathbb{C}), \quad C' = \begin{pmatrix} y_1 & \cdots & y_n \\ 0 & \cdots & 0 \\ \cdots & \cdots & \cdots \\ 0 & \cdots & 0 \end{pmatrix} \in \mathcal{M}_n(\mathbb{C}).
$$

Then

$$
\det(I_n + A) = \det(I_n + B'C') = \det(I_n + C'B')
$$

$$
= \begin{vmatrix} 1 + \sum_{k=1}^{n} x_ky_k & 0 & \cdots & 0 \\ 0 & 1 & \cdots & 0 \\ \cdots & \cdots & \cdots & \cdots \\ 0 & 0 & \cdots & 1 \end{vmatrix} = 1 + \sum_{k=1}^{n} x_ky_k = 1 + \mathrm{tr}(A).
$$

The other equality can be obtained by changing A with $\lambda^{-1}A$. Indeed,

$$
\det(I_n + \lambda^{-1}A) = 1 + \mathrm{tr}(\lambda^{-1}A)
$$

$$
\Rightarrow \frac{1}{\lambda^n} \det(\lambda I_n + A) = 1 + \frac{1}{\lambda} \mathrm{tr}(A)
$$

$$
\Rightarrow \det(\lambda I_n + A) = \lambda^n + \lambda^{n-1} \mathrm{tr}(A).
$$

9. The matrix A is invertible, so

$$\mathrm{rank}(AB) = \mathrm{rank}(B) = 1.$$

A surprising property is that the sum of all entries of the matrix A is equal to $\mathrm{tr}(AB)$. Indeed, with the notation $C = AB$, we have

$$c_{ij} = \sum_{k=1}^{n} a_{ik} b_{kj} = \sum_{k=1}^{n} a_{ik},$$

for all $1 \le i, j \le n$. Thus

$$\mathrm{tr}(AB) = \mathrm{tr}(C) = \sum_{i=1}^{n} c_{ii} = \sum_{i=1}^{n} \left(\sum_{k=1}^{n} a_{ik} \right) = \sum_{i,k=1}^{n} a_{ik},$$

which is the sum of all elements of the matrix A.

By a previous problem, we have $\mathrm{tr}(AB) + 1 = \det(I + AB) = 1$, so $\mathrm{tr}(AB) = 0$.

10. First assume that A is invertible. In this case,

$$\det(A + X) \cdot \det(A - X) = (\det A)^2 \det(I_n + A^{-1}X) \det(I_n - A^{-1}X),$$

with

$$\mathrm{rank}(A^{-1}X) = \mathrm{rank}(X) = 1.$$

Consequently,

$$(\det A)^2 \det(I_n + A^{-1}X) \det(I_n - A^{-1}X)$$
$$= (\det A)^2 [1 + \mathrm{tr}(A^{-1}X)][1 - \mathrm{tr}(A^{-1}X)]$$
$$= (\det A)^2 [1 - (\mathrm{tr}(A^{-1}X))^2]$$
$$= (\det A)^2 - (\det A)^2 (\mathrm{tr}(A^{-1}X))^2 \le (\det A)^2.$$

Now we can see that the result remains true under the weaker condition $\mathrm{rank}(X) = 1$. In case $\det A = 0$, replace A by $A + \varepsilon I_n$ and let $\varepsilon \to 0$.

11. If $m = n$, there is nothing to prove (just choose $C = A^{-1}B^{-1}$), so we consider further that $m < n$.

Let P be an $n \times n$ permutation matrix such that the determinant of the submatrix of AP with entries at the intersections of its m rows and first m columns is nonzero. Let M be the $(n - m) \times n$ matrix consisting of two blocks as follows:

$$M = \left(O_{n-m,\,m} \; I_{n-m} \right)$$

and let A_1 be the $n \times n$ matrix

$$A_1 = \begin{pmatrix} AP \\ M \end{pmatrix} ;$$

using Binet's rule for computing determinants, one sees that $\det A_1 \neq 0$; hence A_1 is invertible in $\mathcal{M}_n(\mathbb{R})$.

Similarly, because B has rank m, there exists an $n \times n$ permutation matrix Q such that QB has a nonsingular submatrix with entries at the intersections of its first m rows and its m columns. Let

$$N = \begin{pmatrix} O_{m,\,n-m} \\ I_{n-m} \end{pmatrix}$$

and

$$B_1 = \left(QB \; N \right) ;$$

one sees that B_1 is an invertible $n \times n$ matrix.

We consider $C_1 = A_1^{-1} B_1^{-1}$; thus we have

$$I_n = A_1 C_1 B_1 = \begin{pmatrix} AP \\ M \end{pmatrix} C_1 \left(QB \; N \right) = \begin{pmatrix} APC_1QB & APC_1N \\ MC_1QB & MC_1N \end{pmatrix},$$

whence (by reading the equality for the upper left $m \times m$ corner) $I_m = APC_1QB$ follows. Now, for $C = PC_1Q$ (which is an $n \times n$ matrix), we get $ACB = I_m$ and the proof is complete.

This is problem 335 Proposed by Vasile Pop in *Gazeta Matematică, seria A*, 1-2/2011.

12. We consider invertible $n \times n$ matrices M, N, P, Q such that $A = MJN$ and $B = PJQ$, where J is the $n \times n$ matrix having the identity matrix I_r (of order r, the rank of A and B) in the upper left corner and all the other entries equal to zero:

$$J = \begin{pmatrix} I_r & O_{r,\,n-r} \\ O_{n-r,\,r} & O_{n-r} \end{pmatrix} .$$

From $BAB = B$, we conclude $PJQMJNPJQ = PJQ$; hence $JQMJNPJ = J$. We write QM and NP in the form

$$QM = \begin{pmatrix} R & S \\ T & U \end{pmatrix}$$

and

$$NP = \begin{pmatrix} V & X \\ Y & Z \end{pmatrix},$$

where R and V are $r \times r$ matrices, S and X are $r \times (n - r)$ matrices, T and Y are $(n - r) \times r$ matrices, and U and Z are $(n - r) \times (n - r)$ matrices. Then the equality $JQMJNPJ = J$ becomes

$$\begin{pmatrix} I_r & O_{r,\,n-r} \\ O_{n-r,\,r} & O_{n-r} \end{pmatrix} \begin{pmatrix} R & S \\ T & U \end{pmatrix} \begin{pmatrix} I_r & O_{r,\,n-r} \\ O_{n-r,\,r} & O_{n-r} \end{pmatrix} \cdot$$

$$\cdot \begin{pmatrix} V & X \\ Y & Z \end{pmatrix} \begin{pmatrix} I_r & O_{r,\,n-r} \\ O_{n-r,\,r} & O_{n-r} \end{pmatrix} = \begin{pmatrix} I_r & O_{r,\,n-r} \\ O_{n-r,\,r} & O_{n-r} \end{pmatrix}$$

and it yields $RV = I_r$. Of course, this implies $VR = I_r$, too, and this is equivalent to $JNPJQMJ = J$, or $MJNPJQMJN = MJN$, that is, $ABA = A$.

13. **Solution I.** For $X \in \mathcal{M}_n(\mathbb{C})$, we also denote by X the linear transform (from $\mathbb{C}^n$ to $\mathbb{C}^n$) having matrix X; then $\ker(X)$ and $\mathrm{im}(X)$ represent the kernel (null space) and the range of this linear transform, respectively. Since $A^2B = A$, $\ker(B)$ is contained in $\ker(A)$; having equal dimensions (because of the equality of the ranks), these subspaces of $\mathbb{C}^n$ (of which one is a subspace of the other) must be equal. From the equality $A(AB - I_n) = O_n$, it follows that $\mathrm{im}(AB - I_n)$ is included in $\ker(A)$; therefore it is also included in $\ker(B)$; this yields $B(AB - I_n) = O_n$, which means that $BAB = B$. According to the result of the previous problem, $ABA = A$ follows.

We can rewrite this as $A(BA - I_n) = O_n$ and thus conclude that $\mathrm{im}(BA - I_n)$ is included in $\ker(A) = \ker(B)$; therefore $B(BA - I_n) = O_n$, which is equivalent to $B^2A = B$, and finishes the proof.

Solution II. We have $\mathrm{rank}(A) = \mathrm{rank}(A^2B) \le \mathrm{rank}(A^2) \le \mathrm{rank}(A)$; hence $\mathrm{rank}(A^2) = \mathrm{rank}(A)$. Because the null space of A is contained in the null space of A^2, it follows that these spaces are equal. Also, $\ker(B) = \ker(A)$, as we saw in the first solution; hence $\ker(B) = \ker(A) = \ker(A^2)$.

Now $A^2B = A$ implies $A^2BA = A^2$, or $A^2(BA - I_n) = O_n$, showing that $\mathrm{im}(BA - I_n)$ is contained in $\ker(A^2) = \ker(B)$. Consequently, $B(BA - I_n) = O_n$, which is equivalent to $B^2A = B$.

This is problem 11239 proposed by Michel Bataille in *The American Mathematical Monthly*, 6/2006. The second solution (by John W. Hagood) was published in the same magazine, number 9/2008.

14. Note first that if $a_1, \ldots, a_n$ are complex numbers that sum to 0, there exist *distinct* complex numbers $x_1, \ldots, x_n$ and a permutation σ of $\{1, \ldots, n\}$ such that $a_k = x_k - x_{\sigma(k)}$ for every $1 \le k \le n$. Indeed, this is clearly true for $n = 1$ (when there is nothing to prove) and for $n = 2$ (when we can write $a_1 = x_1 - x_2$

and $a_2 = x_2 - x_1$, with $x_1 \neq x_2$ if $a_1 \neq 0$, while if $a_1 = 0$, then $a_2 = 0$, too, and we can take $a_1 = 0 = x_1 - x_1$ and $a_2 = 0 = x_2 - x_2$, with x_1 and x_2 chosen to be different). Further, we use induction on n (a quick look over the case $n = 2$ shows what we need to do). Suppose that $n \geq 2$ and the assertion is proved for any m numbers, where $m < n$. Let $a_1, \ldots, a_n$ be complex numbers such that $a_1 + \cdots + a_n = 0$. Suppose, in the first place, that $\{a_1, \ldots, a_n\}$ has no proper subset that sums to 0. We can find $x_1, \ldots, x_n$ such that $a_k = x_k - x_{k+1}$ for every $1 \leq k \leq n$ (with $x_{n+1} = x_1$), and the supplementary assumption assures that all x_k are distinct. (Just put, in an arbitrary way, $a_1 = x_1 - x_2$, then $a_2 = x_2 - x_3, \ldots, a_{n-1} = x_{n-1} - x_n$, for $x_3, \ldots, x_n$ that are precisely defined; $a_n = x_n - x_1$ follows from $a_1 + \cdots + a_n = 0$; if $x_k = x_l$, with $1 \leq k < l \leq n$, then $a_k + \cdots + a_{l-1} = 0$ follows, but this is not allowed in this case.) So we are done by considering the cyclic permutation σ defined by $\sigma(k) = k + 1$ for all $k < n$ and $\sigma(n) = 1$. Otherwise, there exists such a proper subset of $\{a_1, \ldots, a_n\}$ with sum of elements 0. We can assume, without loss of generality, that $a_1 + \cdots + a_m = 0$, where $1 \leq m < n$. But we also have $a_1 + \cdots + a_n = 0$; hence $a_{m+1} + \cdots + a_n = 0$, too. Now m and $n - m$ are both less than n; therefore, we can use the induction hypothesis to get $a_k = y_k - y_{\pi(k)}$ for all $1 \leq k \leq m$ (with $y_1, \ldots, y_m$ mutually distinct and π a permutation of $\{1, \ldots, m\}$) and $a_k = z_k - z_{\tau(k)}$ for all $m + 1 \leq k \leq n$ (with $z_{m+1}, \ldots, z_n$ mutually distinct and τ a permutation of $\{m + 1, \ldots, n\}$). If we further choose some t different from all numbers $y_i - z_j$ ($1 \leq i \leq m$, $m + 1 \leq j \leq n$), then we have $a_k = x_k - x_{\sigma(k)}$ for all $1 \leq k \leq n$, and $x_1, \ldots, x_n$ are all distinct, if we put $x_i = y_i$ and $\sigma(i) = \pi(i)$ for all $1 \leq i \leq m$ and $x_j = z_j + t$ and $\sigma(j) = \tau(j)$ for every $m + 1 \leq j \leq n$.

Now, in our problem, we are given a matrix $A = (a_{ij})_{1 \leq i,j \leq n}$ with $\mathrm{tr}(A) = 0$, that is, $a_{11} + \cdots + a_{nn} = 0$. By the above observation, there exist distinct $x_1, \ldots, x_n$ and a permutation σ of $\{1, \ldots, n\}$ such that $a_{ii} = x_i - x_{\sigma(i)}$ for every $1 \leq i \leq n$. Thus we can write the matrix A in the form $A = B - C$, with

$$
B = \begin{pmatrix} x_1 & 0 & \ldots 0 \\ a_{21} & x_2 & \ldots 0 \\ \ldots & \ldots & \ldots \ldots \\ a_{n1} & a_{n2} & \ldots x_n \end{pmatrix} \quad \text{and} \quad C = \begin{pmatrix} x_{\sigma(1)} & -a_{12} & \ldots & -a_{1n} \\ 0 & x_{\sigma(2)} & \ldots & -a_{2n} \\ \ldots & \ldots & & \ldots \ldots \\ 0 & 0 & & \ldots x_{\sigma(n)} \end{pmatrix}.
$$

Because B and C are triangular, their eigenvalues are the elements on their main diagonals; therefore both B and C have the *distinct* eigenvalues $x_1, \ldots, x_n$. It is well-known (and very easy to prove, if one uses, for instance, the canonical Jordan form) that B and C are similar, that is, there exists an invertible matrix P such that $C = PBP^{-1}$. Then we have $A = B - PBP^{-1} = (BP^{-1})P - P(BP^{-1})$, which is exactly what we wanted to prove.

However we won't stop here, because we intend to effectively prove the similarity of B and C with the help of elementary operations (and elementary matrices). We first show that B can be transformed by similarity into a diagonal matrix having on its main diagonal the same entries as B. To this end, the important observation is that $W_{ij}(\alpha)BW_{ij}(\alpha)^{-1} = W_{ij}(\alpha)BW_{ij}(-\alpha)$ is a matrix that, when compared to B, has only the entries on row i and column j changed; moreover, for $i > j$, the entry in position (i, j) is replaced by $\alpha(x_j - x_i) + a_{ij}$. If we choose $\alpha = -a_{ij}(x_j - x_i)^{-1}$ (which is possible because $x_i \neq x_j$), the new matrix $W_{ij}(\alpha)BW_{ij}(\alpha)^{-1}$ has the same elements as B except for those on row i and column j, and it has 0 in position (i, j). Now we proceed as in the proof of theorem 1, only we use similarities at each step. Namely, we first transform all elements that are on the first column under x_1 into zeros, beginning with a_{21} and finishing with a_{n1}: we first consider $W_{21}(\alpha)BW_{21}(\alpha)^{-1}$ which is a matrix with the same elements as B, except for the elements in the second row and first column. Note, however, that the entries in the second row do not change, due to the zeros from the first row, except for the entry in position $(2, 1)$ that becomes 0 for an appropriate α; yet, the element in position $(1, 1)$ remains unchanged, because of the zeros from the first row. Then multiply this new matrix to the left with $W_{31}(\alpha)$ and to the right with its inverse, in order to obtain a matrix that differs from B only by the elements on the first column situated *under* the second row and has one more zero in position $(3, 1)$ for suitable α, and so on. After we finish with the first column, we pass to the second, where we change all entries under x_2 into zeros by using the same type of similarities (by using, in order, $W_{32}, \ldots, W_{n2}$), and we continue in this manner until we transform $a_{n,n-1}$ into 0. (Every time we apply some W_{ij} to the left and W_{ij}^{-1} to the right $(i > j)$, only the elements in positions $(i, j), \ldots, (n, j)$ change, and the entry in position (i, j) becomes 0 if we choose well the value of α.) At this moment, we transformed (by similarities) B into a diagonal matrix B' that has on the main diagonal precisely the same elements as B.

In an analogous manner, we can show that C is similar with a diagonal matrix C' having the same entries as C on its main diagonal. Because B' and C' have the same elements on the main diagonal (possibly in a different order), and they have all the other entries equal to 0, they are similar—use matrices of the form U_{ij} in order to see that $(U_{ij}B'U_{ij}^{-1} = U_{ij}B'U_{ij}$ only interchanges the entries in position (i, i) and (j, j), and, since any permutation is a product of transpositions, the conclusion follows). Finally, by the transitivity of similarity, we conclude that B and C are similar, and thus we have a complete proof (that needs no Jordan form or any other advanced tools) of this fact and consequently, we have an elementary solution to our problem.

Remark. For the sake of (some kind of) completeness, here is the sketch of a little more advanced proof of the similarity of two (not necessarily triangular) matrices with the same *distinct* eigenvalues. Let $x_1, \ldots, x_n$ be the distinct eigenvalues and let $U_1, \ldots, U_n$ be eigenvectors of B corresponding to $x_1, \ldots, x_n$ respectively, and let $V_1, \ldots, V_n$ be eigenvectors of C corresponding to $x_1, \ldots, x_n$ respectively (column vectors, of course). Thus we have $BU_i = x_i U_i$ and $CV_i = x_i V_i$ for every $1 \leq i \leq n$. Because $x_1, \ldots, x_n$ are distinct, the eigenvectors $U_1, \ldots, U_n$ are linearly independent and $V_1, \ldots, V_n$ are also linearly independent; hence, the matrices $(U_1 \ldots U_n)$ and $(V_1 \ldots V_n)$ are invertible. Let P be the matrix that changes the basis from $\{U_1, \ldots, U_n\}$ to $\{V_1, \ldots, V_n\}$, namely, such that $P(U_1 \ldots U_n) = (V_1 \ldots V_n)$. Then P is invertible too, and $PB = CP$, that is, $C = PBP^{-1}$.

Chapter 6
Equivalence Relations on Groups and Factor Groups

Let $(G, \cdot)$ be a group with identity denoted by e and let $\emptyset \neq H \subseteq G$ satisfy the implications:

a) $x, y \in H \Rightarrow xy \in H$;
b) $x \in H \Rightarrow x^{-1} \in H$.

Such a subset H is called a *subgroup* of G, which will be denoted by $H \leq G$. For every $x, y \in G$, we put

$$x\mathcal{R}^l y (\bmod H) \Leftrightarrow x^{-1}y \in H$$

and

$$x\mathcal{R}^r y (\bmod H) \Leftrightarrow xy^{-1} \in H.$$

We will prove that the relations $\mathcal{R}^l$ and $\mathcal{R}^r$ are equivalence relations on G, also called the left equivalence relation modulo H, and the right equivalence relation modulo H, *respectively*. We will write simply $x\mathcal{R}^l y$ and $x\mathcal{R}^r y$ when no confusion is possible.

Indeed, $x\mathcal{R}^l x$ because $x^{-1}x = e \in H$. If $x\mathcal{R}^l y$, then

$$x^{-1}y \in H \Rightarrow (x^{-1}y)^{-1} \in H \Rightarrow y^{-1}x \in H \Rightarrow y\mathcal{R}^l x,$$

so $\mathcal{R}^l$ is reflexive and symmetric. For transitivity, let $x, y, z \in G$ be such that

$$\begin{cases} x\mathcal{R}^l y \\ y\mathcal{R}^l z \end{cases} \Leftrightarrow \begin{cases} x^{-1}y \in H \\ y^{-1}z \in H \end{cases}.$$

H is a subgroup, so

$$(x^{-1}y)(y^{-1}z) \in H \Leftrightarrow x^{-1}(yy^{-1})z \in H \Leftrightarrow x^{-1}z \in H \Leftrightarrow x\mathcal{R}^l z.$$

© Springer Science+Business Media LLC 2017
T. Andreescu et al., *Mathematical Bridges*, DOI 10.1007/978-0-8176-4629-5_6

For $x \in G$, we denote by $\widehat{x}^l$ the equivalence class of the element x,

$$\widehat{x}^l = \{y \in H \mid y\mathcal{R}^l x\}.$$

Analogously, we put $\widehat{x}^r = \{y \in H \mid y\mathcal{R}^r x\}$. When the difference between the two relations is not important, we denote simply $\widehat{x}$ one of the equivalence classes $\widehat{x}^l$ or $\widehat{x}^r$. For every $x \in G$, we have $\widehat{x}^l = xH$, $\widehat{x}^r = Hx$, where $xH = \{xh \mid h \in H\}$ and $Hx = \{hx \mid h \in H\}$.

Theorem. *Let G be a group and let H be a subgroup of G. Then for every $x, y \in G$, the sets xH, yH, Hx, Hy have the same cardinality.*

Proof. We prove that the sets xH and Hx are cardinal equivalent with H. Indeed, the maps $\alpha : H \to xH$, $\beta : H \to Hx$ given by the laws $\alpha(h) = xh$, $\beta(h) = hx$, $h \in H$ are bijective. A classical proof can be given or we can simply note that the inverses $\alpha^{-1} : xH \to H$, and $\beta^{-1} : Hx \to H$ are given by $\alpha^{-1}(xh) = h$, and $\beta^{-1}(hx) = h$, respectively.

Denote by

$$G/\mathcal{R}^l = \{xH \mid x \in G\}, \quad G/\mathcal{R}^r = \{Hx \mid x \in G\}$$

the set of equivalence classes with respect to $\mathcal{R}^l$ and $\mathcal{R}^r$. The sets $G/\mathcal{R}^l$ and $G/\mathcal{R}^r$ have the same cardinality; the map

$$\phi : G/\mathcal{R}^l \to G/\mathcal{R}^r, \quad \phi(xH) = Hx^{-1}$$

is bijective. First we prove that ϕ is well defined. More precisely, we have to prove the implication

$$xH = x'H \Rightarrow Hx^{-1} = Hx'^{-1}.$$

Indeed, we have

$$xH = x'H \Rightarrow x^{-1}x' \in H \Leftrightarrow (x^{-1})(x'^{-1})^{-1} \in H \Leftrightarrow Hx^{-1} = Hx'^{-1}.$$

By its definition law, ϕ is surjective. For injectivity, let $x, x' \in G$ be such that $\phi(xH) = \phi(x'H)$. Then

$$Hx^{-1} = Hx'^{-1} \Leftrightarrow (x^{-1})(x'^{-1})^{-1} \in H \Leftrightarrow x^{-1}x' \in H \Leftrightarrow xH = x'H. \ \square$$

A subgroup $N \leq G$ is called a *normal subgroup* of G, and we denote this by $N \trianglelefteq G$, if one of the following equivalent assertions is true:

(n') for every $x \in G$ and $n \in N$, $xnx^{-1} \in N$
(n'') for every $x \in G$, $xN = Nx$
(n''') for every $x \in G$ and $n \in N$, there exists $n' \in N$ such that $xn = n'x$.

We prove now that the assertions $(n') - (n''')$ are equivalent.

$(n') \Rightarrow (n'')$. We prove that $xN \subseteq Nx$, because the reverse inclusion is similar. Let $xn \in xN$ be an arbitrary element, with $n \in N$. Then

$$xn = (xnx^{-1})x \in Nx,$$

taking into account that $xnx^{-1} \in N$.

$(n'') \Rightarrow (n''')$. For $x \in G$ and $n \in N$, $xn \in xN$. But $xN = Nx$, so $xn \in Nx$. There exists an element $n' \in N$ such that $xn = n'x$.

$(n''') \Rightarrow (n')$. For $x \in G$ and $n \in N$, let $n' \in N$ such that $xn = n'x$. It follows $xnx^{-1} = n' \in N$. $\square$

Example. If G is commutative, then any subgroup of G is normal.

Example. Let G, G' be groups and let $f : G \to G'$ be a morphism. The kernel

$$\ker f = \left\{ x \in G \mid f(x) = e' \right\}$$

is a normal subgroup of G (e' is the identity of G').

First, if $x, y \in \ker f$, then $f(x) = f(y) = e'$ and

$$f(xy^{-1}) = f(x)f(y^{-1}) = f(x)(f(y))^{-1} = e' \cdot e' = e'.$$

Hence $xy^{-1} \in \ker f$, which means that $\ker f$ is a subgroup of G. Further, for every $x \in G$ and $n \in \ker f$, we have $xnx^{-1} \in \ker f$. Indeed,

$$f(xnx^{-1}) = f(x)f(n)f(x^{-1}) = f(x) \cdot e' \cdot f(x^{-1})$$
$$= f(x)f(x^{-1}) = f(xx^{-1}) = f(e) = e'. \square$$

If N is a normal subgroup of G, $N \trianglelefteq G$, then the left equivalence class and the right equivalence class are identical. Consequently, the sets of equivalence classes $G/\mathcal{R}^l$ and $G/\mathcal{R}^r$ coincide. We denote by G/N the factor set $G/\mathcal{R}^l = G/\mathcal{R}^r$. Now we also denote

$$\widehat{x} = xN = Nx.$$

For all classes x, $y \in G/N$, we define the operation

$$\widehat{x} \cdot \widehat{y} = \widehat{xy}.$$

First note that this operation is well defined. If $\widehat{x} = \widehat{x'}$ and $\widehat{y} = \widehat{y'}$, we must have $\widehat{xy} = \widehat{x'y'}$. Indeed,

$$\begin{cases} \widehat{x} = \widehat{x'} \\ \widehat{y} = \widehat{y'} \end{cases} \Rightarrow \begin{cases} x^{-1}x' \in N \\ y'y^{-1} \in N \end{cases}.$$

On the other hand,

$$(xy)^{-1}(x'y') = y^{-1}x^{-1}x'y' = y^{-1}(x^{-1}x')(y'y^{-1})y = y^{-1}ny \in N,$$

where $n = (x^{-1}x')(y'y^{-1}) \in N$. Hence $\widehat{xy} = \widehat{x'y'}$.

The element $\widehat{e} = eN = Ne = N \in G/N$ is the identity since

$$\widehat{e} \cdot \widehat{n} = \widehat{en} = \widehat{n}, \quad \widehat{n} \cdot \widehat{e} = \widehat{ne} = \widehat{n}$$

and the inverse of $\widehat{x}$ is $\widehat{x^{-1}}$:

$$\widehat{x} \cdot \widehat{x^{-1}} = \widehat{x \cdot x^{-1}} = \widehat{e}, \quad \widehat{x^{-1}} \cdot \widehat{x} = \widehat{x^{-1}x} = \widehat{e}.$$

The group $(G/N, \cdot)$ is called the factor group with respect to the normal subgroup N.

Theorem (Lagrange). *Let G be a finite group. Then for every subgroup H of G, we have ord H which divides ord G, where ord G is the cardinality of G.*

Proof. Let us consider the left equivalence relation $\mathcal{R}$ with respect to H.

In general, any two different equivalence classes are disjoint. Because G is finite, there are only a finite number of equivalence classes, which define a partition of G, say

$$G = x_1 H \cup x_2 H \cup \cdots \cup x_n H,$$

with $x_k \in G$, $1 \leq k \leq n$. As we proved,

$$|x_1 H| = |x_2 H| = \cdots = |x_n H| = \text{ord } H,$$

so by taking the cardinality, we derive

$$\text{ord } G = |x_1 H| + |x_2 H| + \cdots + |x_n H| = n \cdot |H|.$$

Hence ord $G = n \cdot$ ord H. $\square$

The number n of the equivalence classes is called the *index* of H in G, denoted by $n = [G : H]$. Therefore the relation

$$|G| = [G : H] \cdot |H|$$

will be called Lagrange's relation.

The fundamental theorem of isomorphism. *Let G and G' be groups and $f : G \to G'$ be a morphism. Then the factor group $G/\ker f$ is isomorphic to $\text{Im} f$.*

Proof. The proof is constructive, in the sense that we will indicate the requested isomorphism. We are talking about the map

$$\rho : G/\ker f \to \operatorname{Im} f$$

given by $\rho(\widehat{x}) = f(x)$, where $x \in G$.

First we prove that ρ is well defined. To do this, let us consider $x, x' \in G$ with $\widehat{x} = \widehat{x'}$. Thus

$$\widehat{x} = \widehat{x'} \Leftrightarrow x^{-1}x' \in \ker f \Leftrightarrow f(x^{-1}x') = e'$$

$$\Leftrightarrow f(x^{-1})f(x') = e' \Leftrightarrow (f(x))^{-1}f(x') = e'.$$

By multiplying to the left with $f(x)$, we deduce that $f(x) = f(x')$.

Now we use the fact that f is a morphism to prove that the map ρ is a morphism,

$$\rho(\widehat{x} \cdot \widehat{y}) = \rho(\widehat{xy}) = f(xy) = f(x)f(y) = \rho(\widehat{x})\rho(\widehat{y}).$$

If $x, y \in G$ are so that $\rho(\widehat{x}) = \rho(\widehat{y})$, then

$$f(x) = f(y) \Leftrightarrow f(x)(f(y))^{-1} = e'$$

$$\Leftrightarrow f(x)f(y^{-1}) = e' \Leftrightarrow f(xy^{-1}) = e' \Leftrightarrow xy^{-1} \in \ker \Leftrightarrow \widehat{x} = \widehat{y},$$

so f is injective. By its definition, f is also surjective. $\square$

We end this chapter with some important and nontrivial results that can be also obtained using equivalence relations.

Theorem (Cauchy). *Let G be a group of order n and let p be a prime divisor of n. Then there exists an element of order p in G.*

Proof. Consider S the set of p-tuples $(x_1, x_2, \ldots, x_p)$ such that $x_1 x_2 \cdots x_p = e$. Clearly, this set has n^{p-1} elements, because for any choice of $x_1, x_2, \ldots, x_{p-1}$, the element x_p is uniquely determined. Moreover, we can define a function f from S to S by $f(x_1, x_2, \ldots, x_p) = (x_2, x_3, \ldots, x_p, x_1)$, because if $x_1 x_2 \cdots x_p = e$, we also have $x_2 x_3 \cdots x_p x_1 = e$ (due to the fact that $x_1 = (x_2 \cdots x_p)^{-1}$). This function f is manifestly injective and thus it is a permutation of S. Also, we have $f^p(x) = x$ for all $x \in S$, (where f^p denotes the composition taken p times). Now, define an equivalence relation by xRy if and only if there exists an integer k such that $x = f^k(y)$. Because $f^p(x) = x$ for all x, any equivalence class has either 1 or p elements. But the sum of the cardinalities of the equivalence classes is the cardinality of S, which is a multiple of p. Because the class of $(e, e, \ldots, e)$ has one element, it follows that there is some $(x_1, x_2, \ldots, x_p) \in S$, different from $(e, e, \ldots, e)$ and whose equivalence class also has one element. But this implies $(x_1, x_2, \ldots, x_p) = f(x_1, x_2, \ldots, x_p)$, that is $x_1 = x_2 = \cdots = x_p = x$ and this element x satisfies $x^p = e$ because $(x_1, \ldots, x_p) \in S$ and $x \neq e$. Thus, x has order p. $\square$

And here is an application:

Problem. Prove that any commutative group whose order is square-free is cyclic.

Solution. Let G be such a group, whose order is $p_1 p_2 \ldots p_n$ for some distinct prime numbers $p_1, p_2, \ldots, p_n$. By Cauchy's theorem, there are elements x_i whose order equals p_i for all i. We claim that $y = x_1 x_2 \ldots x_n$ has order $p_1 p_2 \ldots p_n$. Indeed, if $y^k = e$, then $y^{kp_2 \cdots p_n} = e$, and because G is abelian, it follows that $x_1^{kp_2 \cdots p_n} = e$. But the order of x_1 is p_1, which is relatively prime to $p_2 p_3 \ldots p_n$. Thus p_1 divides k. Similarly, all numbers $p_1, p_2, \ldots, p_n$ are divisors of k and so k is a multiple of $p_1 p_2 \ldots p_n$. This shows that G has an element whose order is at least equal to the order of the group, so G is indeed cyclic. $\square$

It is very easy to prove, using Lagrange's theorem, that any group whose order is a prime number is cyclic. The following result is however much more difficult:

Problem. Prove that any group whose order is the square of a prime number is abelian.

Solution. Let G be a group of order p^2, where p is a prime number. Also, consider the following equivalence relation on G: xRy if and only if there exists $g \in G$ such that $y = gxg^{-1}$. For any x, it is easy to check that the set $C(x) = \{g | gx = xg\}$ is a subgroup of G. We claim that the cardinality of the equivalence class of x equals $\frac{p^2}{|C(x)|}$. Indeed, it is not difficult to check that the function $f : G/C(x) \to [x]$ (here $[x]$ denotes the equivalence class of x) defined by $f(gC(x)) = gxg^{-1}$ is well defined and bijective: if $g_1 C(x) = g_2 C(x)$, then $g_1 = g_2 c$ for some $c \in C(x)$ and so

$$g_1 x g_1^{-1} = g_2 c x c^{-1} g_2^{-1} = g_2 x g_2^{-1}$$

because $cxc^{-1} = x$. Also, if $gxg^{-1} = hxh^{-1}$, we have $h^{-1}g \in C(x)$ and so $gC(x) = hC(x)$. Thus f is injective. Obviously, f is surjective. This proves the claim that $|[x]| = \frac{p^2}{|C(x)|}$. Now, G is the disjoint union of the equivalence classes, so p^2 equals the sum of the cardinalities of all classes. By the above argument, any class has a cardinality equal to $1, p$ or p^2 and a class has cardinality 1 if and only if $C(x) = G$, that is, x commutes with all elements of the group. Let $Z(G)$ the set of such x. The previous remark implies that $p \mid |Z(G)|$. We need to show that $Z(G) = G$, so it is enough to show that we cannot have $|Z(G)| = p$. Let us assume that this is the case. Then $G/Z(G)$ is a group (clearly, $Z(G)$ is normal) with p elements, thus cyclic. Let $p(a)$ be a generator of this group, where p is the natural projection from G to $G/Z(G)$. Then we know that for any $x, y \in G$, there are integers k, m such that $p(x) = p(a^k)$ and $p(y) = p(a^m)$. It follows that $a^{-k}x$ and $a^{-m}y$ are in $Z(G)$ and this easily implies that $xy = yx$. Thus, any two elements of G commute and $Z(G) = G$, contradicting the assumption that $|Z(G)| = p$. This contradiction shows that we must have $Z(G) = G$ and G is therefore abelian. $\square$

Proposed Problems

1. Let G be a finite group, H a subgroup of G, and K a subgroup of H, $K \leq H \leq G$. Prove that

$$[G : K] = [G : H] \cdot [H : K].$$

2. Let G be a group and let H be a subgroup of index 2 in G, $[G : H] = 2$. Prove that H is a normal subgroup of G.

3. Let G be a group, H a subgroup of G, and let $x, y \in G$. Find a bijective function from xH onto yH.

4. Let G be a finite group. Prove that for every $x \in G \setminus \{e\}$ with $\operatorname{ord} x = k$, we have k divides $\operatorname{ord} G$.

5. Let G be a group and N a normal subgroup of G. Assume that an element $a \in G$ has finite order, $\operatorname{ord} a = n$. Prove that the least positive integer k for which $a^k \in N$ is a divisor of n. Prove that this remains true if N is any subgroup of G (not necessarily a normal subgroup).

6. Let G be a cyclic group, $G = \{x^k \mid k \in \mathbb{Z}\}$, for some element $x \in G$. Prove that G is isomorphic to the additive group of the integers $(\mathbb{Z}, +)$ or to an additive group $\mathbb{Z}/n\mathbb{Z}$, for some integer $n \geq 2$.

7. Let G be a group with identity e such that $f : G \to G$, given by $f(x) = x^n$ is a morphism for some positive integer n. Prove that

$$H = \{x \in G \mid x^n = e\}, \quad K = \{x^n \mid x \in G\}$$

are normal subgroups of G. Moreover, prove that if G is finite, then $\operatorname{ord} K = [G : H]$.

8. Consider the additive group $(\mathbb{R}, +)$ of real numbers with the (normal) subgroup $(\mathbb{Z}, +)$. Prove that the factor group $\mathbb{R}/\mathbb{Z}$ is isomorphic to the multiplicative group $\mathbb{T}$ of complex numbers of module 1.

9. Consider the additive group $(\mathbb{C}, +)$ of complex numbers with the (normal) subgroup $(\mathbb{R}, +)$. Prove that the factor group $\mathbb{C}/\mathbb{R}$ is isomorphic to the additive group $(\mathbb{R}, +)$ of real numbers.

10. Let G be a group with the property that $x^2 = e$, for all $x \in G$. Prove that G is abelian. Moreover, prove that if G is finite, then $\operatorname{ord} G$ is a power of 2.

11. Let G be a group. Denote by $(\operatorname{Inn} G, \circ)$ the group of inner automorphisms,

$$\operatorname{Inn} G = \{\tau_g \mid g \in G\},$$

where $\tau_g : G \to G$ is the automorphism

$$\tau_g(x) = gxg^{-1},$$

for all $x \in G$. Prove that the group $(\text{Inn } G, \circ)$ is isomorphic to $G/Z(G)$, where $Z(G) \leq G$ is the center of G :

$$Z(G) = \{x \in G \mid xy = yx, \ \forall \ y \in G\} .$$

12. Let G be a group and let N be a subgroup of $Z(G)$. Prove that:

a) N is a normal subgroup of G;
b) if G/N is cyclic, then G is abelian.

13. Let $p \geq 3$ be prime. Prove that the set of integers $0 \leq a \leq p - 1$ for which the congruence

$$x^2 \equiv a(\mathrm{mod} \, p)$$

has solutions in the set of integers consists of $\dfrac{p+1}{2}$ elements.

14. Let us consider the group $(\mathbb{Z}, +)$ as a subgroup of the group $(\mathbb{Q}, +)$ of the rational numbers. Prove that every finite subgroup of the factor group $\mathbb{Q}/\mathbb{Z}$ is cyclic.

15. How many morphisms $f : (\mathbb{Z}_2 \times \mathbb{Z}_2, +) \to (S_3, \circ)$ are there?

16. Let $(G, \cdot)$ be a finite group with n elements, and let S be a nonempty subset of G. Prove that if we denote by S^k the set of all products $s_1 \cdots s_k$, with $s_i \in S$ for all $1 \leq i \leq k$, then S^n is a subgroup of G.

17. Let $(G, \cdot)$ be a finite group whose order is square-free, let x be an element of G, and let m and n be relatively prime positive integers. Prove that if there exist y and z in G such that $x = y^m = z^n$, then there also exists t in G such that $x = t^{mn}$.

18. Let A be a nonsingular square matrix with integer entries. Suppose that for every positive integer k, there is a matrix X_k with integer entries such that $X_k^k = A$. Prove that A must be the identity matrix.

Solutions

1. By Lagrange's relation,

$$[G : H] \cdot [H : K] = \frac{|G|}{|H|} \cdot \frac{|H|}{|K|} = \frac{|G|}{|K|} = [G : K].$$

2. We have $[G : H] = 2$, so there are only two left (or right) equivalence classes which define a partition of G. One of the sets of the partition is always H, so the partition is $G = H \cup (G \setminus H)$. Then H is a normal subgroup of G, because

$$G/\mathcal{R}^l = G/\mathcal{R}^r = \{H, G \setminus H\} .$$

3. We will prove that the application $\omega : xH \rightarrow yH$ given by the law $\omega(xh) = yh$, $h \in H$ is bijective. By its definition, ω is surjective. For injectivity, if $\omega(xh) = \omega(xh')$, then

$$\omega(xh) = \omega(xh') \Rightarrow yh = yh'$$

$$\Rightarrow y^{-1}(yh) = y^{-1}(yh') \Rightarrow h = h' \Rightarrow xh = xh'.$$

4. Let us consider the set

$$H = \left\{e, x, x^2, \ldots, x^{k-1}\right\}.$$

Any two elements of H are different. Indeed, if there exist integers $0 \le i < j \le k - 1$ such that $x^i = x^j$, then $x^{j-i} = e$, with $0 < j - i < k$. But the last relation contradicts the fact that $\operatorname{ord} x = k$. We also have the representation $H = \left\{x^i \mid i \in \mathbb{Z}\right\}$. Hence for any elements x^i, $x^j \in H$, it follows

$$x^i(x^j)^{-1} = x^i \cdot x^{-j} = x^{i-j} \in H,$$

which means that H is a subgroup of G. According to Lagrange's theorem,

$$\operatorname{ord} H \mid \operatorname{ord} G \Leftrightarrow k \mid \operatorname{ord} G.$$

5. Let us consider the factor group G/N. We have $a^n = e \in N$, so $\widehat{a}^n = \widehat{e}$ in G/N. We also have $a^k \in N$, so $\widehat{a}^k = \widehat{e}$ in G/N, where k is the minimal value with this property. In other words, k is the order of the element $\widehat{a}$ in G/N. We also have $\widehat{a}^n = \widehat{e}$, so k divides n.

   Let's now assume that the subgroup N is not normal. For the sake of contradiction, let us put $n = kq + r$, with $1 \le r \le k - 1$. We have

$$a^r = a^{n-kq} = a^n \cdot (a^k)^{-q} = (a^k)^{-q} \in N,$$

because $a^k \in N$. We obtained $a^r \in N$, with $r < k$, which contradicts the minimality of k. Hence $k = 0$, which means that k divides n.

6. The application $f : (\mathbb{Z}, +) \rightarrow G$ given by $f(k) = x^k$, $k \in \mathbb{Z}$ is a surjective morphism,

$$f(k + k') = x^{k+k'} = x^k \cdot x^{k'} = f(k)f(k').$$

According to the fundamental theorem of isomorphism, $\mathbb{Z}/\ker f \simeq G$, where $\operatorname{Im} f = G$. If f is injective, then $\ker f = \{0\}$; thus

$$\mathbb{Z}/\ker f \simeq \mathbb{Z}/\{0\} \simeq \mathbb{Z},$$

so $\mathbb{Z} \simeq G$. If f is not injective, then let $k < k'$ be integers with

$$f(k) = f(k') \Leftrightarrow x^k = x^{k'} \Leftrightarrow x^{k'-k} = e.$$

If n is the least positive integer with $x^n = e$, then $\ker f = n\mathbb{Z}$. Hence

$$G \simeq \mathbb{Z}/\ker f \simeq \mathbb{Z}/n\mathbb{Z}.$$

7. In general, the kernel of a morphism is a normal subgroup, and here, $H = \ker f \trianglelefteq G$. For any elements x^n, $y^n \in K$, we have

$$x^n(y^n)^{-1} = x^n(y^{-1})^n = (xy^{-1})^n \in K,$$

so K is a subgroup of G. It is also normal, because for every $x^n \in K$ and $y \in G$, we have

$$yx^ny^{-1} = (yxy^{-1})(yxy^{-1}) \cdots (yxy^{-1}) = (yxy^{-1})^n \in G.$$

For the second part of the problem, we use the fundamental theorem of isomorphism,

$$G/\ker f \simeq \operatorname{Im} f$$

and taking into account that $\ker f = H$, $\operatorname{Im} f = K$, we obtain $G/H \simeq K$. Hence

$$|G/H| = |K| \Leftrightarrow \operatorname{ord} K = [G : H].$$

8. Let us consider the application $f : (\mathbb{R}, +) \to (\mathbb{C}^*, \cdot)$ given by $f(x) = e^{2i\pi x}$. Clearly, f is a morphism and $x \in \ker(f)$ if and only if $e^{2i\pi x} = 1$, that is, x is an integer. On the other hand, it is clear that $\operatorname{Im} f = \mathbb{T}$. By the fundamental theorem of isomorphism, we deduce that

$$\mathbb{R}/\ker f \simeq \operatorname{Im} f \Leftrightarrow \mathbb{R}/\mathbb{Z} \simeq \mathbb{T}.$$

9. Let us consider the application $f : (\mathbb{C}, +) \to (\mathbb{R}, +)$ given by $f(z) = \operatorname{Im} z$, where $\operatorname{Im} z$ denotes the imaginary part of the complex number z. Clearly, f is a morphism and $\ker f = \mathbb{R}$. Obviously, $\operatorname{Im} f = \mathbb{R}$. By the fundamental theorem of isomorphism, we deduce that

$$\mathbb{C}/\ker f \simeq \operatorname{Im} f \Leftrightarrow \mathbb{C}/\mathbb{R} \simeq \mathbb{R}.$$

10. For every $x, y \in G$, we have $x^2 = e$, $y^2 = e$, $(xy)^2 = e$. Thus

$$(xy)^2 = x^2y^2 \Leftrightarrow xyxy = xxyy.$$

By multiplying with x^{-1} to the left and with y^{-1} to the right, we obtain

$$x^{-1}(xyxy)y^{-1} = x^{-1}(xxyy)y^{-1} \Leftrightarrow yx = xy,$$

so G is abelian.

For the second part, we will use Cauchy's theorem: let G be a group of order n in which any element has order at most 2. If $p > 2$ is a prime divisor of n, by Cauchy's theorem, G has an element of order p, which is a contradiction. Therefore any prime divisor of n is 2 and n is thus a power of 2.

11. Let us consider the application $\phi : G \to \operatorname{Inn} G$ given by the law $\phi(g) = \tau_g$. For every $g, g' \in G$, we have

$$\phi(g) \circ \phi(g') = \phi(gg').$$

Indeed, for all $x \in G$, we have

$$\tau_g(\tau_{g'}(x)) = \tau_g(g'xg'^{-1}) = g(g'xg'^{-1})g^{-1} = (gg')x(gg')^{-1} = \tau_{gg'}(x),$$

thus

$$\tau_g \circ \tau_{g'} = \tau_{gg'} \Leftrightarrow \phi(g) \circ \phi(g') = \phi(gg').$$

Hence ϕ is a (surjective) morphism. If $g \in \ker \phi$, then

$$\phi(g) = 1_G \Leftrightarrow \tau_g = 1_G.$$

It follows that for every $x \in G$,

$$\tau_g(x) = x \Leftrightarrow gxg^{-1} = x \Leftrightarrow gx = xg,$$

which is equivalent to $g \in Z(G)$. Then with $\ker \phi = Z(G)$,

$$G/\ker \phi \simeq \operatorname{Inn} G \Leftrightarrow G/Z(G) \simeq \operatorname{Inn} G.$$

12. a) Every element of N is also an element of $Z(G)$, so every element of N commutes with all elements of the group G. Then for every $x \in G$ and $n \in N$,

$$xnx^{-1} = nxx^{-1} = n \in N,$$

so N is a normal subgroup of G.

b) Let us assume that $G/N =< \hat{a} >$, for some $a \in G$. Now let $x \in G$ be arbitrary. We have $\hat{x} \in < \hat{a} >$, so there is an integer k such that

$$\hat{x} = \hat{a}^k \Leftrightarrow \hat{x} = \widehat{a^k} \Leftrightarrow xa^{-k} \in N.$$

Hence $xa^{-k} = n \Leftrightarrow x = a^k n$, for some $n \in N$. Finally, for every $x, y \in G$, let $k, p \in \mathbb{Z}$ and $n, q \in N$ such that $x = a^k n$, $y = a^p q$. It follows that

$$xy = a^k n \cdot a^p q = a^k a^p nq = a^{k+p} nq$$

and

$$yx = a^p q \cdot a^k n = a^p a^k qn = a^{p+k} qn,$$

so $xy = yx$.

13. Let us define the application $f : \mathbb{Z}_p^* \to \mathbb{Z}_p^*$ by the formula $f(x) = x^2$. Easily, f is a morphism from the group $(\mathbb{Z}_p^*, \cdot)$ to itself with $\ker f = \left\{\widehat{1}, \widehat{p-1}\right\}$. Indeed, if $\widehat{x} \in \ker f$, then $\widehat{x}^2 = \widehat{1}$, so p divides $x^2 - 1$. The number p is prime, so p divides $x - 1$ or p divides $x + 1$. Thus $x = 1$, respectively $x = p - 1$. Now, according to the fundamental theorem of isomorphism,

$$\mathrm{Im} f \simeq \mathbb{Z}_p^* / \ker f.$$

In particular,

$$\mathrm{card}\, \mathrm{Im} f = \mathrm{card}(\mathbb{Z}_p^* / \ker f) = \frac{\mathrm{card}\, \mathbb{Z}_p^*}{\mathrm{card}\, \ker f} = \frac{p-1}{2}.$$

If we add the solution $a = 0$, we obtain

$$\frac{p-1}{2} + 1 = \frac{p+1}{2}$$

elements.

14. Let H be a finite subgroup of the group $\mathbb{Q}/\mathbb{Z}$ with $\mathrm{ord}\, H = n$. We can assume that

$$H = \{\widehat{a_1}, \widehat{a_2}, \ldots, \widehat{a_n}\},$$

for some rational numbers $a_1, a_2, \ldots, a_n \in [0, 1)$. Then for every $\widehat{h} \in H$ with $h \in \mathbb{Q} \cap [0, 1)$, we have

$$\mathrm{ord}\, H \cdot \widehat{h} = \widehat{0} \Leftrightarrow n \cdot \widehat{h} = \widehat{0} \Leftrightarrow nh \in \mathbb{Z}.$$

It follows that $h = \frac{k}{n}$, for some integer $1 \le k \le n - 1$. We proved that

$$H \subseteq \left\{\widehat{0}, \frac{\widehat{1}}{n}, \ldots, \frac{\widehat{n-1}}{n}\right\}.$$

The other inclusion is true, because $\mathrm{ord}\, H = n$. In conclusion, $H = \left\langle \frac{\widehat{1}}{n} \right\rangle$.

15. The answer is 10. Each element of the group $\mathbb{Z}_2 \times \mathbb{Z}_2$ is of order at most 2, so $(f(x))^2 = e$, for all $x \in \mathbb{Z}_2 \times \mathbb{Z}_2$. It follows that

$$\mathrm{Im} f \subseteq \{e, (12), (23), (31)\}.$$

If for some $x, y \in \mathbb{Z}_2 \times \mathbb{Z}_2, f(x) = (12), f(y) = (23)$, then

$$f(x + y) = f(x)f(y) = (12)(23) = (231) \notin \mathrm{Im} f,$$

a contradiction. Hence $f(x) = e$ or $\mathrm{Im} f = \{e, \tau\}$, where $\tau \in S_3$ is transposition. According to the fundamental theorem of isomorphism,

$$\mathrm{Im} f \simeq (\mathbb{Z}_2 \times \mathbb{Z}_2) / \ker f.$$

In particular,

$$2 = \mathrm{card}\, \mathrm{Im} f = \mathrm{card}\, [(\mathbb{Z}_2 \times \mathbb{Z}_2) / \ker f] = \frac{\mathrm{card}\, (\mathbb{Z}_2 \times \mathbb{Z}_2)}{\mathrm{card}\, \ker f},$$

so $\ker f$ has 2 elements. One of them is $(0, 0)$ and the other can be chosen in three ways.

In conclusion, there are $3 \times 3 = 9$ such morphisms, and if we add the null morphism, we obtain ten morphisms.

16. Fix some $s \in S$. We have $sS^k \subseteq S^{k+1}$; hence $|S^k| = |sS^k| \leq |S^{k+1}|$ for all $k \in \mathbb{N}^*$, where $|X|$ denotes the number of elements of the (finite) set X. Thus we have the nondecreasing set of cardinalities

$$|S| \leq |S^2| \leq |S^3| \leq \cdots.$$

Note that if, for some j, we have $|S^j| = |S^{j+1}|$, then $|S^k| = |S^{k+1}|$ for all $k \geq j$. Indeed, we have $sS^j \subseteq S^{j+1}$ and $|sS^j| = |S^j| = |S^{j+1}|$; thus $sS^j = S^{j+1}$, which implies $sS^{j+1} = sS^j S = S^{j+1} S = S^{j+2}$, yielding $|S^{j+1}| = |sS^{j+1}| = |S^{j+2}|$, and the conclusion follows inductively.

Now, if in the sequence of inequalities $|S| \leq |S^2| \leq \cdots \leq |S^n|$ we have equality between two consecutive cardinalities, then $|S^k| = |S^{k+1}|$ for all $k \geq j$, with $j \leq n - 1$. Otherwise (when all inequalities are strict), we get $|S^n| \geq n$; but $|S^n| \leq |G| = n$; thus $|S^n| = n$, and this yields $|S^k| = n$ for all $k \geq n$. Either way, we have

$$|S^n| = |S^{n+1}| = |S^{n+2}| = \cdots.$$

As we have seen before, this also leads to $sS^k = S^{k+1}$ for all $k \geq n$, and, by iterating, we obtain $s^n S^n = S^{2n}$, that is, $S^n = S^{2n} = S^n S^n$ (as $s^n = e$, the unit element of G, as a consequence of Lagrange's theorem). Thus S^n is closed under the multiplication of G, and the fact that it contains the identity is obvious: $e = s^n \in S^n$; that is, S^n is a subgroup of G.

This appeared as (a folklore) problem B in *Nieuw Archief voor Wiskunde*, issue 3/2007. The above solution is from the same magazine, 1/2008.

17. Let $o(g)$ denote the order of an element $g \in G$. From $x = y^m$, we infer that

$$o(x) = o(y^m) = \frac{o(y)}{\mu},$$

where $\mu = (o(y), m)$ is the greatest common divisor of the order of y and m. We have that μ divides the order of y, which in turn divides the order $|G|$ of the group; thus, μ divides $|G|$. Also $o(x) = o(y)/\mu$ and m/μ are relatively prime.

Similarly, for $\nu = (o(z), n)$, we have that ν divides $|G|$ and that $o(x)$ and n/ν are relatively prime.

Also, μ and ν are themselves relatively prime, since they are divisors of m and n, respectively; therefore their product divides the order of the group. So, we can write $|G| = \mu\nu\rho$, with integer ρ, and, because $|G|$ is square-free, any two of μ, ν, and ρ are relatively prime.

Now $x^{\nu\rho} = y^{m\nu\rho} = e$ (the identity element of G), because $m\nu\rho = (m/\mu)|G|$ is a multiple of the order of the group, and similarly, $x^{\mu\rho} = z^{n\mu\rho} = e$. It follows that both numbers $\mu\rho$ and $\nu\rho$ are divisible by $o(x)$, the order of x, and therefore, their greatest common divisor ρ is also divisible by $o(x)$. As ρ is prime to both μ and ν, the same is true for the order of x. So, in the end we get that $o(x)$ is prime to any of the numbers m/μ, n/ν, μ, and ν, hence it is prime to their product mn. Again, we can find integers p and q such that $1 = po(x) + qmn$; thus we have

$$x = x^{po(x)} x^{qmn} = (x^q)^{mn} = t^{mn}$$

for $t = x^q \in G$.

Remark. Let $(M, \cdot)$ be a monoid. We say that M has property (P) if, whenever m and n are positive integers, and $x \in M$ has the property that there are y and z in M such that $x = y^m = z^n$, there also exists $t \in M$ for which $x = t^{[m,n]}$ (where $[m, n]$ is the least common multiple of m and n). The reader will easily prove that commutative finite groups have property (P). Also, by using the result of this problem, we can prove that finite groups with square-free order have property (P). We say that a ring has property (P) if its multiplicative monoid has it. Then factorial rings, like the ring of integers, have property (P), and with a more elaborate argument, one can prove that any factor ring of a principal ring has property (P). The proof is essentially the same as in the case of factor rings $\mathbb{Z}/n\mathbb{Z}$, and this is the starting point for the problem above, too. Actually to say that $\mathbb{Z}/n\mathbb{Z}$ has property (P) represents a rewording of the fact that if an infinite arithmetic progression of integers contains a power with exponent m, and a power with exponent n, then it also has a term that is a power with exponent [m, n]. This is a known folklore problem that appeared a few times in various mathematical contests. A particular case was the subject of problem 11182 from *The American Mathematical Monthly* from November 2005 (proposed by Shahin Amrahov).

18. Let p be a prime number that does not divide the determinant of A (which is nonzero). In the group G of invertible matrices with entries in the field $\mathbb{Z}/p\mathbb{Z}$, the equation $X_k^k = A$ is also valid for every positive integer k (we don't use different notations for the respective reduced modulo p matrices). In particular, we have $X_{|G|}^{|G|} = A$, but every matrix from G has the $|G|$th power the identity matrix, according to Lagrange's theorem. Thus $A = I$ in this group, meaning that all elements of $A - I$ (this time regarded as an integer matrix) are divisible by p. Since this happens for infinitely many primes p, it follows that $A - I$ is the zero matrix; thus $A = I$, the identity matrix, as claimed.

This is problem 11401, proposed by Marius Cavachi in *The American Mathematical Monthly* 10/2008 and solved by Microsoft Research Problem Group in the same *Monthly*, 10/2010.

Chapter 7
Density

We say that a set $A \subseteq \mathbb{R}$ is *dense in* $\mathbb{R}$ if any open interval of real numbers contains elements from A. One of the practicalities of dense sets follows from the fact that two continuous functions from $\mathbb{R}$ to $\mathbb{R}$ are equal if they are equal on a dense subset of $\mathbb{R}$. We have the following characterization of dense sets:

Theorem. *Let $A \subseteq \mathbb{R}$. The following assertions are equivalent:*

a) *A is dense in $\mathbb{R}$.*
b) *for every real number x, there exists a sequence of elements of A, converging to x.*

Proof. If A is dense, then for every real number x and every positive integer n, we have

$$\left(x - \frac{1}{n}, x + \frac{1}{n} \right) \cap A \neq \emptyset.$$

Therefore, for each positive integer n, we can choose an element

$$a_n \in \left(x - \frac{1}{n}, x + \frac{1}{n} \right) \cap A.$$

We obtained a sequence $(a_n)_{n \geq 1}$ of elements of A, which clearly converges to x. For the converse, let $(a, b) \subseteq \mathbb{R}$ and let $x \in (a, b)$ be fixed. Then there exists a sequence $(a_n)_{n \geq 1}$ with elements in A, converging to x. But (a, b) is a neighborhood of the limit x; thus we can find an element $a_{n_0} \in (a, b)$. Because $a_{n_0} \in A$, it follows that $(a, b) \cap A \neq \emptyset$. In conclusion, A is dense in $\mathbb{R}$. $\square$

The set $\mathbb{Q}$ of rational numbers is dense in $\mathbb{R}$. We will prove that each nonempty interval contains at least one rational number. Indeed, let (a, b) be an interval, with $a < b$. Looking for two integers m, n, say n positive, so that $a < \frac{m}{n} < b$, we observe that it is enough to prove the existence of a positive integer n such that

© Springer Science+Business Media LLC 2017
T. Andreescu et al., *Mathematical Bridges*, DOI 10.1007/978-0-8176-4629-5_7

$nb - ba > 1$. Indeed, if such an integer exists, the interval (na, nb) has length greater than 1, so it contains an integer m. Then $\frac{m}{n} \in (a, b)$. But it is clear that $n = 1 + \left[\frac{1}{b-a}\right]$ is a solution. Moreover, one can easily prove that any nonempty interval contains infinitely many rational numbers. The inequality $x - \frac{1}{n} < [nx] \le x$ shows that $\lim_{n \to \infty} \frac{[nx]}{n} = x$. In fact, this is another proof that $\mathbb{Q}$ is dense in $\mathbb{R}$. Indeed, for every $x \in \mathbb{R}$, the sequence of rational numbers $\left(\frac{[nx]}{n}\right)_{n \ge 1}$ converges to x. Now, our assertion follows from the theorem. For instance, the sequence

$$\left[\sqrt{2}\right], \frac{\left[2\sqrt{2}\right]}{2}, \frac{\left[3\sqrt{2}\right]}{3}, \frac{\left[4\sqrt{2}\right]}{4}, \ldots, \frac{\left[n\sqrt{2}\right]}{n}, \ldots$$

is a nice example of a sequence of rational numbers which converges to $\sqrt{2}$.

Further, the set $\mathbb{R} \setminus \mathbb{Q}$ of irrational numbers is dense in $\mathbb{R}$. Indeed, let (a, b) be a nonempty interval. As we have proved, the interval $\left(\frac{a}{\sqrt{2}}, \frac{b}{\sqrt{2}}\right)$ contains at least one rational number, say q. Then $q\sqrt{2} \in (a, b)$ and $q\sqrt{2}$ is irrational. So we proved that there are no intervals consisting only of rational numbers. Observe that this is also immediate by a cardinality argument: we have seen that any interval is uncountable, so it cannot be included in the set of rational numbers, which is countable. Finally, here is one more possible argument: we assert that each interval of the form $(0, \varepsilon)$, $\varepsilon > 0$, contains at least one irrational number. Indeed, we have the implication:

$$\sqrt{2} \in (0, 2) \Rightarrow \frac{\sqrt{2}}{n} \in \left(0, \frac{2}{n}\right),$$

for all positive integers n. If q is rational, in (a, b), then let ω be irrational in $(0, b - q)$. Now, the number $\omega + q$ is irrational and $\omega + q \in (q, b) \subset (a, b)$.

Finally, note that the notion of density has nothing to do with the order structure of $\mathbb{R}$—it is a topological notion. It can easily be extended to other spaces, such as $\mathbb{R}^n$. Indeed, if $\|x\|$ is the Euclidean norm of the vector $x \in \mathbb{R}^n$, then we say that a set A is dense if every open ball $B(x, r)$ (defined as the set of points $y \in \mathbb{R}^n$ such that $\|y - x\| < r$) contains at least one element of A. All properties of continuous functions related to dense sets that we discussed remain clearly true in such a larger context.

A very useful and general result is the following:

Theorem (Stone-Weierstrass). *Let K be a compact subset of the set of real numbers and let A be an algebra of continuous functions defined on K and having real values. Suppose that all constant functions belong to A and also that for any distinct points $x, y \in K$, there exists $f \in A$ such that $f(x) \ne f(y)$. Then A is dense in the set of continuous functions on K, with real values, for the uniform convergence norm. That is, for any continuous function f defined on K, with real values and for any $\epsilon > 0$, there exists $g \in A$ such that $|f(x) - g(x)| < \epsilon$ for all $x \in K$.*

The proof of this result is far from being easy and will require several steps. First, observe that if we define $\overline{A}$ to be the closure of A in $(C(K), \|\cdot\|_\infty)$ (i.e., the set of continuous functions on K with $\|f\|_\infty = \max_{x \in K} |f(x)|$), then $\overline{A}$ is also an algebra having the same properties as A. Hence we can directly assume that A is closed. Now, let us show that with this assumption, if $f \in A$, then $|f| \in A$. By working with $\dfrac{f}{\|f\|_\infty}$ instead of f, we may assume that $\|f\|_\infty \leq 1$. Write

$$|f| = \sqrt{1 + (f^2 - 1)} = \sum_{n=0}^{\infty} \binom{1/2}{n} (f^2 - 1)^n,$$

where

$$\binom{1/2}{n} = \frac{\dfrac{1}{2}\left(\dfrac{1}{2} - 1\right) \cdots \left(\dfrac{1}{2} - n + 1\right)}{n!}.$$

This is just a consequence of the Taylor series of $(1 + x)^\alpha$. On the other hand, using Stirling's formula, we obtain that

$$\sum_{n \geq 1} \left| \binom{1/2}{n} \right| < \infty,$$

and since $\|f^2 - 1\|_\infty \leq 1$, it follows that $|f|$ can be approximated with polynomials in f to any degree of accuracy: simply take the partial sums of the above series. Since these polynomials in f belong to $\overline{A} = A$, so does $|f|$. This finishes the proof of the first step.

This shows that $\min(f, g) \in A$ and $\max(f, g) \in A$ if $f, g \in A$. Indeed, it is enough to note that $\min(f, g) = \dfrac{f + g - |f - g|}{2}$ and the hypothesis made on A shows that $\min(f, g) \in A$. The next step will be to prove that for all $x \neq y \in K$ and for all real numbers a, b, there exists $f \in A$ such that $f(x) = a$ and $f(y) = b$. This can be done by taking

$$f(t) = a + (b - a)\frac{g(t) - g(x)}{g(y) - g(x)},$$

where $g \in A$ is such that $g(y) \neq g(x)$. Then $f \in A$ (because A is stable under multiplication, addition, and contains the constant functions) and clearly f(x)=a, $f(y) = b$.

Consider $x, y \in K$ and a continuous function f defined on K. We know (by the previous step) that there exists $g_{x,y} \in A$ such that $g_{x,y}(x) = f(x)$ and $g_{x,y}(y) = f(y)$. Let $\epsilon > 0$ be fixed and also fix a point $x \in K$. By continuity of $f - g_{x,y}$ at y, for all $y \in K$, there exists an open interval I_y centered at y and such that $g_{x,y}(t) < f(t) + \epsilon$ for all $t \in I_y$. Because the intervals I_y are an open cover of the compact set K, there exist $y_1, y_2, \ldots, y_n \in K$ such that K is covered by $I_{y_1}, I_{y_2}, \ldots, I_{y_n}$. The function $g_x = \min(g_{x,y_1}, g_{x,y_2}, \ldots, g_{x,y_n})$ belongs to A (recall the first step of the proof), $g_x(x) = f(x)$, and also, $g_x(t) - f(t) < \epsilon$ for all $t \in K$. By continuity at x, for all $x \in K$, there exists an open interval J_x centered at x such that $f(t) < g_x(t) + \epsilon$ for all $t \in J_x$. Using again the compactness of K, we can extract from the open cover $(J_x)_{x \in K}$ a finite cover $J_{x_1}, J_{x_2}, \ldots, J_{x_s}$. Then the function $g = \max(g_{x_1}, g_{x_2}, \ldots, g_{x_s})$ belongs to A and satisfies $|g(x) - f(x)| < \epsilon$ for all $x \in K$. This finishes the proof of the theorem. $\square$

Here is an application, which can hardly be proved by elementary arguments.

Problem. Let $f : [0, 1] \to \mathbb{R}$ be a continuous function such that for all nonnegative integers n,

$$\int_0^1 f(x) x^n dx = 0.$$

Prove that $f = 0$.

Solution. We deduce that $\int_0^1 f(x) P(x) dx = 0$ for all polynomials P with real coefficients. Now, let $\epsilon > 0$ and let P be a polynomial such that $|P(x) - f(x)| < \epsilon$ for all $x \in [0, 1]$. Such a polynomial exists by the Stone-Weierstrass theorem applied to the algebra of polynomial functions on $[0, 1]$. Let M be such that $|f(x)| \le M$ for all $x \in [0, 1]$ (it exists by the continuity of f). Then

$$\int_0^1 f(x)^2 dx = \int_0^1 f(x)(f(x) - P(x)) dx \le M\epsilon.$$

Because this is true for all ϵ, it follows that $\int_0^1 f(x)^2 dx = 0$; hence $f = 0$, due to the continuity. $\square$

We present now a quite challenging problem, from Gazeta Matematica's contest:

Problem. Find all continuous functions $f : \mathbb{R} \to \mathbb{R}$ such that

$$f(2x - y) + f(2y - x) + 2f(x + y) = 9f(x) + 9f(y) \text{ for all } x, y.$$

Solution. We clearly have $f(0) = 0$. By taking $x = y$, we deduce that $f(2x) = 8f(x)$. Also, with $y = 0$, we deduce that $f(-x) = -f(x)$. Finally, by taking $y = 2x$ we deduce that $f(3x) = 27f(x)$. Using these results, we can immediately prove that $f(x) = x^3 f(1)$ for all $x \in A$, where $A = \{2^m 3^n | m, n \in \mathbb{Z}\}$. We claim that A is dense in $[0, \infty)$. This follows if we prove that $\mathbb{Z} + \mathbb{Z}(\log_2 3)$ is dense in $\mathbb{R}$. This is clearly a noncyclic additive subgroup of $\mathbb{R}$. The conclusion follows from the following general and useful result:

Lemma. *Any additive subgroup of the additive group of real numbers is either cyclic or dense.*

Indeed, let G be such a group and suppose it is not the trivial group. Let α be the greatest lower bound of the set $X = G \cap (0, \infty)$. Suppose that G is not cyclic. We claim that $\alpha \in G$. Otherwise, there exists a decreasing sequence x_n of positive elements of G that converges to α. Then $x_{n-1} - x_n$ is a positive element of G that converges to 0; thus $\alpha = 0 \in G$, which is a contradiction. Therefore, $\alpha \in G$. Now, assume that $\alpha \neq 0$. If $x \in G$ and $x > 0$, then $y = x - \alpha \cdot [\frac{x}{\alpha}] \in G$ and $\alpha > y \geq 0$. Thus $y = 0$ and so any element of G is an integer multiple of α, also a contradiction. This shows that $\alpha = 0$ and there exists a sequence $x_n \in G$ that converges to 0, with $x_n > 0$. Therefore, if (a, b) is any interval and $c \in (a, b)$, for sufficiently large n, we have $[\frac{c}{x_n}] x_n \in (a, b) \cap G$ and so G is dense. $\square$

Coming back to the problem, we deduce that $f(x) = x^3 f(1)$ on a dense set; therefore $f(x) = x^3 f(1)$ everywhere. Thus, any such function is of the form $f(x) = ax^3$ for some a. It is not difficult to check that any such function satisfies the given relation, so these are all solutions of the problem. $\square$

We said in the beginning of the chapter (and we saw in the previous problem) that dense sets are practical when dealing with continuous functions. The next problem, taken from the 2000 Putnam Competition highlights this assertion:

Problem. Let f be a continuous real function such that

$$f(2x^2 - 1) = 2xf(x)$$

for all real numbers x. Prove that $f(x) = 0$ for all $x \in [-1, 1]$.

Solution. Let $F(x) = \dfrac{f(\cos x)}{\sin x}$, defined and continuous except on $\mathbb{Z}\pi$. The given relation implies that $F(2x) = F(x)$. But F is clearly 2π periodic. Hence

$$F(1) = F(2^{k+1}) = F(2^{k+1} + 2n\pi) = F\left(1 + \frac{n\pi}{2^k}\right)$$

for all integers n, k. It is however easy to see that the set $\left\{\dfrac{n\pi}{2^k} \mid n, k \in \mathbb{Z}\right\}$ is dense in $\mathbb{R}$, since for instance

$$\lim_{n \to \infty} \frac{\left[2^n \cdot \dfrac{\alpha}{\pi}\right]\pi}{2^n} = \alpha, \text{ for all } \alpha \in \mathbb{R}.$$

This shows that F is constant on each interval of the form $(k\pi, (k+1)\pi)$. Change x and $-x$ in the given relation and conclude that f is odd; thus $F(x) = F(x + \pi)$. This, combined with the previous result shows that F is constant on $\mathbb{R} \setminus \mathbb{Z}\pi$ and this constant is 0, because F is odd. Thus $f(x) = 0$ for $x \in (-1, 1)$ and by continuity, this also holds for $x \in \{-1, 1\}$. $\square$

The next problem, given on the Putnam Competition in 1989, uses the density of the rational numbers in a very unexpected way:

Problem. Does there exist an uncountable set of subsets of $\mathbb{N}$ such that any two distinct subsets have finite intersection?

Solution. Surprisingly, the answer is yes and the construction is not intricate. Consider for each $x \in (0, 1)$ a sequence $(x_n)_{n \geq 1}$ of rationals converging to x.

Let A_x be the set consisting of the numbers x_n. Because the limit of a convergent sequence is unique, $A_x \cap A_y$ is finite if $x \neq y$. Also, we have uncountably many sets A_x, because $(0, 1)$ is uncountable. Now, it is enough to consider a bijection $f : \mathbb{Q} \to \mathbb{N}$ (an earlier chapter shows that f exists) and to consider the sets $f(A_x)$. They satisfy the conditions of the problem. $\square$

Proposed Problems

1. Prove that the set $A = \left\{ \dfrac{m}{2^n} \mid m \in \mathbb{Z} \,,\, n \in \mathbb{N} \right\}$ is dense in $\mathbb{R}$.

2. Let $(a_n)_{n \in \mathbb{N}}$ be a sequence of nonzero real numbers, converging to zero. Prove that the set $A = \{ma_n \mid m \in \mathbb{Z} \,,\, n \in \mathbb{N}\}$ is dense in $\mathbb{R}$.

3. Prove that the following sets are dense in $\mathbb{R}$:

   a) $A = \left\{ p + q\sqrt{2} \mid p, q \in \mathbb{Z} \right\}$;

   b) $B = \left\{ a\sqrt[3]{4} + b\sqrt[3]{2} + c \mid a, b, c \in \mathbb{Z} \right\}$.

4. Prove that the set $A = \left\{ \dfrac{m}{2^n} - \dfrac{n}{2^m} \mid m, n \in \mathbb{N} \right\}$ is dense in $\mathbb{R}$.

5. Prove that the set $A = \{m \sin n \mid m, n \in \mathbb{Z}\}$ is dense in $\mathbb{R}$.

6. Let x be a real number. Prove that the set of all numbers of the form $\{nx\}$, $n \in \mathbb{N}$, is dense in $[0, 1]$ if and only if x is irrational (here, $\{\cdot\}$ denotes the fractional part function).

7. Prove that the following sets are dense in $[1, \infty)$:

   a) $A = \left\{ \sqrt[m]{n} \mid m, n \in \mathbb{N} \,,\, m \geq 2 \right\}$;

   b) $A = \left\{ \sqrt[m]{1 + \dfrac{1}{2} + \dfrac{1}{3} + \cdots + \dfrac{1}{n}} \mid m, n \in \mathbb{N} \,,\, m \geq 2 \right\}$.

8. Let $(a_n)_{n \in \mathbb{N}}$ be a sequence of real numbers with limit $+\infty$, such that

$$\lim_{n \to \infty} (a_{n+1} - a_n) = 0.$$

Prove that the set $A = \{a_m - a_n \mid m, n \in \mathbb{N}\}$ is dense in $\mathbb{R}$.

9. Is $\sqrt{2}$ the limit of a sequence of numbers of the form $\sqrt[3]{m} - \sqrt[3]{n}$, with positive integers m, n?

10. Is $\sqrt{3}$ the limit of a sequence of numbers of the form

$$\frac{1}{n} + \frac{1}{n+1} + \cdots + \frac{1}{m},$$

with positive integers $m > n$?

11. Let $f, g : \mathbb{R} \to \mathbb{R}, f$ continuous, g monotone. Prove that if

$$f(x) = g(x),$$

for all rational numbers x, then $f = g$.

12. Let $f : \mathbb{R} \to \mathbb{R}$ be a continuous function such that

$$f(x) = f\left(x + \frac{1}{n}\right),$$

for all reals x and for all nonnegative integers n. Prove that f is constant.

13. Let $f : \mathbb{R} \to \mathbb{R}$ be a continuous function such that

$$f(x) = f(x + \sqrt{2}) = f(x + \sqrt{3}),$$

for all real numbers x. Prove that f is constant.

14. Define a function $f : \mathbb{N} \to [0, 2)$, by

$$f(n) = \{\sqrt{n}\} + \{\sqrt{n+1}\}.$$

Prove that $\mathrm{Im} f$ is dense in $[0, 2]$.

15. Let M be the set of real numbers of the form $\dfrac{m+n}{\sqrt{m^2 + n^2}}$, with m, n positive integers. Prove that if $u, v \in M$, $u < v$, then there exists $w \in M$ so that $u < w < v$.

16. Let $(a_n)_{n \in \mathbb{N}}$ be a sequence decreasing to zero, and let $(b_n)_{n \in \mathbb{N}}$ be a sequence increasing to infinity, such that the sequence $(b_{n+1} - b_n)_{n \in \mathbb{N}}$ is bounded. Prove that the set $A = \{a_m b_n \mid m, n \in \mathbb{N}\}$ is dense in $[0, \infty)$.

17. Let $(b_n)_{n \in \mathbb{N}}$ be a sequence increasing to infinity, such that

$$\lim_{n \to \infty} \frac{b_{n+1}}{b_n} = 1.$$

Prove that the set

$$A = \left\{ \frac{b_m}{b_n} \mid m, n \in \mathbb{N} \right\}$$

is dense in $[0, \infty)$.

18. Let $(a_n)_{n \geq 1} \subset (1, \infty)$ be a sequence convergent to 1, such that

$$\lim_{n \to \infty} (a_1 a_2 \cdots a_n) = \infty.$$

Prove that the set $A = \{a_{n+1} a_{n+2} \cdots a_m \mid n, m \in \mathbb{N}, \ n < m\}$ is dense in $[1, \infty)$.

19. Are there positive integers $m > n$ such that

$$\frac{\dfrac{1}{\sqrt{n+1}} + \dfrac{1}{\sqrt{n+2}} + \cdots + \dfrac{1}{\sqrt{m}}}{1 + \dfrac{1}{\sqrt{2}} + \cdots + \dfrac{1}{\sqrt{n}}} \in \left(\sqrt{2007} - \frac{1}{2007}, \sqrt{2007} + \frac{1}{2007} \right) ?$$

20. Is there a dense set of the space that does not contain four coplanar points?

Solutions

1. For each real α, we have

$$\lim_{n \to \infty} \frac{[2^n \alpha]}{2^n} = \alpha.$$

Indeed, using the inequality $x - 1 < [x] \leq x$, we derive

$$\frac{2^n \alpha - 1}{2^n} < \frac{[2^n \alpha]}{2^n} \leq \alpha$$

or

$$\alpha - \frac{1}{2^n} < \frac{[2^n \alpha]}{2^n} \leq \alpha.$$

In conclusion, for every real number α, there exists the sequence $\left(\frac{[2^n \alpha]}{2^n} \right)_{n \geq 1}$ with elements from A, which converges to α. Hence A is dense in $\mathbb{R}$.

Otherwise, let (a, b) be a nonempty interval of real numbers. The question is whether there exists an element of A lying in (a, b). In other words, can we find integers $m, n, \ n \geq 1$ so that

$$a < \frac{m}{2^n} < b?$$

This inequality is equivalent to

$$2^n a < m < 2^n b,$$

which we now prove. Indeed, first let us choose a positive integer $n >$ $\log_2(\frac{1}{b-a})$. Thus

$$2^n > \frac{1}{b-a} \Rightarrow 2^n b - 2^n a > 1.$$

Now the interval $(2^n a, 2^n b)$ contains at least one integer, because its length is greater than 1. Finally, if m is that integer, then

$$2^n a < m < 2^n b \Rightarrow a < \frac{m}{2^n} < b \Rightarrow \frac{m}{2^n} \in (a,b).$$

2. Let us consider real numbers $a < b$ and prove that the interval (a,b) contains at least one element of A. Because $\lim\limits_{n \to \infty} \dfrac{b-a}{|a_n|} = \infty$, we can find n_0 such that $\dfrac{b-a}{|a_{n_0}|} > 1$. The length of the interval $\left(\dfrac{a}{|a_{n_0}|}, \dfrac{b}{|a_{n_0}|} \right)$ is greater than 1, so it contains an integer m_0 :

$$\frac{a}{|a_{n_0}|} < m_0 < \frac{b}{|a_{n_0}|}.$$

This means that

$$a < m_0 |a_{n_0}| < b,$$

so A is dense in $\mathbb{R}$. In particular, for $a_n = \dfrac{1}{2^n}$, we deduce that the set

$$\left\{ \frac{m}{2^n} \mid m \in \mathbb{Z}, \ n \in \mathbb{N} \right\}$$

is dense in $\mathbb{R}$.

3. a) The sequence $a_n = (\sqrt{2} - 1)^n$ converges to zero; therefore the set

$$\left\{ m \cdot (\sqrt{2} - 1)^n \mid m \in \mathbb{Z}, \ n \in \mathbb{N} \right\}$$

is dense in $\mathbb{R}$, by the previous exercise. Now, there are integers a_n, b_n such that

$$(\sqrt{2} - 1)^n = a_n \sqrt{2} + b_n.$$

Indeed, using the binomial theorem, we have

$$(\sqrt{2} - 1)^n = \sum_{k=0}^{n} (-1)^k \binom{n}{k} (\sqrt{2})^{n-k}.$$

Hence

$$m(\sqrt{2} - 1)^n = ma_n\sqrt{2} + mb_n.$$

Because

$$\left\{p\sqrt{2} + q \mid p, q \in \mathbb{Z}\right\} \supset \left\{m(\sqrt{2} - 1)^n \mid m \in \mathbb{Z},\ n \in \mathbb{N}\right\},$$

the conclusion follows. Note that we could also use the fact that A is an additive subgroup of $\mathbb{R}$, which is not cyclic because $\sqrt{2} \notin \mathbb{Q}$. The conclusion follows from a result discussed in the theoretical part of the text.

b) The sequence $a_n = (\sqrt[3]{2} - 1)^n$ converges to zero. Therefore the set

$$\left\{m \cdot (\sqrt[3]{2} - 1)^n \mid m \in \mathbb{Z},\ n \in \mathbb{N}\right\}$$

is dense in $\mathbb{R}$. Now there are integers a_n, b_n, c_n such that

$$(\sqrt[3]{2} - 1)^n = a_n\sqrt[3]{4} + b_n\sqrt[3]{2} + c_n.$$

Indeed, using the binomial theorem, we have

$$(\sqrt[3]{2} - 1)^n = \sum_{k=0}^{n}(-1)^k \binom{n}{k} (\sqrt[3]{2})^{n-k}.$$

Hence

$$m(\sqrt[3]{2} - 1)^n = ma_n\sqrt[3]{4} + mb_n\sqrt[3]{2} + mc_n.$$

Because

$$\left\{a\sqrt[3]{4} + b\sqrt[3]{2} + c \mid a, b, c \in \mathbb{Z}\right\} \supset \left\{m(\sqrt[3]{2} - 1)^n \mid m \in \mathbb{Z},\ n \in \mathbb{N}\right\},$$

the conclusion follows.

4. If $x \in A$, then $-x \in A$. For $\alpha > 0$, note that

$$\lim_{n\to\infty} \left(\frac{[2^n\alpha]}{2^n} - \frac{n}{2^{[2^n\alpha]}}\right) = \alpha.$$

Thus for every real number $\alpha > 0$, the sequence

$$x_n = \frac{[2^n\alpha]}{2^n} - \frac{n}{2^{[2^n\alpha]}}, \quad n \geq 1$$

of elements from A converges to α. Since $0 \in A$, we are done.

5. We prove that for each $\varepsilon > 0$, there is a positive integer t such that

$$|t - 2k\pi| < \varepsilon,$$

for some integer k. Indeed, let us consider a positive integer N with $1/N < \varepsilon$. Two of the $N + 1$ numbers $\{2s\pi\}$ (where $\{x\}$ denotes the fractional part of the real number x), $1 \le s \le N + 1$, must be in the same interval of the form

$$\left[\frac{j}{N}, \frac{j+1}{N} \right), \quad 0 \le j \le N - 1.$$

If these numbers are $\{2p\pi\}$ and $\{2q\pi\}$, with $1 \le p < q \le N + 1$, then the absolute value of their difference is less than the length $1/N$ of each interval:

$$|\{2p\pi\} - \{2q\pi\}| < \frac{1}{N} < \varepsilon.$$

Thus we have $|t - 2k\pi| < \varepsilon$ for either $t = [2p\pi] - [2q\pi]$ and $k = p - q$, or $t = [2q\pi] - [2p\pi]$ and $k = q - p$ (depending on which of these choices gives a positive t).

Note that actually there are infinitely many pairs (t, k) with positive integer t and integer k such that $|t - 2k\pi| < \varepsilon$ (because we can repeat the above reasoning for N bigger and bigger) and that in these pairs t cannot assume only finitely many values (we encourage the reader to think of the complete proof of these statements); hence we can find such pairs (t, k) (having all the above properties) with t as large as we want.

Now we can find such a pair (t_1, k_1) for $\varepsilon = 1$, that is, with the property that

$$|t_1 - 2k_1\pi| < 1.$$

We then consider $\varepsilon = 1/2$, and we pick a pair (t_2, k_2), with $t_2 > t_1$, and such that

$$|t_2 - 2k_2\pi| < \frac{1}{2}.$$

In general, if (t_j, k_j) were found such that $t_1 < t_2 < \cdots < t_n$ and

$$|t_j - 2k_j\pi| < \frac{1}{j}$$

for every $1 \le j \le n$, we can pick a pair (t_{n+1}, k_{n+1}) of integers (t_{n+1} being positive) such that

$$|t_{n+1} - 2k_{n+1}\pi| < \frac{1}{n+1}$$

and $t_{n+1} > t_n$. Thus there exists a sequence $(t_n)_{n \geq 1}$ of positive integers such that

$$|t_n - 2k_n\pi| < \frac{1}{n},$$

with integers k_n. Consequently,

$$\lim_{n \to \infty} \sin t_n = 0$$

and finally, the set

$$\{m \sin t_n \mid m \in \mathbb{Z}, \; n \in \mathbb{N}^*\}$$

is dense in $\mathbb{R}$, by exercise 2. The conclusion follows because this is a subset of the set $\{m \sin n \mid m, n \in \mathbb{Z}\}$.

6. Let us suppose that x is rational, $x = \dfrac{p}{q}$, where p, q are relatively prime integers. In this case,

$$\{\{nx\} \mid n \in \mathbb{N}\} = \left\{0, \frac{1}{q}, \frac{2}{q}, \dots, \frac{p-1}{q}\right\}$$

so this set cannot be dense in $[0, 1]$.

So assume that x is an irrational number. We need to prove that the set of fractional parts of the numbers nx with positive integer n is dense in $[0, 1]$. We first show that if $\mu = \inf\{\{nx\} \mid n \in \mathbb{N}^*\}$, then $\mu = 0$.

Assume, to get a contradiction, that $\mu > 0$; thus we can find some positive integer N such that

$$\frac{1}{N+1} \leq \mu < \frac{1}{N}.$$

We conclude that $\mu < (\mu + 1)/(N + 1) < 1/N$; hence there exists a positive integer k such that

$$\frac{1}{N+1} \leq \mu \leq \{kx\} < \frac{\mu+1}{N+1} < \frac{1}{N}.$$

It follows that $1 \leq (N + 1)\{kx\} < 1 + 1/N$; therefore the integral part of $(N + 1)\{kx\}$ is 1. Thus we have

$$\begin{aligned}
\{(N + 1)kx\} &= \{(N + 1)kx - (N + 1)[kx]\} \\
&= \{(N + 1)\{kx\}\} \\
&= (N + 1)\{kx\} - [(N + 1)\{kx\}] \\
&= (N + 1)\{kx\} - 1 < \mu \\
&= \inf\{\{nx\} \mid n \in \mathbb{N}^*\},
\end{aligned}$$

which is clearly a contradiction. Note that this part of the proof does not require the assumption that x is irrational (and, indeed, we have already seen that $\mu = 0$ in the case of rational x).

Now let $0 \le a < b \le 1$ and consider a positive integer p such that p is small enough in order that $(b - a)/\{px\} > 1$. (Here is where we use the irrationality of x, which implies that $\{nx\}$ is never 0; hence we can find p with not only $\{px\} < b - a$ but also with $\{px\} > 0$.) Then, having length greater than 1, the interval from $a/\{px\}$ to $b/\{px\}$ contains at least one integer m, for which we have

$$0 \le a < m\{px\} < b \le 1.$$

Thus $0 < m\{px\} < 1$; hence m is positive and $\{mpx\} = m\{px\}$ is between a and b, finishing our proof. (Alternatively, one can proceed as in problem 2, by considering a sequence of positive fractional parts $\{n_k x\}$ tending to 0, etc.)

This is actually a celebrated theorem of Kronecker (1884). Note that it also implies the (already proved) result saying that if x is irrational, then $\mathbb{Z} + x\mathbb{Z}$ is dense in $\mathbb{R}$. In fact, we can prove that, for a positive irrational number x, the set $\{mx - n \,|\, m, n \in \mathbb{N}^*\}$ is dense in $\mathbb{R}$. Indeed, let us consider arbitrary real numbers a and b, with $a < b$. Assume first that $0 < a < b$, and choose a positive integer $p > b$; we then have $0 < a/p < b/p < 1$, and Kronecker's theorem implies the existence of some positive integer q with the property that $a/p < \{qx\} < b/p$, so that $a < pqx - p[qx] < b$, that is $mx - n \in (a, b)$ for the positive integers $m = pq$, and $n = p[qx]$ (n is positive because $x > 0$ and q can be chosen as big as we want, for instance $q > 1/x$). Now if a and b are arbitrary real numbers with $a < b$, there is a positive integer k such that $0 < a + k < b + k$; thus (according to what we just proved) there are positive integers m and n such that $a + k < mx - n < b + k$; thus $mx - (n + k)$ is in (a, b), which we had to prove.

7. Let $1 < a < b$. Because

$$\lim_{m \to \infty} (b^m - a^m) = \infty,$$

we can find m_0 such that

$$b^{m_0} - a^{m_0} > 1.$$

Further, the interval (a^{m_0}, b^{m_0}) contains at least one integer,

$$n_0 \in (a^{m_0}, b^{m_0}) \cap \mathbb{Z};$$

thus

$$a^{m_0} < n_0 < b^{m_0} \Leftrightarrow a < \sqrt[m_0]{n_0} < b.$$

Moreover, because $a^{m_0} > 1$, we can denote by k the smallest integer so that

$$1 + \frac{1}{2} + \cdots + \frac{1}{k} > a^{m_0}.$$

But

$$b^{m_0} > a^{m_0} + 1,$$

so

$$a^{m_0} < 1 + \frac{1}{2} + \cdots + \frac{1}{k} < b^{m_0}$$

or

$$a < \sqrt[m_0]{1 + \frac{1}{2} + \cdots + \frac{1}{k}} < b.$$

8. Let $a, b \in \mathbb{R}$, $a < b$. Let n_0 be so that $m > n_0$ implies

$$a_{m+1} - a_m < b - a.$$

We have

$$\lim_{m \to \infty} (a_m - a_{n_0}) = \infty,$$

so let $m_0 > n_0 + 1$ be the smallest integer with the property that

$$a_{m_0} - a_{n_0} > a.$$

Therefore $a_{m_0-1} - a_{n_0} \leq a$ and

$$a_{m_0} - a_{n_0} = (a_{m_0} - a_{m_0-1}) + (a_{m_0-1} - a_{n_0}) < (b - a) + a = b.$$

In conclusion, $a < a_{m_0} - a_{n_0} < b$.
9. The answer is yes. The sequence

$$a_n = \sqrt[3]{n}, \quad n \geq 1$$

has limit ∞ and

$$\lim_{n \to \infty} (a_{n+1} - a_n) = 0.$$

Indeed,

$$\lim_{n \to \infty} (a_{n+1} - a_n) = \lim_{n \to \infty} \left(\sqrt[3]{n+1} - \sqrt[3]{n} \right)$$

$$= \lim_{n \to \infty} \frac{1}{\sqrt[3]{(n+1)^2} + \sqrt[3]{n(n+1)} + \sqrt[3]{n^2}} = 0.$$

As we have proved, the set

$$A = \{a_m - a_n \mid m, n \in \mathbb{N}\} = \{\sqrt[3]{m} - \sqrt[3]{n} \mid m, n \in \mathbb{N}\}$$

is dense in $\mathbb{R}$.

10. The answer is yes. The sequence

$$a_n = 1 + \frac{1}{2} + \cdots + \frac{1}{n}, \quad n \geq 1$$

is increasing, unbounded and

$$\lim_{n \to \infty} (a_{n+1} - a_n) = 0.$$

For all $m > n$, we have

$$a_m - a_n = \frac{1}{n+1} + \cdots + \frac{1}{m}.$$

As we have proved, the set

$$A = \{a_m - a_n \mid m, n \in \mathbb{N}\} = \left\{ \frac{1}{n+1} + \cdots + \frac{1}{m} \mid m, n \in \mathbb{N}, \ m > n \right\}$$

is dense in $\mathbb{R}_+$.

11. For each real number x, consider two sequences of rational numbers $(a_n)_{n \geq 1}$ and $(b_n)_{n \geq 1}$ converging to x such that

$$a_n \leq a_{n+1} \leq x \leq b_{n+1} \leq b_n,$$

for all positive integers n. The function g is monotone, say increasing; thus the sequence $(g(a_n))_{n \geq 1}$ is increasing and bounded above:

$$g(a_n) \leq g(x).$$

Let $l = \lim_{n \to \infty} g(a_n)$. On the other hand,

$$g(a_n) = f(a_n) \to f(x),$$

because of continuity of f at x; thus $l = f(x)$. Analogously, we deduce

$$\lim_{n \to \infty} g(a_n) = \lim_{n \to \infty} g(b_n) = f(x).$$

By taking $n \to \infty$ in the relations

$$g(a_n) \leq g(x) \leq g(b_n),$$

we derive

$$f(x) \le g(x) \le f(x),$$

so $f = g$.

12. We have

$$f\left(x + \frac{2}{n}\right) = f\left(x + \frac{1}{n}\right) = f(x)$$

and by replacing x with $x - \frac{2}{n}$,

$$f(x) = f\left(x - \frac{1}{n}\right) = f\left(x - \frac{2}{n}\right).$$

By induction,

$$f(x) = f\left(x + \frac{m}{n}\right)$$

for all integers $m, n \ne 0$. For $x = 0$,

$$f\left(\frac{m}{n}\right) = f(0),$$

for all $m \in \mathbb{Z}$, $n \in \mathbb{N}^*$. The continuous function f is equal to the constant function $f(0)$ on $\mathbb{Q}$, so

$$f(x) = f(0),$$

for all reals x.

There is also another argument: we proved that a continuous, periodic, nonconstant function has a minimum period. It is not this case, when f has periods of the form $\frac{1}{n}$, for every positive integer n.

13. First, with $x \to x + \sqrt{2}$ and so on, we obtain

$$f(x) = f(x + \sqrt{2}) = f(x + 2\sqrt{2}) = \ldots,$$

and with $x \to x + \sqrt{3}$ and so on, we obtain

$$f(x) = f(x + \sqrt{3}) = f(x + 2\sqrt{3}) = \ldots$$

so

$$f(x) = f(x + m\sqrt{2}) = f(x + n\sqrt{3}),$$

for all integers m, n. Moreover,

$$f(x + m\sqrt{2} + n\sqrt{3}) = f(x + m\sqrt{2}) = f(x)$$

and for $x = 0$,

$$f\left(m\sqrt{2} + n\sqrt{3}\right) = f(0),$$

for all $m, n \in \mathbb{Z}$. It follows that f is constant because the set

$$\{m\sqrt{2} + n\sqrt{3} \mid m, n \in \mathbb{Z}\}$$

is dense in $\mathbb{R}$ (using the same argument as in exercise 3a).

14. Let $0 < a < b < 2$. For each positive integer k, with

$$k > \frac{1}{b - a}\left(2 - \frac{b^2 - a^2}{4}\right),$$

the length of the interval

$$\left(\left(k + \frac{a}{2}\right)^2, \left(k + \frac{b}{2}\right)^2 - 1\right)$$

is greater than 1, so at least one integer n lies in this interval. For such an n, we have

$$\left(k + \frac{a}{2}\right)^2 < n < \left(k + \frac{b}{2}\right)^2 - 1$$

$$\Rightarrow k + \frac{a}{2} < \sqrt{n} < \sqrt{n + 1} < k + \frac{b}{2}.$$

Hence

$$\frac{a}{2} < \{\sqrt{n}\} < \frac{b}{2}$$

and

$$\frac{a}{2} < \{\sqrt{n + 1}\} < \frac{b}{2}.$$

In conclusion,

$$a < \{\sqrt{n}\} + \{\sqrt{n + 1}\} < b.$$

15. The function

$$f : [1, \infty) \cap \mathbb{Q} \to \mathbb{R},$$

given by the formula

$$f(x) = \frac{x + 1}{\sqrt{x^2 + 1}}$$

is decreasing. The set M is the image of the function f. Let $u, v \in M$, $u < v$ be of the form

$$u = \frac{m_1 + n_1}{\sqrt{m_1^2 + n_1^2}}, \quad v = \frac{m_2 + n_2}{\sqrt{m_2^2 + n_2^2}},$$

with $m_1 > n_1$ and $m_2 > n_2$. We can write

$$u = \frac{\dfrac{m_1}{n_1} + 1}{\sqrt{\dfrac{m_1^2}{n_1^2} + 1}}, \quad v = \frac{\dfrac{m_2}{n_2} + 1}{\sqrt{\dfrac{m_2^2}{n_2^2} + 1}},$$

or

$$u = f\left(\frac{m_1}{n_1}\right), \quad v = f\left(\frac{m_2}{n_2}\right).$$

Hence from

$$\frac{m_2}{n_2} \leq \frac{\dfrac{m_1}{n_1} + \dfrac{m_2}{n_2}}{2} \leq \frac{m_1}{n_1}$$

we deduce that

$$w = f\left(\frac{\dfrac{m_1}{n_1} + \dfrac{m_2}{n_2}}{2}\right) \in M$$

lies between u and v, because of the monotony of f.

16. Assume that $b_{n+1} - b_n < M$. Let $0 < a < b$. Choose an integer m such that

$$a_m < \frac{b - a}{M}.$$

We can find an integer n such that

$$\frac{a}{a_m} < b_n < \frac{b}{a_m},$$

because the situation

$$b_k < \frac{a}{a_m} < \frac{b}{a_m} < b_{k+1}$$

is not possible. Indeed, this would imply

$$b_{k+1} - b_k > \frac{b-a}{a_m} > M,$$

which is a contradiction. In conclusion, $a < a_m b_n < b$.

17. With the notation $a_n = \ln b_n$, we have $\lim a_n = \infty$ and

$$\lim_{n\to\infty} (a_{n+1} - a_n) = \lim_{n\to\infty} \ln \frac{b_{n+1}}{b_n} = 0.$$

As we have proved, the set

$$\{a_m - a_n \mid m, n \in \mathbb{N}\}$$

is dense in $\mathbb{R}$, so the set

$$\left\{ \frac{b_m}{b_n} \mid m, n \in \mathbb{N} \right\}$$

is dense in $[0, \infty)$.

18. Let us define the sequence $(b_n)_{n\geq 1}$ by the formula

$$b_n = a_1 a_2 \cdots a_n.$$

We have

$$\frac{b_{n+1}}{b_n} = \frac{a_1 a_2 \cdots a_n a_{n+1}}{a_1 a_2 \cdots a_n} = a_{n+1} > 1,$$

so the sequence $(b_n)_{n\geq 1}$ is increasing. Then

$$\lim_{n\to\infty} \frac{b_{n+1}}{b_n} = \lim_{n\to\infty} a_{n+1} = 1;$$

hence we have the hypothesis of the previous problem. Thus the set

$$A = \left\{ \frac{b_m}{b_n} \mid m, n \in \mathbb{N}, \ n \geq 1 \right\}$$

is dense in $[0, \infty)$ and the conclusion follows if we take into account that

$$\frac{b_m}{b_n} = \frac{a_1 a_2 \cdots a_m}{a_1 a_2 \cdots a_n} = a_{n+1} a_{n+2} \cdots a_m.$$

19. The answer is yes. We will use problem 17 with the sequence

$$b_n = 1 + \frac{1}{\sqrt{2}} + \cdots + \frac{1}{\sqrt{n}}, \quad n \geq 1.$$

As we know, the sequence $(b_n)_{n \geq 1}$ is unbounded and is clearly increasing. Moreover,

$$\frac{b_{n+1}}{b_n} = 1 + \frac{1}{b_n \sqrt{n+1}},$$

so $\dfrac{b_{n+1}}{b_n}$ converges to 1. Hence the set

$$A = \left\{ \frac{b_m}{b_n} \mid m, n \in \mathbb{N}, \ m, n \geq 1 \right\}$$

is dense in $[0, \infty)$. But for $m > n$, we have

$$\frac{b_m}{b_n} = \frac{\left(1 + \dfrac{1}{\sqrt{2}} + \cdots + \dfrac{1}{\sqrt{n}} \right) + \left(\dfrac{1}{\sqrt{n+1}} + \dfrac{1}{\sqrt{n+2}} + \cdots + \dfrac{1}{\sqrt{m}} \right)}{1 + \dfrac{1}{\sqrt{2}} + \cdots + \dfrac{1}{\sqrt{n}}}$$

$$= 1 + \frac{\dfrac{1}{\sqrt{n+1}} + \dfrac{1}{\sqrt{n+2}} + \cdots + \dfrac{1}{\sqrt{m}}}{1 + \dfrac{1}{\sqrt{2}} + \cdots + \dfrac{1}{\sqrt{n}}}.$$

Finally, if $m > n$ are so that

$$\frac{b_m}{b_n} \in \left(1 + \sqrt{2007} - \frac{1}{2007}, 1 + \sqrt{2007} + \frac{1}{2007} \right),$$

then by subtracting unity, we obtain

$$\frac{\dfrac{1}{\sqrt{n+1}}+\dfrac{1}{\sqrt{n+2}}+\cdots+\dfrac{1}{\sqrt{m}}}{1+\dfrac{1}{\sqrt{2}}+\cdots+\dfrac{1}{\sqrt{n}}} \in \left(\sqrt{2007}-\frac{1}{2007}, \sqrt{2007}+\frac{1}{2007}\right).$$

20. Surprisingly, the answer is yes. Indeed, take a_n to be the sequence of points in space with all coordinates rational numbers and consider the balls B_n centered at a_n and having radius $\frac{1}{n}$. It is clear that any sequence $(x_n)_{n \geq 1}$ with $x_n \in B_n$ for any $n \geq 1$ defines a dense subset of the space. But since a ball cannot be covered by a finite number of planes, by induction, we can construct a sequence $x_n \in B_n$ such that x_n is not on any plane determined by three of the points $x_1, x_2, \ldots, x_{n-1}$. This shows the existence of such a set.

Chapter 8
The Nested Intervals Theorem

It is well known that every monotone sequence of real numbers has a limit, finite or infinite. If $(x_n)_{n\geq 1}$ is increasing, then

$$\lim_{n\to\infty} x_n = \sup_{n\geq 1} x_n$$

(where it is possible for the supremum to be ∞) and if $(x_n)_{n\geq 1}$ is decreasing, then

$$\lim_{n\to\infty} x_n = \inf_{n\geq 1} x_n$$

(where the infimum can be $-\infty$). The monotone convergence theorem gives further information, namely, that every bounded monotone sequence of real numbers is convergent, i.e., it surely has a finite limit. The proof is based on the least upper bound axiom, asserting that any nonempty bounded above set of real numbers has a least upper bound. (Actually this statement is equivalent to the nested intervals theorem that follows, and each of them expresses the completeness of the system of real numbers.)

As a direct consequence of the monotone convergence theorem, we give the following result. We note that if $I = [a, b]$ is a closed interval, then $l(I) = b - a$ denotes the length of I.

The Nested Intervals Theorem. *Let*

$$I_1 \supseteq I_2 \supseteq \cdots \supseteq I_n \supseteq \cdots,$$

be a decreasing sequence of closed intervals. Then

$$\bigcap_{n\geq 1} I_n \neq \emptyset.$$

© Springer Science+Business Media LLC 2017
T. Andreescu et al., *Mathematical Bridges*, DOI 10.1007/978-0-8176-4629-5_8

Moreover, $\bigcap_{n\geq 1} I_n$ *is a singleton in case* $\lim_{n\to\infty} l(I_n) = 0$.

Proof. If $I_n = [a_n, b_n]$, $n \geq 1$, then from the inclusions

$$[a_1, b_1] \supseteq [a_2, b_2] \supseteq \cdots \supseteq [a_n, b_n] \supseteq \cdots$$

we deduce the inequalities

$$a_1 \leq a_2 \leq \cdots \leq a_n \leq \cdots \leq b_n \leq \cdots \leq b_2 \leq b_1.$$

Hence $(a_n)_{n\geq 1}$ is increasing and $(b_n)_{n\geq 1}$ is decreasing. Let us denote

$$a = \lim_{n\to\infty} a_n, \quad b = \lim_{n\to\infty} b_n.$$

From $a_n \leq b_n$, it follows $a \leq b$. Further, we will prove that $\bigcap_{n\geq 1} I_n = [a, b]$. If $z \in \bigcap_{n\geq 1} I_n$, then $z \in I_n \Rightarrow a_n \leq z \leq b_n$ for all integers $n \geq 1$. By taking $n \to \infty$ in the last inequality, we derive $a \leq z \leq b$, thus $z \in [a, b]$.

Conversely, if $z \in [a, b]$, then $a_n \leq a \leq z \leq b \leq b_n$ so $a_n \leq z \leq b_n$, for all integers $n \geq 1$ and then $z \in \bigcap_{n\geq 1} I_n$.

For the second part, we have

$$\lim_{n\to\infty} (b_n - a_n) = 0,$$

so

$$\lim_{n\to\infty} a_n = \lim_{n\to\infty} b_n = a.$$

Finally, $\bigcap_{n\geq 1} I_n = \{a\}$. Indeed, if $x \in \bigcap_{n\geq 1} I_n$, then $a_n \leq x \leq b_n$ for all $n \geq 1$, and for $n \to \infty$, we deduce $a \leq x \leq a$, which is $x = a$. $\square$

This theorem is an important tool to establish some basic results in mathematical analysis, as we can see below.

Problem. Let $f : [a, b] \to \mathbb{R}$ be a continuous function. If $f(a)f(b) \leq 0$, prove that there exists $c \in [a, b]$ such that $f(c) = 0$.

Solution. Let us consider the decomposition

$$[a, b] = \left[a, \frac{a+b}{2}\right] \cup \left[\frac{a+b}{2}, b\right]$$

and denote by $I_1 = [a_1, b_1]$ one of the intervals $\left[a, \dfrac{a+b}{2}\right]$ or $\left[\dfrac{a+b}{2}, b\right]$ such that $f(a_1)f(b_1) \leq 0$.

If the

$$I_k = [a_k, b_k], \; 1 \leq k \leq n-1$$

are already defined, then let $I_n = [a_n, b_n]$ be one of the intervals

$$\left[a_{n-1}, \frac{a_{n-1} + b_{n-1}}{2}\right] \quad \text{or} \quad \left[\frac{a_{n-1} + b_{n-1}}{2}, b_{n-1}\right]$$

such that $f(a_n)f(b_n) \leq 0$. Inductively, we defined the decreasing sequence $(I_n)_{n\geq 1}$ of closed intervals with $l(I_n) = \dfrac{b-a}{2^n}$ and

$$f(a_n)f(b_n) \leq 0, \tag{8.1}$$

for all $n \geq 1$. Let $\{c\} = \bigcap_{n\geq 1} I_n$. As we proved, $a_n \to c$ and $b_n \to c$. If we take $n \to \infty$ in (8.1), we get that $(f(c))^2 \leq 0$, because of the continuity of f.

In conclusion, $f(c) = 0$. $\square$

In fact, this is basically a proof of the following:

Theorem (Cauchy-Bolzano). *Let $I \subseteq \mathbb{R}$ be an interval. Then any continuous function $f : I \to \mathbb{R}$ has the intermediate value property.*

Proof. We have to show that for any $a, b \in I$ and for each λ between $f(a)$ and $f(b)$, we can find c between a and b such that $f(c) = \lambda$.

We assume that $a < b$, without loss of generality. The function $g(x) = f(x) - \lambda$ is continuous on $[a, b]$, and, from the bounds on λ, $g(a)g(b) \leq 0$. Consequently, $g(c) = 0$ for some $c \in [a, b]$. $\square$

Cauchy-Bolzano's theorem has many other interesting proofs. We present here a proof of this theorem using the compactness of $[a, b]$.

Assume, by way of contradiction, that the continuous function $g : [a, b] \to \mathbb{R}$ takes the values $g(a)$ and $g(b)$ of opposite signs and $g(x) \neq 0$, for all x in $[a, b]$. For each $x \in [a, b]$, we have $g(x) \neq 0$, so there is an open interval $I_x \ni x$ such that g has the same sign as $g(x)$ on $I_x \cap [a, b]$.

The family $(I_x)_{x\in[a,b]}$ is an open cover for the compact set $[a, b]$. Consequently, there are

$$x_1, x_2, \ldots, x_n \in [a, b]$$

such that

$$I_{x_1} \cap I_{x_2} \cap \cdots \cap I_{x_n} \supseteq [a, b].$$

Assume that

$$x_1 < x_2 < \cdots < x_n,$$

so the intervals

$$I_{x_1}, I_{x_2}, \ldots, I_{x_n}$$

are denoted from left to right. In particular, $a \in I_{x_1}$ and $b \in I_{x_n}$.

The function g keeps the same sign on $I_{x_1} \cap [a, b]$ and $I_{x_2} \cap [a, b]$.

Because $I_{x_1} \cap I_{x_2} \neq \emptyset$, g will keep the same sign on $(I_{x_1} \cup I_{x_2}) \cap [a, b]$. Inductively, g keeps the same sign on

$$(I_{x_1} \cup I_{x_2} \cup \cdots \cup I_{x_k}) \cap [a, b],$$

for each $1 \leq k \leq n$. In particular, g keeps the same sign on

$$(I_{x_1} \cup I_{x_2} \cup \cdots \cup I_{x_n}) \cap [a, b] = [a, b],$$

which contradicts the fact that $g(a)$ and $g(b)$ have opposite signs.

Moreover, we know that continuous functions transform connected sets into connected sets. The interval $[a, b]$ is connected, so $g([a, b])$ is connected. This remark can be another proof of the Cauchy-Bolzano theorem. $\square$

The next application is the following fixed point result:

Theorem **(Knaster).** *Any increasing function $f : [a, b] \to [a, b]$ has at least one fixed point.*

   Proof. We have $f(a) \geq a$ and $f(b) \leq b$. If

$$f\left(\frac{a+b}{2}\right) \leq \frac{a+b}{2},$$

then put $a_1 = a$, $b_1 = \dfrac{a+b}{2}$. In case

$$f\left(\frac{a+b}{2}\right) > \frac{a+b}{2},$$

we put $a_1 = \dfrac{a+b}{2}$, $b_1 = b$. Either way, for the interval $I_1 = [a_1, b_1]$, we have

$$f(a_1) \geq a_1, \quad f(b_1) \leq b_1.$$

By induction, we can define the decreasing sequence $I_n = [a_n, b_n]$, $n \geq 1$ of closed intervals with

$$l(I_n) = \frac{b - a}{2^n}$$

and

$$f(a_n) \geq a_n, \quad f(b_n) \leq b_n, \tag{8.2}$$

for all $n \geq 1$. Let c be the common limit of $(a_n)_{n\geq 1}$ and $(b_n)_{n\geq 1}$. If we take $n \to \infty$ in

$$a_n \leq f(a_n) \leq f(b_n) \leq b_n,$$

we derive

$$\lim_{n\to\infty} f(a_n) = \lim_{n\to\infty} f(b_n) = c.$$

Further, by the monotonicity of f and from the inequalities

$$a_n \leq c \leq b_n,$$

we deduce

$$f(a_n) \leq f(c) \leq f(b_n).$$

Finally, for $n \to \infty$, we have $c \leq f(c) \leq c$, that is, $f(c) = c$. $\square$

Let us try to extend the lemma of the nested intervals to higher dimensions. Let us recall first some basic facts about the topology of $\mathbb{R}^n$. By definition, a set $A \subseteq \mathbb{R}^n$ is *open* if for all $x \in A$, there exists $r > 0$ such that $B(x, r) \subset A$. Here, $B(x, r)$ is the open ball centered at x and having radius r, that is, the set

$$\{y \in \mathbb{R}^n \mid \|x - y\| < r\},$$

where $\|x\| = \sqrt{x_1^2 + \cdots + x_n^2}$ is the Euclidean norm. A is called *closed* if $\mathbb{R}^n \setminus A$ is open. The *interior* of a set A is the largest open set contained in A. If $Int(A)$ denotes the interior of A, then $x \in Int(A)$ if and only if there exists $r > 0$ such that $B(x, r) \subset A$. Finally, if A is bounded, the *diameter* of A is defined by

$$diam(A) = \sup_{x,y \in A} \|x - y\|.$$

A famous theorem asserts that for a subset A of $\mathbb{R}^n$, the following statements are equivalent:

1) A is compact, that is if $A \subset \bigcup_{i \in I} O_i$, where O_i are open sets and I is an arbitrary

   set, then there exists a finite subset J of I such that $A \subset \bigcup_{i \in J} O_i$.

2) A is closed and bounded;

3) Any sequence whose terms are in A has a subsequence converging to an element of A.

We will not prove this theorem here, but will use it to generalize the lemma of the nested intervals:

Theorem. *Let $(K_n)_{n \geq 1}$ be a decreasing sequence of nonempty compact sets. Then $\bigcap_{n \geq 1} K_n$ is nonempty and moreover, if $\lim_{n \to \infty} diam(K_n) = 0$, then $\bigcap_{n \geq 1} K_n$ is a singleton.*

Let us briefly present the proof. Choose a sequence $x_n \in K_n$ and observe that all its terms lie in the compact set K_1. Thus it has a subsequence $(x_{n_k})_{k \geq 1}$ which converges to a certain $x \in K_1$. Observe that $x_{n_k} \in K_p$ for all $k \geq p$ (because $n_k \geq k \geq p$, thus $x_{n_k} \in K_{n_k} \subset K_p$), thus $x \in K_p$, too (because K_p is closed). Because p was arbitrary, $x \in \bigcap_{p \geq 1} K_p$, and the first part is proved. The second part is obvious: if $a \neq b \in \bigcap_{n \geq 1} K_n$, then

$$diam(K_n) \geq \|a - b\|, \quad \forall \, n \geq 1$$

thus passing to the limit, $a = b$, a contradiction. $\square$

We end this theoretical part with a very useful result called Baire's theorem. It has a very easy statement, but the consequences are striking.

Theorem (Baire).

1) Let $(F_n)_{n \geq 1}$ be a sequence of closed sets in $\mathbb{R}^d$, each of them having empty interior. Then $\bigcup_{n \geq 1} F_n$ also has empty interior.

2) Let $(O_n)_{n \geq 1}$ be a sequence of dense open sets in $\mathbb{R}^d$. Then $\bigcap_{n \geq 1} O_n$ is also dense.

Observe that it is enough to prove just the second assertion because the first one is obtained by considering $O_n = \mathbb{R}^d \setminus F_n$. So, let us fix an open ball $B(x, r)$ and let us prove that $\bigcap_{n \geq 1} O_n$ intersects this ball. Because O_1 is dense in $\mathbb{R}^d$, it intersects $B(x, r)$ and because it is open, there is a closed ball $\overline{B(x_1, r_1)} \subset B(x, r) \cap O_1$. Because O_2 is dense in $\mathbb{R}^d$ and open, there is $r_2 \leq \frac{r_1}{2}$ and x_2 such that the closed ball $\overline{B(x_2, r_2)} \subset B(x_1, r_1) \cap O_2$. Inductively, we construct x_n and $r_{n+1} \leq \frac{r_n}{2}$ such that $\overline{B(x_{n+1}, r_{n+1})} \subset B(x_n, r_n) \cap O_{n+1}$.

Now, $\|x_{n+1} - x_n\| \leq r_n \leq \dfrac{r_1}{2^{n-1}}$, thus $\displaystyle\sum_{n \geq 1}(x_{n+1} - x_n)$ is absolutely convergent,
thus convergent. This implies that x_n converges to a certain $\rho \in \mathbb{R}^d$. It is immediate
to see that $\rho \in \left(\displaystyle\bigcap_{n \geq 1} O_n\right) \cap B(x, r)$, which finishes the proof of the theorem. $\square$

Actually, using exactly the same idea (except for the argument showing that x_n
converges) one can show that we can replace $\mathbb{R}^d$ by any complete metric space. Here
is a beautiful application, for which you can find another solution by studying the
proposed problems.

Problem. Prove that there is no function $f : \mathbb{R} \to \mathbb{R}$ which is continuous exactly
at the rational numbers.

Solution. We will prove first that the set of points where a function is continuous
is a countable intersection of open sets. Indeed, let us define

$$\omega(f, x) = \inf_{r > 0} \ \sup_{a,b \in (x-r,x+r)} |f(a) - f(b)|$$

and $O_n = \left\{x \in \mathbb{R} \mid \omega(f, x) < \dfrac{1}{n}\right\}$.

We claim that f is continuous precisely on $\displaystyle\bigcap_{n \geq 1} O_n$. Indeed, it is pretty clear that f
is continuous at x if and only if $\omega(f, x) = 0$ (just use the definition of continuity and
the triangle inequality).

Now, we prove that each O_n is open. Let $x \in O_n$ and let $r > 0$ be such that

$$\sup_{a,b \in (x-r,x+r)} |f(a) - f(b)| < \frac{1}{n}.$$

Clearly, there is $\delta > 0$ such that if $|x - y| < \delta$, then

$$\left(y - \frac{r}{2}, y + \frac{r}{2}\right) \subset (x - r, x + r),$$

thus

$$\sup_{a,b \in (y-\frac{r}{2},y+\frac{r}{2})} |f(a) - f(b)| \leq \sup_{a,b \in (x-r,x+r)} |f(a) - f(b)| < \frac{1}{n}$$

and so $\omega(f, y) < \dfrac{1}{n}$. This shows that $(x - \delta, x + \delta) \subset O_n$ and so O_n is open.

Finally, suppose that $\mathbb{Q} = \bigcap_{n \geq 1} O_n$, where O_n are open sets. Then

$$\mathbb{R} \setminus \mathbb{Q} = \bigcup_{n \geq 1} (\mathbb{R} \setminus O_n) \Rightarrow \mathbb{R} = \bigcup_{x \in \mathbb{Q}} \{x\} \cup \bigcup_{n \geq 1} (\mathbb{R} \setminus O_n).$$

Because $\{x\}$ and $\mathbb{R} \setminus O_n$ are closed sets, using Baire's theorem, we deduce that either some $\{x\}$ or some $\mathbb{R} \setminus O_n$ has nonempty interior. The first case is clearly impossible. The second case would imply that $\mathbb{R} \setminus \mathbb{Q}$ has nonempty interior, that is, $\mathbb{Q}$ is not dense in $\mathbb{R}$, which is again impossible. $\square$

Proposed Problems

1. Let $(x_n)_{n \geq 1}$ be a sequence of real numbers such that x_{n+2} lies between x_n and x_{n+1}, for each integer $n \geq 1$. If $x_{n+1} - x_n \to 0$, as $n \to \infty$, prove that $(x_n)_{n \geq 1}$ is convergent.

2. Prove that the interval $[0, 1]$ is not a countable set, using the lemma of the nested intervals.

3. Consider a continuous function $g : [a, b] \to \mathbb{R}$, with $g(a) \leq g(b)$ and $f : [a, b] \to [g(a), g(b)]$ increasing. Prove that there exists $c \in [a, b]$ such that $f(c) = g(c)$.

4. Is there a function $f : [0, 1] \to \mathbb{R}$ with the property that

$$\lim_{x \to a} |f(x)| = \infty,$$

for all rational numbers a from $[0, 1]$?

5. Let $f : [0, 1] \to [0, \infty)$ be an integrable function such that $f \geq 0$ and

$$\int_0^1 f(x) \, dx = 0.$$

Prove that there is c in $[0, 1]$ with $f(c) = 0$. Deduce that the set of zeroes of the function f is dense in $[0, 1]$.

6. Let $f, g : [a, b] \to \mathbb{R}$ be two functions whose continuity sets are dense in $[0, 1]$. Prove that there exists z in $[a, b]$ such that f and g are both continuous at z. (By *the continuity set* of a function f, denoted C_f we mean the set of all points x such that f is continuous at x).

7. It is well known that the Riemann (or Thomae) function is continuous at every irrational point and it is discontinuous at every rational point.

   (a) Does there exist a function $s : [0, 1] \to \mathbb{R}$ which is continuous at every rational point from $[0, 1]$ and discontinuous at every irrational point from $[0, 1]$?

   (b) Is there any continuous function $f : \mathbb{R} \to \mathbb{R}$ that maps the rational numbers to irrationals and vice versa?

8. (a) Find a function $f : \mathbb{R} \to \mathbb{R}_+$ with the property:

$$\text{any } x \in \mathbb{Q} \text{ is a point of strict local minimum of } f. \qquad (*)$$

(b) Find a function $f : \mathbb{Q} \to \mathbb{R}_+$ with the property that every point is a point of strict local minimum and f is unbounded on any set $\mathbb{Q} \cap I$, with I an interval (that is not reduced to one single point).

(c) Let $f : \mathbb{R} \to \mathbb{R}_+$ be a function unbounded on any set $\mathbb{Q} \cap I$ (with I an interval). Prove that f does not have the property $(*)$.

9. Let $f : \mathbb{R} \to \mathbb{R}$ be a continuous function with the property that $\lim\limits_{n \to \infty} f(nx) = 0$ for all $x \in \mathbb{R}$. Prove that $\lim\limits_{x \to \infty} f(x) = 0$.

10. Let $f : [0, 1] \to [0, 1]$ be a continuous function with the property that

$$0 \in \{x, f(x), f(f(x)), f(f(f(x))), \ldots\}$$

for all $x \in [0, 1]$. Prove that f^n is identically 0 for some n.

11. Prove Lagrange's mean value theorem by using the nested intervals theorem.

Solutions

1. Let I_n be the closed interval with extremities at x_n and x_{n+1}. The length of the interval I_n is equal to $|x_{n+1} - x_n|$, so

$$\lim_{n \to \infty} l(I_n) = \lim_{n \to \infty} |x_{n+1} - x_n| = 0.$$

From the fact that x_{n+2} lies between x_n and x_{n+1}, it follows that $I_{n+1} \subseteq I_n$, so the sequence $(I_n)_{n \geq 1}$ of closed intervals is decreasing. We can therefore find a real α such that

$$\bigcap_{n \geq 1} I_n = \{\alpha\}.$$

Moreover, the sequence $(x_n)_{n \geq 1}$ is convergent to α, as the sequence of the extremities of intervals I_n, $n \geq 1$.

2. Let us assume, by way of contradiction, that

$$[0, 1] = \{x_n \mid n \in \mathbb{N}, \ n \geq 1\}.$$

Let I_1 be one of the intervals

$$\left[0, \frac{1}{3}\right], \quad \left[\frac{1}{3}, \frac{2}{3}\right], \quad \left[\frac{2}{3}, 1\right]$$

for which $x_1 \notin I_1$. Further, we divide the interval I_1 into three equal closed intervals and denote by I_2 one of them so that $x_2 \notin I_2$. In a similar way, we can inductively define a decreasing sequence of closed intervals

$$I_1 \supseteq I_2 \supseteq I_3 \supseteq \ldots \supseteq I_n \supseteq \ldots$$

such that $x_n \notin I_n$, for all integers $n \geq 1$. The length of the intervals tends to zero as $n \to \infty$,

$$l(I_n) = \frac{1}{3^n},$$

so all intervals have a unique common point, say

$$\bigcap_{n \geq 1} I_n = \{c\}.$$

Because $c \in [0, 1]$, we can find an integer $k \geq 1$ so that $c = x_k$. But $x_k \notin I_k$, so

$$x_k \notin \bigcap_{n \geq 1} I_n,$$

which is a contradiction. In conclusion, $[0, 1]$ is not countable.

3. We have $f(a) \geq g(a)$ and $f(b) \leq g(b)$. If

$$f\left(\frac{a+b}{2}\right) \leq g\left(\frac{a+b}{2}\right),$$

then put

$$a_1 = a, \quad b_1 = \frac{a+b}{2}.$$

In case

$$f\left(\frac{a+b}{2}\right) > g\left(\frac{a+b}{2}\right),$$

we put

$$a_1 = \frac{a+b}{2}, \quad b_1 = b.$$

Either way, for the interval $I_1 = [a_1, b_1]$ we have

$$f(a_1) \geq g(a_1), \quad f(b_1) \leq g(b_1).$$

By induction, we can define the decreasing sequence $I_n = [a_n, b_n]$, $n \geq 1$ of closed intervals with

$$l(I_n) = \frac{b - a}{2^n}$$

and

$$f(a_n) \geq g(a_n), \quad f(b_n) \leq g(b_n), \tag{8.2}$$

for all $n \geq 1$. Let c be the common limit of $(a_n)_{n \geq 1}$ and $(b_n)_{n \geq 1}$. If we take $n \to \infty$ in

$$g(a_n) \leq f(a_n) \leq f(b_n) \leq g(b_n),$$

we derive

$$\lim_{n \to \infty} f(a_n) = \lim_{n \to \infty} f(b_n) = g(c).$$

Further, by the monotonicity of f and from the inequalities $a_n \leq c \leq b_n$, we deduce

$$f(a_n) \leq f(c) \leq f(b_n).$$

Finally, for $n \to \infty$, we have $g(c) \leq f(c) \leq g(c)$, which is $f(c) = g(c)$.

4. The answer is no. Because

$$\lim_{x \to 1/2} |f(x)| = \infty,$$

we can find a closed neighborhood I_1 of $1/2$ for which $|f(x)| > 1$, for all $x \in I_1 \setminus \{\frac{1}{2}\}$. Now, if we take a rational number r_1 in the interior of $I_1 \setminus \{\frac{1}{2}\}$, then from

$$\lim_{x \to r_1} |f(x)| = \infty,$$

we deduce the existence of a closed neighborhood I_2 of r_1, say $I_2 \subset I_1$ and

$$l(I_2) < \frac{1}{2} l(I_1)$$

such that $|f(x)| > 2$, for all $x \in I_2 \setminus \{r_1\}$. Inductively, we can construct a decreasing sequence of closed intervals $(I_n)_{n \geq 1}$ with

$$l(I_n) < \frac{1}{2^{n-1}} l(I_1)$$

such that for each $n \geq 1$, $|f(x)| > n$, for all $x \in I_n \setminus \{r_n\}$. If c is the common point of all intervals I_n, $n \geq 1$, then we must have

$$|f(c)| > n,$$

for all $n \geq 1$, which is impossible.

5. For each $\varepsilon > 0$, there is a closed interval denoted $I(\varepsilon)$ for which $f(x) < \varepsilon$, for all $x \in I(\varepsilon)$. Assuming the contrary, then there exists $\varepsilon_0 > 0$ with the following property: for every interval $I \subseteq [0, 1]$, we can find $x \in I$ such that $f(x) \geq \varepsilon_0$. Then for every partition $(x_k)_{0 \leq k \leq n}$ of $[0, 1]$, we choose $\xi_k \in [x_{k-1}, x_k]$ so that $f(\xi_k) \geq \varepsilon_0$. Now we can find a sequence of partitions with norm converging to zero for which the Riemann sums

$$\sigma_\Delta(f, \xi_k) = \sum_{k=1}^{n} f(\xi_k)(x_k - x_{k-1}) \geq \sum_{k=1}^{n} \varepsilon_0(x_k - x_{k-1}) = \varepsilon_0.$$

Therefore

$$\int_0^1 f(x)\, dx \geq \varepsilon_0 > 0,$$

which is a contradiction. Thus, the assertion we made is true. Now let $I_1 \subseteq [0, 1]$, $I_1 = [a_1, b_1]$ be such that $f(x) < 1$, for all $x \in I_1$. Hence

$$0 \leq \int_{a_1}^{b_1} f(x)\, dx \leq \int_0^1 f(x)\, dx = 0,$$

so

$$\int_{a_1}^{b_1} f(x)\, dx = 0.$$

We can apply the assertion for the restriction $f|_{I_1}$. Indeed, there exists a closed interval $I_2 \subseteq I_1$ so that $f(x) < \frac{1}{2}$, for all $x \in I_2$. By induction, we can find a decreasing sequence $(I_n)_{n \geq 1}$ of closed intervals for which $f(x) < \frac{1}{n}$, for all $x \in I_n$. Finally, if c is a common point of all intervals I_n, $n \geq 1$, then

$$0 \leq f(c) < \frac{1}{n},$$

for all $n \geq 1$, so $f(c) = 0$.

Next, let $[a, b] \subseteq [0, 1]$ and define the restriction

$$f|_{[a,b]} : [a, b] \to \mathbb{R}$$

of the function f. Because $f|_{[a,b]} \geq 0$, it follows that

$$0 \leq \int_a^b f(x)\,dx \leq \int_0^1 f(x)\,dx = 0,$$

so

$$\int_a^b f(x)\,dx = 0.$$

According to the first part of the problem, there exists $c \in [a, b]$ so that $f(c) = 0$. In conclusion, f has a zero in every interval $[a, b] \subseteq [0, 1]$.

6. We have to prove that $C_f \cap C_g \neq \emptyset$. Let $x_0 \in C_f \cap (a, b)$. From the continuity of f at x_0, we can find a closed interval $x_0 \in I_0 \subset (0, 1)$, with $l(I_0) > 0$ for which

$$x, y \in I_0 \Rightarrow |f(x) - f(y)| < 1.$$

Indeed, I_0 can be taken with the property that

$$x \in I_0 \Rightarrow |f(x) - f(x_0)| < \frac{1}{2}.$$

In this way, for any $x, y \in I_0$, we have

$$|f(x) - f(y)| \leq |f(x) - f(x_0)| + |f(y) - f(x_0)| < \frac{1}{2} + \frac{1}{2} = 1.$$

Further, let $x_1 \in C_g \cap I_0$. Similarly, we can find a closed interval $x_1 \in I_1 \subset I_0$, $l(I_1) > 0$ for which

$$x, y \in I_1 \Rightarrow |g(x) - g(y)| < 1.$$

We also have

$$|f(x) - f(y)| < 1, \quad |g(x) - g(y)| < 1,$$

for all $x, y \in I_1$. By induction, we can define a decreasing sequence $(I_n)_{n \geq 1}$ of closed intervals, with $l(I_n) \to 0$ as $n \to \infty$, such that

$$|f(x) - f(y)| < \frac{1}{2^{n-1}}, \quad |g(x) - g(y)| < \frac{1}{2^{n-1}},$$

for all $x, y \in I_n$. If $\{c\} = \bigcap_{n \geq 1} I_n$, then obviously $c \in C_f \cap C_g$.

7. We denote by r both the Riemann (or Thomae) function, given by $r(x) = 0$ if x is irrational, and $r(x) = 1/q$ if x is rational, written $x = p/q$ in its lowest terms (that is, p and q are relatively prime integers, and $q > 0$), and its restriction to the interval $[0, 1]$. The restriction is used in the first part and the function defined on the entire set of the reals—in the second.

For (a) the answer is no. If such a function s exists, then the continuity set of s is dense in $[0, 1]$, $C_s = \mathbb{Q} \cap [0, 1]$. Consider here also the Riemann function r with the continuity set

$$C_r = (\mathbb{R} \setminus \mathbb{Q}) \cap [0, 1].$$

Now we have the functions $r, s : [0, 1] \to \mathbb{R}$ with the continuity sets dense in $[0, 1]$,

$$\overline{C}_r = \overline{C}_s = [0, 1].$$

According to the previous problem, there exists a point $z \in C_r \cap C_s$. This is impossible, because the following relations are contradictory:

$$z \in C_r \Leftrightarrow z \in (\mathbb{R} \setminus \mathbb{Q}) \cap [0, 1]$$

and

$$z \in C_s \Leftrightarrow z \in \mathbb{Q} \cap [0, 1].$$

For (b) the answer is still no. Suppose, on the contrary, that $f : \mathbb{R} \to \mathbb{R}$ is continuous and has the property that $f(x)$ is an irrational (respectively rational) number whenever x is rational (respectively irrational). Let r be Riemann's function defined on the entire set of reals. Thus r is continuous (and takes value 0) exactly at the irrational points. Let $g = r \circ f$ and note that g is also continuous at every rational point q (as q is transformed by f into an irrational point, at which r is continuous). On the other hand, let i be any irrational number, and assume that g is continuous at i. If $(q_n)_{n \geq 1}$ is a sequence of rational numbers with limit i, we must have

$$g(i) = \lim_{n \to \infty} g(q_n) \Leftrightarrow r(f(i)) = \lim_{n \to \infty} r(f(q_n)) = 0$$

(because each $f(q_n)$ is irrational; hence $f(i)$ has to be irrational—which is not true. Thus, the function g is not continuous at i—consequently, g is continuous precisely at the rational points. Now the existence of r and g (actually of their restrictions to $[0, 1]$) is contradictory—the contradiction being the same as in part (a) of the problem; therefore such a function f cannot exist.

8. (Claudiu Raicu) For (a), we could define

$$f(x) = \begin{cases} 0, & x = 0 \\ 1 - \dfrac{1}{q}, & x = \dfrac{p}{q} \in \mathbb{Q}^* \\ 1, & x \in \mathbb{R} \setminus \mathbb{Q}. \end{cases},$$

where we write p/q for a fraction in its lowest terms, as defined in the statement of problem 7. Actually, one sees that this is $1 - r$, for r defined in that problem, too.

Similarly, for (b) it is enough to choose

$$f\left(\frac{p}{q}\right) = \begin{cases} q, p \neq 0 \\ 0, p = 0 \end{cases}.$$

Let us now solve (c). Suppose, by way of contradiction, that such a function exists. Let

$$I_0 \supseteq I_1 \supseteq I_2 \supseteq \ldots \supseteq I_{n-1},$$

such that $f(I_k) \subset [k, \infty)$ for $k = 1, 2, \ldots, n - 1$. Then there is some $x_n \in \mathbb{Q} \cap I_{n-1}$ with $f(x_n) \geq n$ and consider a segment $I_n \subseteq I_{n-1}$ such that $f(x_n) \leq f(x)$ for all $x \in I_n$. Such I_n exists because x_n is a point of local minimum. Then, $f(I_n) \subset [n, \infty)$. We can continue inductively to construct a sequence

$$I_0 \supseteq I_1 \supseteq I_2 \supseteq \ldots$$

such that $f(I_k) \subset [k, \infty)$ for all k. If

$$x \in \bigcap_{k=0}^{\infty} I_k,$$

then $f(x) \geq k$ for all k, a contradiction.

9. Suppose, by way of contradiction, that the conclusion is false. Then there is a sequence $x_n \to \infty$ such that

$$\lim_{n \to \infty} f(x_n) = l \neq 0.$$

Suppose, without loss of generality, that $l > 0$ and let $0 < \epsilon < l$. Then there is $N \in \mathbb{N}$ such that $f(x_n) > \epsilon$ for all $n \geq N$. By continuity, it follows that for every $n \geq N$, there is a closed interval $I_n = [\alpha_n, \beta_n] \ni x_n$ such that $f(x) > \epsilon$ for all $x \in I_n$.

We will prove that there is $x \in \mathbb{R}$ such that $\{nx : x \in \mathbb{R}\} \cap I_k \neq \emptyset$ for infinitely many $k \in \mathbb{N}$. Clearly we can assume, without loss of generality, that $\beta_n - \alpha_n \to 0$.

We will build inductively a sequence of closed intervals $(I_{k_n})_{n \geq 1}$ as follows. Choose k_1 arbitrarily. Consider k_n fixed. It is easily checked using $\beta_n - \alpha_n \to 0$ and $\alpha_n, \beta_n \to \infty$ that for large enough k_{n+1}, there is $M_n \in \mathbb{N}$ such that $[M_n \alpha_{k_n}, M_n \beta_{k_n}] \supset [\alpha_{k_{n+1}}, \beta_{k_{n+1}}]$. So, let us choose k_{n+1} with this property. By the construction,

$$[\alpha_{k_1}, \beta_{k_1}] \supset \left[\frac{\alpha_{k_2}}{M_1}, \frac{\beta_{k_2}}{M_1}\right] \supset \left[\frac{\alpha_{k_3}}{M_1 M_2}, \frac{\beta_{k_3}}{M_1 M_2}\right] \supset \cdots$$

$$\Rightarrow \bigcap_{n=1}^{\infty} \left[\frac{\alpha_{k_n}}{\prod_1^{n-1} M_i}, \frac{\beta_{k_n}}{\prod_1^{n-1} M_i}\right] \neq \emptyset.$$

Then, for x from this set, we see that, in fact, $\{nx : x \in \mathbb{R}\} \cap I_k \neq \emptyset$ for infinitely many k, which contradicts the fact that $\lim_{n \to \infty} f(nx) = 0$ since $f(nx) > \epsilon$ for infinitely many $n \in \mathbb{N}$. Hence, our supposition was false and the problem is solved.

Note that this problem (called Croft's lemma) was also discussed in Chapter 1. The reader might be interested to compare two wordings of (basically) the (same) solution.

10. It is clear that $f(0) = 0$ and $f(x) < x$ for all $x > 0$. There are 2 possible cases.

First Case. There is some $\eta > 0$ such that $f(x) = 0$ for all $x \in [0, \eta]$. Then, for all x, there is some open interval $V(x) \ni x$ and some $n(x)$ such that

$$f^{n(x)}(V(x)) \subset [0, \eta].$$

Let $x_1, x_2, \ldots, x_p$ such that

$$V(x_1) \cup V(x_2) \cup \ldots \cup V(x_p) \supset [0, 1].$$

Then it is clear that for

$$n = 1 + \max_{1 \leq i \leq p} n(x_i), \qquad f^n \equiv 0.$$

Second Case. We build inductively a sequence $I_n = [a_n, b_n]$ with the properties:

$$f(x) > 0, \forall x \in I_n \text{ and } b_{n+1} < \max_{x \in I_n} f(x)$$

The construction is clear due to the hypothesis of the second case. Let us define

$$K_n = \left\{x \in I_1 : f(x) \in I_2, \ldots, f^{n-1}(x) \in I_n\right\}.$$

We obviously have $K_n \neq \emptyset$ and $K_n \supset K_{n+1}$. Since K_n is compact, we have

$$x \in \bigcap_n K_n.$$

For such x,

$$0 \notin \{x, f(x), f(f(x)), f(f(f(x))), \ldots\},$$

which is a contradiction, hence the second case is impossible.

11. Let $f : [a, b] \to \mathbb{R}$ be a continuous function, differentiable on (a, b). We intend to prove that there exists $c \in (a, b)$ such that $R(a, b) = f'(c)$, where we define, for all $x, y \in [a, b]$, $x \neq y$,

$$R(x, y) = \frac{f(x) - f(y)}{x - y}.$$

We also let

$$R(x, x) = f'(x)$$

for every $x \in (a, b)$. To begin, we note the following properties of the function R.

First, R is continuous at any point where it is defined, by the continuity of f and by the definition of derivative at points (x, x), $x \in (a, b)$. It follows that, if we fix some $\alpha \in (a, b)$, the function $x \mapsto R(\alpha, x)$ is continuous at every point of $[a, b]$.

We also have that, for all $x < y < z$ in $[a, b]$, $R(x, z)$ is between $R(x, y)$ and $R(y, z)$. This is because

$$R(x, z) = \frac{y - x}{z - x} R(x, y) + \frac{z - y}{z - x} R(y, z) = (1 - t)R(x, y) + tR(y, z)$$

with $t = (z - y)/(z - x) \in (0, 1)$. This is an almost trivial identity (isn't it?) but it becomes critical in this proof of Lagrange's theorem (due, as far as we know, to the Romanian mathematician Dimitrie Pompeiu). Note that if two of $R(x, y)$, $R(y, z)$, and $R(x, z)$ are equal, then all three are equal. Of course, $R(x, y) = R(y, x)$ holds for every distinct $x, y \in [a, b]$.

And now for the proof. If we have $R(a, x) = R(x, b)$ for all $x \in (a, b)$, then $R(a, x) = R(x, b) = R(a, b)$ and $f(x) = R(a, b)(x - a) + f(a)$, hence $f'(x) = R(a, b)$ for all $x \in (a, b)$, and there is nothing left to prove. Thus we can assume that there is such $u \in (a, b)$ with $R(a, u) \neq R(u, b)$, say with $R(u, a) = R(a, u) < R(u, b)$. Because the function $x \mapsto R(u, x)$ is continuous and $R(a, u) < R(a, b) < R(u, b)$, there exists $v \in (a, b)$ such that $R(a, b) = R(u, v)$. If $v = u$, this means $R(a, b) = R(u, u) = f'(u)$ and the proof ends here. Otherwise, we have an interval $I_1 = [a_1, b_1]$ (with $\{a_1, b_1\} = \{u, v\}$) such that $a < a_1 < b_1 < b$ and $R(a, b) = R(a_1, b_1)$.

Now let $a_2' = (a_1 + b_1)/2$ and look at $R(a_1, a_2')$ and $R(a_2', b_1)$. If they are equal, then (as we noticed above), we also have $R(a_1, b_1) = R(a_1, a_2') = R(a_2', b_1)$; in this case, we consider the interval $I_2 = [a_2, b_2]$ to be one of the intervals $[a_1, a_2']$, $[a_2', b_1]$ (it doesn't matter which). Observe that I_2 is included in I_1, has length half the length of I_1, and has the property that $R(a, b) = R(a_1, b_1) = R(a_2, b_2)$. If not, then $R(a_1, b_1)$ is *strictly* between $R(a_1, a_2')$ and $R(a_2', b_1)$, and, since the function $x \mapsto R(a_2', x)$ is continuous, there must be some $b_2' \in (a_1, b_1)$ such that $R(a_1, b_1) = R(a_2', b_2')$. If $a_2' = b_2'$, the proof ends here, because $R(a, b) = R(a_1, b_1) = R(a_2', a_2') = f'(a_2')$. Otherwise, we just built an interval $I_2 = [a_2, b_2]$ which is included in I_1, has length at most half the length of I_1, and which is such that $R(a, b) = R(a_1, b_1) = R(a_2, b_2)$. Of course, a_2 and b_2 are (in the second case) a_2' and b_2' in increasing order.

Clearly, this process can be iterated in order to get a sequence of nested compact intervals $I_1 \supseteq I_2 \supseteq \cdots$ such that each $I_{n+1} = [a_{n+1}, b_{n+1}]$ is included in $I_n = [a_n, b_n]$, has length at most half the length of I_n, and also, we have $R(a, b) = R(a_1, b_1) = \cdots = R(a_n, b_n) = \cdots$. (If at some moment we get $a_n = b_n$, the proof ends because this means $R(a, b) = f'(a_n)$. Otherwise the process continues indefinitely.) Moreover, the first interval is included in (a, b) (and, consequently, all of them are). This means that the common point c of all intervals I_n is in (a, b), too. As $l(I_{n+1}) \leq (1/2)l(I_n)$ for all n, the sequence of lengths of intervals goes to zero; therefore we have

$$\lim_{n \to \infty} a_n = \lim_{n \to \infty} b_n = c$$

and, finally,

$$R(a, b) = \lim_{n \to \infty} R(a_n, b_n) = R(c, c) = f'(c).$$

Note that this (rather complicated) proof shows that Lagrange's theorem can be obtained with very little knowledge on derivatives (basically, only the definition of the derivative is needed). Thus we can obtain other theorems (such as Rolle's theorem) using a method different from the usual one, and we can obtain the *equivalence* of the important theorems in the analysis on the real line.

Chapter 9
The Splitting Method and Double Sequences

We use here the splitting method to establish some useful convergence results. The splitting method is a useful tool to compute the limits of certain sequences of real numbers, whose general form is a sum s_n of n terms, which do not behave in the same way. In fact, this method consists of decomposing the sum s_n into two sums, which are analyzed separately, using different methods in general, adapted to the behavior of the terms composing them. We give first some examples of problems which use the splitting method and then some general results, with lots of practical applications.

The next problem is typical for this type of argument and was discussed in the Jury of the Romanian Mathematical Olympiad, 2002.

Problem. Let $f : [0, \infty) \to [0, 1)$ be continuous, with $\lim\limits_{x \to \infty} f(x) = 0$. Prove that

$$\lim_{n \to \infty} \left[\int_0^1 f^n(x)dx + \int_1^2 f^{n-1}(x)dx + \cdots + \int_{n-1}^n f(x)dx \right] = 0.$$

(Here, $f^k(x)$ denotes $(f(x))^k$.)

Solution. Let $M \in (0, 1)$ satisfy $f(x) \leq M$, for all real numbers $x \in [0, \infty)$. Let us see why such an M exists: there exists some n such that if $x > n$, then $f(x) < \frac{1}{2}$. On the other hand, f has a maximum $M_1 < 1$ on $[0, n]$ because it is continuous. Thus $M = \max(M_1, 0.5)$ is a possible choice. Let us take $\varepsilon > 0$, actually $\varepsilon < 1/2$, and consider a rank k for which $f(x) < \epsilon$ for all $x \in [k, \infty)$. We have the decomposition

$$z_n = \int_0^1 f^n(x)dx + \int_1^2 f^{n-1}(x)dx + \cdots + \int_{n-1}^n f(x)dx$$

$$= \left(\int_0^1 f^n(x)dx + \int_1^2 f^{n-1}(x)dx + \cdots + \int_{k-1}^k f^{n+1-k}(x)dx \right)$$

© Springer Science+Business Media LLC 2017
T. Andreescu et al., *Mathematical Bridges*, DOI 10.1007/978-0-8176-4629-5_9

$$+ \left(\int_k^{k+1} f^{n-k}(x)dx + \int_{k+1}^{k+2} f^{n-k-1}(x)dx + \cdots + \int_{n-1}^n f(x)dx \right).$$

First,

$$\int_0^1 f^n(x)dx + \int_1^2 f^{n-1}(x)dx + \cdots + \int_{k-1}^k f^{n+1-k}(x)dx$$

$$\leq M^n + M^{n-1} + \cdots + M^{n+1-k} < \frac{M^{n+1-k}}{1-M}.$$

Further, $\displaystyle\lim_{n\to\infty} \frac{M^{n+1-k}}{1-M} = 0$, so we can find a rank n_1 for which

$$\int_0^1 f^n(x)dx + \int_1^2 f^{n-1}(x)dx + \cdots + \int_{k-1}^k f^{n+1-k}(x)dx < \varepsilon, \quad \forall\, n \geq n_1.$$

On the other hand,

$$\int_k^{k+1} f^{n-k}(x)dx + \int_{k+1}^{k+2} f^{n-k-1}(x)dx + \cdots + \int_{n-1}^n f(x)dx$$

$$< \varepsilon^{n-k} + \cdots + \varepsilon^2 + \varepsilon < \frac{\varepsilon}{1-\varepsilon} < 2\varepsilon,$$

so

$$\int_k^{k+1} f^{n-k}(x)dx + \int_{k+1}^{k+2} f^{n-k-1}(x)dx + \cdots + \int_{n-1}^n f(x)dx < 2\varepsilon.$$

Finally, by adding the previous inequalities, we obtain $z_n < 3\varepsilon$, $\forall\, n \geq n_1$, so the problem is solved. $\square$

The reader will immediately be convinced that the following problem is not an easy one. However, it becomes much more natural in the framework of the splitting method.

Problem. Prove that

$$\lim_{n\to\infty} \frac{1^n + 2^n + 3^n + \cdots + n^n}{n^n} = \frac{e}{e-1}.$$

Solution. If we write the sequence starting with the last term, we obtain

$$a_n = \left(\frac{n}{n}\right)^n + \left(\frac{n-1}{n}\right)^n + \left(\frac{n-2}{n}\right)^n + \cdots + \left(\frac{2}{n}\right)^n + \left(\frac{1}{n}\right)^n$$

or

$$a_n = 1 + \left(1 - \frac{1}{n}\right)^n + \left(1 - \frac{2}{n}\right)^n + \cdots + \left(1 - \frac{n-2}{n}\right)^n + \left(1 - \frac{n-1}{n}\right)^n.$$

Now we can see that the sequence $(a_n)_{n \geq 1}$ is closely related to the limit

$$\lim_{n \to \infty} \left(1 + e^{-1} + e^{-2} + \cdots + e^{-n}\right) = \frac{e}{e-1}.$$

More precisely, using first the inequality

$$\left(1 - \frac{k}{n}\right)^n < e^{-k},$$

we deduce that

$$a_n < 1 + e^{-1} + e^{-2} + \cdots + e^{-n+1} < \frac{1}{1 - e^{-1}},$$

or $a_n < \dfrac{e}{e-1}, \; \forall \, n \geq 1$.

On the other hand, for all integers $n > k \geq 1$, we have

$$a_n \geq 1 + \left(1 - \frac{1}{n}\right)^n + \left(1 - \frac{2}{n}\right)^n + \cdots + \left(1 - \frac{k}{n}\right)^n.$$

If we consider now the inferior limit with respect to n, we obtain

$$\liminf_{n \to \infty} a_n \geq \liminf_{n \to \infty} \left[1 + \left(1 - \frac{1}{n}\right)^n + \left(1 - \frac{2}{n}\right)^n + \cdots + \left(1 - \frac{k}{n}\right)^n\right]$$

$$= 1 + e^{-1} + e^{-2} + \cdots + e^{-k},$$

so

$$\liminf_{n \to \infty} a_n \geq 1 + e^{-1} + e^{-2} + \cdots + e^{-k},$$

for all positive integers k. Further, by taking the limit as $k \to \infty$ in the last inequality, we infer that

$$\liminf_{n \to \infty} a_n \geq \frac{e}{e-1}.$$

Now the conclusion follows from the inequalities:

$$\frac{e}{e-1} \leq \liminf_{n \to \infty} a_n, \quad a_n < \frac{e}{e-1}. \quad \square$$

Here is another typical application of the splitting method:

Problem. Let $(a_n)_{n \geq 1}$ be a decreasing sequence of real numbers with

$$\lim_{n \to \infty} a_n = 1.$$

Prove that the sequence with general term $b_n = \left(1 + \dfrac{a_1}{n}\right)\left(1 + \dfrac{a_2}{n}\right)\cdots\left(1 + \dfrac{a_n}{n}\right)$
converges to e.

Solution. Let us consider the term

$$b_{k+m} = \left(1 + \frac{a_1}{k+m}\right)\left(1 + \frac{a_2}{k+m}\right)\cdots\left(1 + \frac{a_k}{k+m}\right)\cdots$$
$$\cdots\left(1 + \frac{a_{k+1}}{k+m}\right)\cdots\left(1 + \frac{a_{k+m}}{k+m}\right).$$

The product of the first k factors can be estimated as follows:

$$\left(1 + \frac{a_1}{k+m}\right)\left(1 + \frac{a_2}{k+m}\right)\cdots\left(1 + \frac{a_k}{k+m}\right)$$
$$\leq \left(1 + \frac{a_1}{k+m}\right)^k \leq e^{\frac{ka_1}{k+m}}$$

Assume next that for a given $\varepsilon > 0$, k is chosen such that $1 < a_{k+m} < 1 + \varepsilon$, for all positive integers m. Hence

$$\left(1 + \frac{a_{k+1}}{k+m}\right)\cdots\left(1 + \frac{a_{k+m}}{k+m}\right) \leq \left(1 + \frac{a_{k+1}}{m}\right)\cdots\left(1 + \frac{a_{k+m}}{m}\right)$$
$$\leq \left(1 + \frac{1+\varepsilon}{m}\right)^m \leq e^{1+\varepsilon}.$$

In conclusion,

$$b_{k+m} \leq e^{\frac{ka_1}{k+m}+1+\varepsilon} \leq e^{1+2\varepsilon},$$

for fixed (but arbitrary) $\varepsilon > 0$ and sufficiently large m, because for big enough m, we have $\frac{ka_1}{k+m} \leq \varepsilon$. Since ε can be any positive number, this implies $\limsup_{n \to \infty} b_n \leq e$.

On the other hand,

$$\left(1 + \frac{1}{n}\right)^n \leq \left(1 + \frac{a_1}{n}\right)\left(1 + \frac{a_2}{n}\right)\cdots\left(1 + \frac{a_n}{n}\right) = b_n$$

for all $n \geq 1$; this yields $e \leq \liminf_{n \to \infty} b_n$ and finishes the proof. $\square$

Alternatively (but without using the method that we are discussing now), if one considers the sequence with general term $c_n = b_n/(1 + (1/n))^n$, then one notes (like above) that $1 \leq c_n$ for all $n \geq 1$. Yet, by the inequality $1 + x \leq e^x$, $x \in \mathbb{R}$ that we already used, we see that

$$c_n = \left(1 + \frac{a_1 - 1}{n+1}\right)\left(1 + \frac{a_2 - 1}{n+1}\right)\cdots\left(1 + \frac{a_n - 1}{n+1}\right) \leq e^{d_n},$$

where

$$d_n = \frac{(a_1 - 1) + (a_2 - 1) + \cdots + (a_n - 1)}{n+1},$$

and $\lim_{n\to\infty} d_n = 0$ follows from the hypothesis and the Cesàro-Stolz theorem. Thus we have the inequalities

$$\left(1 + \frac{1}{n}\right)^n \leq b_n \leq \left(1 + \frac{1}{n}\right)^n e^{d_n}$$

for all $n \geq 1$. Now the squeeze theorem finishes the proof. $\square$

Finally observe that the condition about the monotonicity of $(a_n)_{n\geq 1}$ is not necessary; we can assume only that $a_n \to 1$ and $a_n \geq 1$ for all n, and both proofs work.

It is not difficult to see that by changing the order of the terms of a convergent series, usually we do not obtain the same sum (a famous theorem of Riemann asserts much more: if (a_n) is a sequence of real numbers and $\sum_{n\geq 1} a_n$ is convergent, but $\sum_{n\geq 1} |a_n| = \infty$, then for any $\alpha \in \mathbb{R}$, one can permute the terms of the series $\sum_{n\geq 1} a_n$ so that the sum of the resulting series is α). However, if the series is absolutely convergent, we have the following beautiful application of the splitting method:

Problem. Let $(a_n)_n$ be a sequence of real numbers such that $\sum_{n\geq 1} |a_n|$ converges. Let σ be a permutation of the set of positive integers. Prove that

$$\sum_{n=1}^{\infty} a_n = \sum_{n=1}^{\infty} a_{\sigma(n)}.$$

Solution. Because $\sum_{n\geq 1} |a_n| < \infty$, $\sum_{n=1}^{\infty} a_n$ converges. Let l be its sum and let us prove that $\sum_{n=1}^{\infty} a_{\sigma(n)}$ converges to l. Let $\varepsilon > 0$ and choose N such that $\sum_{n\geq N} |a_n| < \varepsilon$.

Let $M \geq N$ be such that $\{\sigma(1), \sigma(2), \ldots, \sigma(M)\}$ contains $\{1, 2, \ldots, N\}$ and let $m \geq M + N$. Finally define $A = \{1, 2, \ldots, m\} \setminus \{1, 2, \ldots, N - 1\}$ and $B = \{\sigma(1), \ldots, \sigma(m)\} \setminus \{1, 2, \ldots, N - 1\}$. Then

$$\left| \sum_{i=1}^{m} a_{\sigma(i)} - l \right| \leq \left| \sum_{i=1}^{m} a_i - l \right| + \left| \sum_{i=1}^{m} a_{\sigma(i)} - \sum_{i=1}^{m} a_i \right|$$

$$\leq \left| \sum_{i=1}^{m} a_i - l \right| + \left| \sum_{i \in B} a_i - \sum_{i \in A} a_i \right| \leq \sum_{i \in A} |a_i| + \sum_{i \in B} |a_i| + \left| \sum_{i=1}^{m} a_i - l \right|$$

$$\leq 2\varepsilon + \left| \sum_{i=1}^{m} a_i - l \right|.$$

Because $l = \sum_{i=1}^{\infty} a_i$, for sufficiently large m, we have $\left| \sum_{i=1}^{m} a_i - l \right| < \varepsilon$ and thus $\left| \sum_{i=1}^{m} a_{\sigma(i)} - l \right| < 3\varepsilon$ for sufficiently large m. This shows that $\sum_{i=1}^{\infty} a_{\sigma(i)} = l$. $\square$

The following two problems have shorter solutions, but are far from being obvious.

Problem. Let $(a_n)_{n \geq 1}$ be a sequence of positive real numbers such that $\sum_{n \geq 1} a_n$ converges. Prove that $\sum_{n \geq 1} a_n^{1 - \frac{1}{n}}$ also converges.

Solution. The idea is that if a_n is very small, so is $a_n^{1 - \frac{1}{n}}$, while if a_n is not very small, say $a_n \geq M$, then $a_n^{1 - \frac{1}{n}} \leq \frac{a_n}{\sqrt[n]{M}}$. Therefore we split the sum into two parts: one corresponding to those a_n greater than or equal to 2^{-n}, and the second one corresponding to $a_n < 2^{-n}$. If $a_n \geq 2^{-n}$, then $a_n^{1 - \frac{1}{n}} \leq 2a_n$, so the first sum is bounded by $2 \sum_{n \geq 1} a_n$. If $a_n < 2^{-n}$, then $a_n^{1 - \frac{1}{n}} \leq \frac{1}{2^{n-1}}$, so the second sum is bounded by $\sum_{n \geq 1} \frac{1}{2^{n-1}}$. Therefore the partial sums of $\sum_{n \geq 1} a_n^{1 - \frac{1}{n}}$ are bounded and so the series converges. $\square$

Problem. Find $\lim_{n \to \infty} \dfrac{\sqrt{n} + \sqrt[3]{n} + \cdots + \sqrt[n-1]{n} + \sqrt[n]{n}}{n}$.

Solution. If $a_n = \dfrac{\sqrt{n} + \sqrt[3]{n} + \cdots + \sqrt[n-1]{n} + \sqrt[n]{n}}{n}$, then clearly $a_n \geq \dfrac{n-1}{n}$.

On the other hand, let $M > 0$ and split $\displaystyle\sum_{i=2}^{n} \sqrt[i]{n}$ in $S_1 = \displaystyle\sum_{2 \leq i < M \ln n} \sqrt[i]{n}$ and $S_2 =$

$\displaystyle\sum_{n \geq i \geq M \ln n} \sqrt[i]{n}$. The first sum is at most $M \ln n \sqrt{n} = o(n)$, while the second one is

at most $n \cdot e^{\frac{1}{M}}$, because for $i \geq M \ln n$, we have $\sqrt[i]{n} = n^{\frac{1}{i}} \leq n^{\frac{1}{M \ln n}} = e^{\frac{1}{M}}$. Thus
$\dfrac{S_1 + S_2}{n} \leq \dfrac{M \ln n}{\sqrt{n}} + e^{\frac{1}{M}}$. Now, if $\varepsilon > 0$, there is M with $e^{\frac{1}{M}} < 1 + \dfrac{\varepsilon}{2}$ and for such

a fixed M, we have $\dfrac{M \ln n}{\sqrt{n}} < \dfrac{\varepsilon}{2}$ for large n, thus $a_n < 1 + \varepsilon$ for large n. Therefore

$\displaystyle\lim_{n \to \infty} a_n = 1$. $\square$

Let us give now some important theoretical results concerning double sequences and their convergence. We discuss this in the same chapter because, as the reader will immediately notice, all proofs are based on the splitting method. We will discuss Toeplitz's theorem, which is in fact a useful convergence criterion for a class of sequences of real numbers. We will study some of its consequences and applications, with a particular emphasis on a very useful result, the Cesàro-Stolz theorem.

Theorem (Toeplitz). *Let* $A = (a_{nk})_{n \geq k \geq 1}$ *be an infinite triangular matrix with nonnegative entries,*

$$A = \begin{pmatrix} a_{11} & 0 & 0 & 0 & 0 & \cdots \\ a_{21} & a_{22} & 0 & 0 & 0 & \cdots \\ a_{31} & a_{32} & a_{33} & 0 & 0 & \cdots \\ \cdots & \cdots & \cdots & \cdots & \cdots \\ a_{n1} & a_{n2} & a_{n3} & \cdots & a_{nn} & \cdots \\ \cdots & \cdots & \cdots & \cdots & \cdots \end{pmatrix}.$$

Assume that:

a) *The sum of the elements in each row is equal to 1, i.e.,* $\displaystyle\sum_{k=1}^{n} a_{nk} = 1$, *for all positive integers n.*

b) *Each sequence determined by each column is convergent to zero, i.e.,* $\displaystyle\lim_{n \to \infty} a_{nk} = 0$, *for all positive integers k.*

Then, for any sequence $(t_n)_{n \geq 1}$ *which has a limit (finite or infinite),*

$$\lim_{n \to \infty} \sum_{k=1}^{n} a_{nk} t_k = \lim_{n \to \infty} t_n.$$

Proof. Let $\alpha = \lim_{n \to \infty} t_n$ and let $\widetilde{t}_n = \sum_{k=1}^{n} a_{nk} t_k, \quad n \geq 1$.

First, we consider the case $\alpha \in \mathbb{R}$. By changing $(t_n)_{n \geq 1}$ to $(t_n - \alpha)_{n \geq 1}$, we can assume that $\alpha = 0$ (note that we used here the first hypothesis of the matrix A). For any $\varepsilon > 0$, we can find a positive integer $N = N(\varepsilon)$ such that $|t_n| < \frac{\varepsilon}{2}$, for all integers $n \geq N$. If $n > N$, then

$$\widetilde{t}_n = (a_{n1} t_1 + \cdots + a_{nN} t_N) + (a_{n,N+1} t_{N+1} + \cdots + a_{nn} t_n).$$

From b), we deduce that

$$\lim_{n \to \infty} (a_{n1} t_1 + \cdots + a_{nN} t_N) = 0,$$

so there exists $n_\varepsilon \geq N$ such that

$$|a_{n1} t_1 + \cdots + a_{nN} t_N| < \frac{\varepsilon}{2}, \ \forall n \geq n_\varepsilon.$$

Finally, for all $n \geq n_\varepsilon$, we have

$$|\widetilde{t}_n| < \frac{\varepsilon}{2} + \sum_{k=N+1}^{n} a_{nk} |t_k| \leq \frac{\varepsilon}{2} + \frac{\varepsilon}{2} \sum_{k=1}^{n} a_{nk} = \frac{\varepsilon}{2} + \frac{\varepsilon}{2} = \varepsilon.$$

To complete the proof, we assume now that $t_n \to \infty$, as $n \to \infty$. Let $M > 0$ and $N \geq 1$ be so that $t_n \geq 3M$, for all integers $n \geq N$. We can find $N_1 > N$ with

$$|a_{n1} t_1 + \cdots + a_{nN} t_N| \leq M,$$

for all integers $n \geq N_1$ and let $N_2 \geq N_1$ be such that $a_{n1} + \cdots + a_{nN} \leq \frac{1}{3}$, for all integers $n \geq N_2$. Now

$$\widetilde{t}_n = (a_{n1} t_1 + \cdots + a_{nN} t_N) + (a_{n,N+1} t_{N+1} + \cdots + a_{nn} t_n)$$

$$\geq -M + 3M \sum_{k=N+1}^{n} a_{nk} = -M + 3M \left(1 - \sum_{k=1}^{N} a_{nk}\right)$$

$$\geq -M + 3M \left(1 - \frac{1}{3}\right) = M,$$

for all integers $n \geq N_2$ and consequently $\widetilde{t}_n \to \infty$, as $n \to \infty$. $\square$

A direct consequence that is very useful in practice is the not so well-known Cesàro-Stolz theorem:

Theorem (Cesàro-Stolz). *Let $(x_n)_{n\geq 1}$ and $(y_n)_{n\geq 1}$ be two sequences of real numbers. Assume that $(y_n)_{n\geq 1}$ is increasing, unbounded, and that there exists*

$$\alpha = \lim_{n\to\infty} \frac{x_n - x_{n-1}}{y_n - y_{n-1}}.$$

Then $\lim\limits_{n\to\infty} \dfrac{x_n}{y_n} = \alpha.$

Proof. We have the following identity

$$\frac{x_n}{y_n} = \sum_{k=1}^{n} \frac{y_k - y_{k-1}}{y_n} \cdot \frac{x_k - x_{k-1}}{y_k - y_{k-1}} = \sum_{k=1}^{n} a_{nk} t_k,$$

where

$$a_{nk} = \frac{y_k - y_{k-1}}{y_n}, \quad t_k = \frac{x_k - x_{k-1}}{y_k - y_{k-1}}, \quad k \geq 1$$

and $x_0 = y_0 = 0$. For all $n \geq 1$, we have

$$\sum_{k=1}^{n} a_{nk} = \sum_{k=1}^{n} \frac{y_k - y_{k-1}}{y_n} = \frac{1}{y_n} \cdot (y_n - y_0) = 1$$

and for each $k \geq 1$,

$$\lim_{n\to\infty} a_{nk} = \lim_{n\to\infty} \frac{y_k - y_{k-1}}{y_n} = 0,$$

because $y_n \to \infty$, as $n \to \infty$. Now $t_n = \dfrac{x_n - x_{n-1}}{y_n - y_{n-1}} \to \alpha$, so $\widetilde{t}_n = \dfrac{x_n}{y_n} \to \alpha.$ $\square$

To see the strength of this criterion, consider the following B6 Putnam problem from 2006:

Problem. Let k be an integer greater than 1. Suppose $a_0 > 0$ and define

$$a_{n+1} = a_n + \frac{1}{\sqrt[k]{a_n}}$$

for $n \geq 0$. Evaluate $\lim\limits_{n\to\infty} \dfrac{a_n^{k+1}}{n^k}$.

Solution. It is clear that the sequence is increasing. We claim that it tends to ∞. Indeed, otherwise it converges to a finite limit l and passing to the limit in the given relation, we obtain a contradiction. Now, in order to get rid of n^k and to apply the Cesàro-Stolz theorem, it is better to take the kth root of the sequence whose limit we are asked to compute. Thus, we will use Cesàro-Stolz for the sequence $\dfrac{a_n^{1+1/k}}{n}$.

If we manage to compute the limit of $a_{n+1}^{1+1/k} - a_n^{1+1/k}$, then the previous sequence will have the same limit. However, this boils down to finding the limit as x tends to ∞ of

$$\left(x + \frac{1}{\sqrt[k]{x}}\right)^{1+1/k} - x^{1+1/k}.$$

By changing the variable to $x^{1/k} = t$, we need to compute the limit for $t \to \infty$ of

$$t^{k+1}\left(\left(1 + \frac{1}{t^{k+1}}\right)^{1+1/k} - 1\right).$$

However, it is clear that this limit is $1 + 1/k$, simply because another change of variable ($t^{-k-1} = u$) shows that it is the same as the limit of $\frac{(1+u)^{1+1/k}-1}{u}$ as u tends to 0. Thus the limit of $\frac{a_n^{1+1/k}}{n}$ is $1 + 1/k$, and now it is clear that the answer to the problem is $(1 + 1/k)^k$. $\square$

Using the same splitting method, we will establish probably the most useful result of this chapter, Lebesgue's dominated convergence theorem for sequences. It gives a very easy-to-verify condition for passing to the limit in an infinite sum.

Lebesgue's theorem. *Let $(a_{mn})_{m,n\geq 1}$ be a double sequence of real numbers for which there exists a sequence $(a_n)_{n\geq 1}$ such that $\lim\limits_{m\to\infty} a_{mn} = a_n$ and $|a_{mn}| \leq \alpha_n$, for all positive integers m, n and some sequence $(\alpha_n)_{n\geq 1}$. Then, if the series $\sum\limits_{n=1}^{\infty} \alpha_n$ converges,*

$$\lim_{m\to\infty} \sum_{n=1}^{\infty} a_{mn} = \sum_{n=1}^{\infty} a_n.$$

Proof. First note that $|a_n| \leq \alpha_n$, so the series $\sum\limits_{n=1}^{\infty} a_{mn}$ and $\sum\limits_{n=1}^{\infty} a_n$ are absolutely convergent. With the notations

$$\sigma = \sum_{n=1}^{\infty} a_n, \quad t_m = \sum_{n=1}^{\infty} a_{mn},$$

we have

$$|\sigma - t_m| \leq \sum_{n=1}^{p} |a_n - a_{mn}| + \sum_{n=p+1}^{\infty} |a_n - a_{mn}|$$

$$\leq \sum_{n=1}^{p} |a_n - a_{mn}| + 2\sum_{n=p+1}^{\infty} \alpha_n.$$

For each real $\varepsilon > 0$, let n_0 be such that $\displaystyle\sum_{n=n_0}^{\infty} \alpha_n < \frac{\varepsilon}{4}$. Then for sufficiently large m,

$$|\sigma - t_m| \leq \sum_{n=1}^{n_0} |a_n - a_{mn}| + \frac{\varepsilon}{2} < n_0 \cdot \frac{\varepsilon}{2n_0} + \frac{\varepsilon}{2} = \varepsilon. \quad \square$$

We end this theoretical part with an application of the previous result:

Problem. Let $\alpha > 1$ be a real number. Prove that

$$\lim_{n\to\infty} \frac{\sqrt{n}}{\alpha^n} e^{-n} \sum_{k=0}^{n} \frac{(\alpha n)^k}{k!} = \frac{\alpha}{\sqrt{2\pi}(\alpha - 1)}.$$

Solution. Let

$$x_n = \sum_{k=0}^{n} \frac{(\alpha n)^k}{k!} = \frac{(\alpha n)^n}{n!} \sum_{k=0}^{n} \frac{n!}{(n-k)!} \cdot \frac{1}{(\alpha n)^k}.$$

Let $a_{nk} = \dfrac{n!}{(n-k)!} \cdot \dfrac{1}{(\alpha n)^k}$ if $k \leq n$ and 0 if $k > n$. Then $0 \leq a_{nk} \leq \dfrac{1}{\alpha^k}$ and because $\displaystyle\sum_{k} \frac{1}{\alpha^k}$ converges and $\displaystyle\lim_{n\to\infty} a_{nk} = \frac{1}{\alpha^k}$, we deduce that

$$\lim_{n\to\infty} \sum_{k=0}^{n} \frac{n!}{(n-k)!} \cdot \frac{1}{(\alpha n)^k} = \sum_{k=0}^{\infty} \frac{1}{\alpha^k} = \frac{\alpha}{\alpha - 1}.$$

Therefore

$$x_n = \sum_{k=0}^{n} \frac{(\alpha n)^k}{k!} \sim \frac{\alpha}{\alpha - 1} \cdot \frac{(\alpha n)^n}{n!}.$$

It is enough to use Stirling's formula $n! \sim \left(\dfrac{n}{e}\right)^n \sqrt{2\pi n}$ to deduce that

$$\lim_{n\to\infty} \frac{\sqrt{n}}{(\alpha e)^n} \sum_{k=0}^{n} \frac{(\alpha n)^k}{k!} = \frac{\alpha}{\sqrt{2\pi}(\alpha - 1)}. \quad \square$$

Proposed Problems

1. Let $(a_{nk})_{n,k\geq 1}$ be a triangular matrix with the following properties:
   a) $\lim\limits_{n\to\infty} a_{nk} = 0$, for each $k \geq 1$;
   b) there is $K > 0$ such that

$$\sum_{k=1}^{n} |a_{nk}| \leq K,$$

   for all positive integers n. If the sequence $(t_n)_{n\geq 1}$ converges to zero, prove that the sequence $(\widetilde{t}_n)_{n\geq 1}$ given by the formula

$$\widetilde{t}_n = \sum_{k=1}^{n} a_{nk} t_k,$$

   also converges to zero.

2. Assume that the hypothesis of Toeplitz's theorem hold. Prove that if $(t_n)_{n\geq 1}$ is bounded, then $(\widetilde{t}_n)_{n\geq 1}$ is bounded and

$$\sup_{n\geq 1} |\widetilde{t}_n| \leq \sup_{n\geq 1} |t_n| \, .$$

3. Let $(x_n)_{n\geq 0}$, $(y_n)_{n\geq 0}$ be sequences converging to zero and such that the series $\sum\limits_{n=0}^{\infty} y_n$ is absolutely convergent. Prove that

$$\lim_{n\to\infty} (x_0 y_n + x_1 y_{n-1} + \cdots + x_{n-1} y_1 + x_n y_0) = 0.$$

4. Let $(x_n)_{n\geq 0}$, $(y_n)_{n\geq 0}$ be sequences convergent to x, respectively y. Prove that

$$\lim_{n\to\infty} \frac{x_0 y_n + x_1 y_{n-1} + \cdots + x_{n-1} y_1 + x_n y_0}{n} = x \cdot y.$$

5. If $x_n \to x$ as $n \to \infty$, prove that

$$\lim_{n\to\infty} \frac{\binom{n}{0} x_0 + \binom{n}{1} x_1 + \cdots + \binom{n}{n} x_n}{2^n} = x.$$

6. Let $(x_n)_{n\geq 1}$ be a sequence such that $\lim\limits_{n\to\infty} (x_{n+1} - \lambda x_n) = 0$, for some $|\lambda| < 1$. Prove that $\lim\limits_{n\to\infty} x_n = 0$.

7. Let $(x_n)_{n \geq 1}$ be a sequence such that $\lim_{n \to \infty} (10x_{n+2} + 7x_{n+1} + x_n) = 18$. Prove that $\lim_{n \to \infty} x_n = 1$.

8. Let $(x_n)_{n \geq 1}$ be a bounded sequence such that

$$x_{n+2} \leq \frac{3x_{n+1} - x_n}{2},$$

for all positive integers n. Prove that $(x_n)_{n \geq 1}$ is convergent.

9. Let $(a_n)_{n \geq 0}$ and $(b_n)_{n \geq 0}$ be sequences of real numbers such that

$$b_n = a_n - \alpha a_{n+1}$$

with α a fixed real number, $|\alpha| < 1$. Assume that $(b_n)_{n \geq 0}$ is convergent and

$$\lim_{n \to \infty} a_n \alpha^n = 0.$$

Prove that $(a_n)_{n \geq 0}$ is convergent.

10. Let $(a_{mn})_{m,n \geq 1}$ be a double sequence of positive integers and assume that each positive integer appears at most ten times in the sequence $(a_{mn})_{m,n \geq 1}$. Prove that there exist positive integers m, n such that $a_{mn} > mn$.

11. Let $(a_n)_{n \geq 1}$ be a sequence of positive real numbers such that the series

$$\sum_{n=1}^{\infty} \frac{1}{a_n}$$

converges. Prove that

$$\sum_{m=1}^{\infty} \sum_{n=1}^{\infty} \frac{1}{a_m(a_m + a_n)} = \frac{1}{2} \left(\sum_{n=1}^{\infty} \frac{1}{a_n} \right)^2,$$

then find the sum of the series

$$\sum_{m-1}^{\infty} \sum_{n=1}^{\infty} \frac{m^2 n}{3^m (n \cdot 3^m + m \cdot 3^n)}.$$

12. Let $a_n \in \mathbb{C}$ be a sequence such that $\lim_{n \to \infty} na_n = 0$ and $\lim_{x \nearrow 1} \sum_{n=0}^{\infty} a_n x^n = 0$. Prove that $\sum_{n=0}^{\infty} a_n = 0$.

13. Let $f : [0, 1] \mapsto \mathbb{R}$ be a Lipschitz function and let the series $\sum_{k=0}^{\infty} a_k$ be convergent with sum a. Prove that

$$\lim_{n \to \infty} \sum_{k=0}^{n} a_k f\left(\frac{k}{n}\right) = af(0).$$

14. Let $f : [0, 1] \mapsto \mathbb{R}$ be a continuous function and let the series $\sum_{k=0}^{\infty} a_k$ be absolutely convergent (hence convergent) with sum a. Prove that

$$\lim_{n \to \infty} \sum_{k=0}^{n} a_k f\left(\frac{k}{n}\right) = af(0).$$

Solutions

1. For $\varepsilon > 0$, we consider an integer $n(\varepsilon)$ for which $|t_n| < \frac{\varepsilon}{2K}$ for all integers $n \geq n(\varepsilon)$. For every $n > n(\varepsilon)$, we have

$$|\widetilde{t}_n| \leq \left|\sum_{k=1}^{n(\varepsilon)} a_{nk} t_k\right| + \sum_{k=n(\varepsilon)+1}^{n} |a_{nk} t_k|$$

$$\leq \left|\sum_{k=1}^{n(\varepsilon)} a_{nk} t_k\right| + \left(\sum_{k=n(\varepsilon)+1}^{n} |a_{nk}|\right) \cdot \frac{\varepsilon}{2K} \leq \left|\sum_{k=1}^{n(\varepsilon)} a_{nk} t_k\right| + \frac{\varepsilon}{2}.$$

Let $k(\varepsilon)$ be such that

$$|a_{nm}| < \frac{\varepsilon}{2(|t_1| + \cdots + |t_{n(\varepsilon)}|)},$$

for all integers $n \geq k(\varepsilon)$ and all $m = 1, 2, \ldots, n(\varepsilon)$.
Hence for every $n > n(\varepsilon) + k(\varepsilon)$, we have

$$\left|\sum_{k=1}^{n(\varepsilon)} a_{nk} t_k\right| \leq \sum_{k=1}^{n(\varepsilon)} |a_{nk}| \cdot |t_k| \leq \frac{\varepsilon}{2(|t_1| + \cdots + |t_{n(\varepsilon)}|)} \cdot (|t_1| + \cdots + |t_{n(\varepsilon)}|) = \frac{\varepsilon}{2}.$$

Finally, $|\widetilde{t}_n| < \varepsilon$, for all integers $n > n(\varepsilon) + k(\varepsilon)$.
2. The proof is identical to the previous one.
3. We can use problem 1. Let us put $a_{nk} = y_{n-k}$ for $n, k \geq 1$, $n \geq k$. We have

$$\lim_{n \to \infty} a_{nk} = \lim_{n \to \infty} y_{n-k} = 0,$$

for each positive integer k and

$$\sum_{k=1}^{n} |a_{nk}| = \sum_{k=1}^{n} |y_{n-k}| < K,$$

for some $K > 0$, because of the absolute convergence of the series $\sum_{k=0}^{\infty} y_k$. Now let $t_n = x_{n-1}$ for all $n \geq 1$, hence $(t_n)_{n\geq 1}$ has limit 0. By the result of problem 1, the sequence with general term

$$\widetilde{t}_n = \sum_{k=1}^{n} a_{nk} t_k = \sum_{k=1}^{n} y_{n-k} x_{k-1}$$

also converges to zero—which is the desired conclusion.

We can also give a direct proof. Let $M > 0$ be such that $|x_n| \leq M$ and $|y_n| \leq M$ for all positive integers n. For $\varepsilon > 0$, let $n(\varepsilon)$ be such that

$$|y_{n+1}| + |y_{n+2}| + \cdots + |y_{n+p}| < \frac{\varepsilon}{2M}$$

for all integers $n \geq n(\varepsilon)$, $p \geq 1$, and $|x_n| < \frac{\varepsilon}{2K}$ for all integers $n \geq n(\varepsilon)$, where $K = \sum_{n=1}^{\infty} |y_n|$. Now, for all $n > 2n(\varepsilon)$, we have

$$|x_1 y_{n-1} + x_2 y_{n-2} + \cdots + x_{n-2} y_2 + x_{n-1} y_1|$$

$$\leq \left(|x_1| \cdot |y_{n-1}| + |x_2| \cdot |y_{n-2}| + \cdots + |x_{n-n(\varepsilon)-1}| \cdot |y_{n(\varepsilon)+1}| \right)$$

$$+ \left(|x_{n-n(\varepsilon)}| \cdot |y_{n(\varepsilon)}| + \cdots + |x_{n-1}| \cdot |y_1| \right)$$

$$\leq M \cdot \frac{\varepsilon}{2M} + \frac{\varepsilon}{2K} \cdot K = \varepsilon.$$

4. We can assume $x = 0$, by taking $(x_n - x)_{n\geq 0}$ instead of $(x_n)_{n\geq 0}$. Further, we can assume that $y = 0$ by changing $(y_n)_{n\geq 0}$ to $(y_n - y)_{n\geq 0}$. If $M > 0$ is such that $|y_n| \leq M$, for all nonnegative integers n, then

$$\left| \frac{x_0 y_n + \cdots + x_n y_0}{n} \right| \leq \frac{|x_0| \cdot |y_n| + \cdots + |x_n| \cdot |y_0|}{n} \leq M \cdot \frac{|x_0| + \cdots + |x_n|}{n}.$$

By the Cesàro-Stolz theorem,

$$\lim_{n\to\infty} \frac{|x_0| + \cdots + |x_n|}{n} = 0,$$

so the problem is solved.

Observe also that if the two sequences converge to 0 (we saw that we can always make this assumption), then the result is clear from Cauchy-Schwarz's inequality combined with the Cesàro-Stolz theorem:

$$\left| \frac{x_0 y_n + x_1 y_{n-1} + \cdots + x_n y_0}{n+1} \right| \leq \sqrt{\frac{x_0^2 + x_1^2 + \cdots + x_n^2}{n+1}} \cdot \sqrt{\frac{y_0^2 + y_1^2 + \cdots + y_n^2}{n+1}}.$$

5. We can use Toeplitz's theorem with

$$a_{nk} = \frac{1}{2^n} \binom{n}{k}, \qquad t_n = x_n.$$

First, for each integer $k \geq 1$, we have $\binom{n}{k} < n^k$, so

$$\lim_{n \to \infty} a_{nk} = \lim_{n \to \infty} \frac{1}{2^n} \binom{n}{k} = 0.$$

Finally,

$$\sum_{k=0}^{n} a_{nk} = \sum_{k=0}^{n} \frac{1}{2^n} \binom{n}{k} = 1.$$

We can also give a proof which uses the splitting method. We assume that $(x_n)_{n \geq 0}$ converges to zero. For arbitrary fixed $\varepsilon > 0$, let $n(\varepsilon)$ be an integer for which $|x_n| \leq \frac{\varepsilon}{2}$, for all integers $n \geq n(\varepsilon)$. Let us put $M = \sup\{|x_n| \mid n \in \mathbb{N}\}$. Then for all $n > n(\varepsilon)$, we have

$$\left| \frac{\binom{n}{0} x_0 + \cdots + \binom{n}{n(\varepsilon)} x_{n(\varepsilon)} + \cdots + \binom{n}{n} x_n}{2^n} \right|$$

$$\leq \frac{\binom{n}{0} |x_0| + \cdots + \binom{n}{n(\varepsilon)} |x_{n(\varepsilon)}|}{2^n}$$

$$+ \frac{\binom{n}{n(\varepsilon)+1} |x_{n(\varepsilon)+1}| + \cdots + \binom{n}{n} |x_n|}{2^n}.$$

First,

$$\frac{\binom{n}{0}|x_0| + \cdots + \binom{n}{n(\varepsilon)}|x_{n(\varepsilon)}|}{2^n}$$

$$\leq M \cdot \frac{\binom{n}{0} + \cdots + \binom{n}{n(\varepsilon)}}{2^n} < M \cdot \frac{\varepsilon}{2M} = \frac{\varepsilon}{2},$$

for sufficiently large n, say $n \geq n_1(\varepsilon)$. On the other hand,

$$\frac{\binom{n}{n(\varepsilon)+1}|x_{n(\varepsilon)+1}| + \cdots + \binom{n}{n}|x_n|}{2^n}$$

$$\leq \frac{\varepsilon}{2} \cdot \frac{\binom{n}{n(\varepsilon)+1} + \cdots + \binom{n}{n}}{2^n} < \frac{\varepsilon}{2}.$$

Finally,

$$\left| \frac{\binom{n}{0}x_0 + \cdots + \binom{n}{n}x_n}{2^n} \right| < \varepsilon,$$

for all integers $n > \max\{n(\varepsilon), n_1(\varepsilon)\}$, which solves the problem.

It is interesting to note that in the case when the sequence converges to 0, a simple argument based on Cauchy-Schwarz's inequality and the Cesàro-Stolz theorem does not work. Indeed, it boils down to the fact that the sequence whose general term is $\frac{n}{4^n} \cdot \binom{2n}{n}$ is bounded, which is false (the reader can prove this, as an exercise).

6. We know that the sequence

$$y_n = x_{n+1} - \lambda x_n, \ n \geq 1$$

converges to zero. From the equalities

$$\frac{y_n}{\lambda^{n+1}} = \frac{x_{n+1}}{\lambda^{n+1}} - \frac{x_n}{\lambda^n}, \ n \geq 1,$$

we deduce that

$$x_n = \lambda^{n-1}x_1 + \sum_{k=1}^{n-1} \lambda^{n-k-1}y_k.$$

Now it is sufficient to prove that the sequence

$$\widetilde{x}_n = \sum_{k=1}^{n} \lambda^{n-k}y_k, \quad n \geq 1$$

is convergent to zero. This follows from a corollary of Toeplitz's theorem, more precisely, from problem 1. Indeed, with $a_{nk} = \lambda^{n-k}$, we have

$$\lim_{n \to \infty} a_{nk} = \lim_{n \to \infty} \lambda^{n-k} = 0,$$

for all positive integers k and

$$\sum_{k=1}^{n} |a_{nk}| = \sum_{k=1}^{n} |\lambda|^{n-k} \leq \frac{1}{1 - |\lambda|}.$$

7. First we will prove a more general result. Let a, b, c be real numbers, $a \neq 0$ such that the quadratic equation $ax^2 + bx + c = 0$ has two real solutions in $(-1, 1)$. Then each sequence $(x_n)_{n \geq 1}$ with the property

$$\lim_{n \to \infty} (ax_{n+2} + bx_{n+1} + cx_n) = 0$$

is convergent to zero. Indeed, if we denote by $\lambda, \mu \in (-1, 1)$ the solutions of the quadratic equation, then according to Viète's formulas,

$$\lambda + \mu = -\frac{b}{a}, \quad \lambda\mu = \frac{c}{a}.$$

Now,

$$\lim_{n \to \infty} (ax_{n+2} + bx_{n+1} + cx_n) = 0$$

becomes

$$\lim_{n \to \infty} \left(x_{n+2} + \frac{b}{a}x_{n+1} + \frac{c}{a}x_n \right) = 0$$

or

$$\lim_{n \to \infty} [x_{n+2} - (\lambda + \mu)x_{n+1} + \lambda\mu x_n] = 0$$

which can be also written as

$$\lim_{n\to\infty} [(x_{n+2} - \lambda x_{n+1}) - \mu(x_{n+1} - \lambda x_n)] = 0.$$

In this way, with the notation $u_n = x_{n+1} - \lambda x_n$, $n \geq 1$, we have

$$\lim_{n\to\infty} (u_{n+1} - \mu u_n) = 0$$

and $|\mu| < 1$. We have seen that this implies $\lim_{n\to\infty} u_n = 0$, that is,

$$\lim_{n\to\infty} (x_{n+1} - \lambda x_n) = 0.$$

Using again the result of the previous problem, we conclude that $(x_n)_{n\geq 1}$ converges to zero. In our case, note that the quadratic equation

$$10x^2 + 7x + 1 = 0$$

has two real solutions in $(-1, 1)$. Let us denote $a_n = x_n - 1$, $n \geq 1$. We have

$$\lim_{n\to\infty} (10a_{n+2} + 7a_{n+1} + a_n) = \lim_{n\to\infty} [10(x_{n+2} - 1) + 7(x_{n+1} - 1) + (x_n - 1)]$$

$$= \lim_{n\to\infty} [(10a_{n+2} + 7a_{n+1} + a_n) - 18] = 0.$$

This implies $a_n \to 0$ and then $x_n \to 1$, as $n \to \infty$.

8. Let us define the sequence

$$y_n = x_{n+1} - \frac{1}{2}x_n, \quad n \geq 1.$$

Clearly, if $(x_n)_{n\geq 1}$ is bounded, so is the sequence $(y_n)_{n\geq 1}$ (just note that $|y_n| \leq |x_{n+1}| + \frac{|x_n|}{2}$). The given inequality allows us to prove the monotony of the sequence $(y_n)_{n\geq 1}$. Indeed,

$$y_{n+1} - y_n = \left(x_{n+2} - \frac{1}{2}x_{n+1}\right) - \left(x_{n+1} - \frac{1}{2}x_n\right)$$

$$= x_{n+2} - \frac{3}{2}x_{n+1} + \frac{1}{2}x_n \leq 0,$$

so the sequence $(y_n)_{n\geq 1}$ is decreasing. Hence $(y_n)_{n\geq 1}$ is convergent and let l be its limit. Searching for a real number α so that the sequence $z_n = x_n - \alpha$, $n \geq 1$ satisfies $\lim_{n\to\infty} (z_{n+1} - \frac{1}{2}z_n) = 0$, we immediately obtain $\alpha = 2l$. Thus the sequence $z_n = x_n - 2l$, $n \geq 1$ satisfies

$$\lim_{n \to \infty} \left(z_{n+1} - \frac{1}{2} z_n \right) = 0.$$

and, as we know, this is possible only if $z_n \to 0$, so $x_n \to 2l$.

9. By induction, for $n \geq 1$, we have

$$a_n = -\sum_{k=-1}^{n-1} b_k \left(\frac{1}{\alpha} \right)^{n-k} = -\frac{1}{\alpha^n} \sum_{k=-1}^{n-1} b_k \alpha^k,$$

where $b_{-1} = -a_0 \alpha$. Since $a_n \alpha^n$ tends to zero, the partial sums on the right-hand side in

$$-a_n \alpha^n = \sum_{k=-1}^{n-1} b_k \alpha^k$$

also approach zero. They may therefore be replaced by the negatives of their corresponding remainders

$$-a_n \alpha^n = -\sum_{k=n}^{\infty} b_k \alpha^k \quad \text{or} \quad a_n = \sum_{k=0}^{\infty} b_{n+k} \alpha^k.$$

If b is the limit of the sequence $(b_n)_{n \geq -1}$, then we can estimate the difference

$$\left| a_n - \frac{b}{1-\alpha} \right| = \left| \sum_{k=0}^{\infty} (b_{n+k} - b) \alpha^k \right| \leq \frac{1}{1 - |\alpha|} \cdot \sup_{k \geq n} |b_k - b|,$$

which becomes arbitrarily small for large values of n. Hence $(a_n)_{n \in \mathbb{N}}$ converges to $b/(1 - \alpha)$.

10. Let us assume, by way of contradiction, that $a_{mn} \leq mn$, for all positive integers m, n. For every positive integer k, the set

$$A_k = \{ (i, j) \mid a_{ij} \leq k \}$$

has at most $10k$ elements. On the other hand, A_k contains all pairs (i, j) with $ij \leq k$ and there are

$$\left[\frac{k}{1} \right] + \left[\frac{k}{2} \right] + \cdots + \left[\frac{k}{k} \right].$$

such pairs. Thus, for all positive integers k,

$$10k \geq \left[\frac{k}{1} \right] + \left[\frac{k}{2} \right] + \cdots + \left[\frac{k}{k} \right].$$

This inequality cannot be true for large k, because

$$\lim_{k \to \infty} \frac{\left[\frac{k}{1}\right] + \left[\frac{k}{2}\right] + \cdots + \left[\frac{k}{k}\right]}{k} = \infty.$$

11. By interchanging m and n, we have

$$s = \sum_{m=1}^{\infty} \sum_{n=1}^{\infty} \frac{1}{a_m(a_m + a_n)} = \sum_{m=1}^{\infty} \sum_{n=1}^{\infty} \frac{1}{a_n(a_m + a_n)}.$$

Thus

$$2s = \sum_{m=1}^{\infty} \sum_{n=1}^{\infty} \left(\frac{1}{a_m(a_m + a_n)} + \frac{1}{a_n(a_m + a_n)} \right)$$

$$= \sum_{m=1}^{\infty} \sum_{n=1}^{\infty} \frac{1}{a_m a_n} = \left(\sum_{m=1}^{\infty} \frac{1}{a_m} \right) \left(\sum_{n=1}^{\infty} \frac{1}{a_n} \right) = \left(\sum_{n=1}^{\infty} \frac{1}{a_n} \right)^2.$$

For the second part, let us take $a_n = \frac{3^n}{n}$ to obtain

$$\sum_{m=1}^{\infty} \sum_{n=1}^{\infty} \frac{1}{a_m(a_m + a_n)} = \sum_{m=1}^{\infty} \sum_{n=1}^{\infty} \frac{m^2 n}{3^m(n \cdot 3^m + m \cdot 3^n)} = \frac{1}{2} \left(\sum_{n=1}^{\infty} \frac{n}{3^n} \right)^2 = \frac{9}{32}.$$

Indeed, if we differentiate

$$\frac{1}{1 - x} = \sum_{n \geq 0} x^n$$

for $|x| < 1$, we obtain $\dfrac{x}{(1 - x)^2} = \displaystyle\sum_{n \geq 1} n x^n$; thus $\displaystyle\sum_{n \geq 1} \frac{n}{3^n} = \frac{3}{4}$.

12. Let $f(x) = \displaystyle\sum_{n=0}^{\infty} a_n x^n$, which is defined for $|x| < 1$ because a_n is bounded. Then
for all $0 \leq x < 1$,

$$\left| \sum_{n=0}^{N} a_n \right| \leq |f(x)| + \left| \sum_{n>N} a_n x^n + \sum_{n=0}^{N} a_n (x^n - 1) \right|$$

$$\leq |f(x)| + \sum_{n>N} |a_n| x^n + \sum_{n=0}^{N} |a_n|(1 - x^n)$$

$$\le |f(x)| + \frac{1}{N} \sup_{n \ge N} |na_n| \sum_{n>N} x^n + (1-x) \sum_{n=0}^{N} |a_n|(1 + x + \cdots + x^{n-1})$$

$$\le |f(x)| + \frac{1}{N} \sup_{n \ge N} |na_n| \frac{x^{N+1}}{1-x} + (1-x) \sum_{n=0}^{N} n|a_n|$$

$$\le |f(x)| + \frac{1}{N(1-x)} \sup_{n \ge N} |na_n| + N(1-x)\frac{1}{N} \sum_{n=0}^{N} n|a_n|$$

Choose $x = 1 - \frac{1}{N}$, then

$$\left| \sum_{n=0}^{N} a_n \right| \le \left| f\left(1 - \frac{1}{N}\right) \right| + \sup_{n \ge N} |na_n| + \frac{1}{N} \sum_{n=0}^{N} n|a_n|.$$

By making $N \to \infty$ in the previous inequality, we obtain

$$\lim_{N \to \infty} \sum_{n=0}^{N} a_n = 0,$$

because

$$\lim_{N \to \infty} f\left(1 - \frac{1}{N}\right) = 0, \qquad \lim_{N \to \infty} \sup_{n \ge N} |na_n| = 0$$

(because $\lim_{n \to \infty} n|a_n| = 0$) and

$$\lim_{N \to \infty} \frac{1}{N} \sum_{n=0}^{N} n|a_n| = 0$$

by the Cesàro-Stolz theorem.

13. **Solution I.** The fact that f is Lipschitz means that there exists a constant $L > 0$ such that $|f(x)-f(y)| \le L|x-y|$ for all $x, y \in [0, 1]$; consequently, $|f((k-1)/n)- f(k/n)| \le L/n$ for all positive integers n and $k \le n$. We have

$$\sum_{k=0}^{n} a_k f\left(\frac{k}{n}\right) - af(0) = \sum_{k=1}^{n} (a_0 + a_1 + \cdots + a_{k-1} - a)\left(f\left(\frac{k-1}{n}\right) - f\left(\frac{k}{n}\right)\right)$$

$$+ (a_0 + a_1 + \cdots + a_n - a)f\left(\frac{n}{n}\right);$$

therefore, using the triangle inequality and the Lipschitz condition, we obtain

$$\left| \sum_{k=0}^{n} a_k f\left(\frac{k}{n}\right) - af(0) \right| \leq L \cdot \frac{1}{n} \sum_{k=1}^{n} |a_0 + a_1 + \cdots + a_{k-1} - a|$$

$$+ |a_0 + a_1 + \cdots + a_n - a||f(1)|.$$

Now, the hypothesis tells us that $\lim_{n\to\infty} |a_0 + a_1 + \cdots + a_n - a| = 0$, and the Cesàro-Stolz theorem ensures that

$$\lim_{n\to\infty} \frac{1}{n} \sum_{k=1}^{n} |a_0 + a_1 + \cdots + a_{k-1} - a| = 0,$$

as well. Thus, the sequence on the right hand side of the above inequality has limit 0, which proves that the left hand side also has limit 0, finishing the proof.

Solution II. We need the following simple (and interesting by itself) lemma: if $(x_n)_{n\geq 0}$ is a convergent sequence of real numbers, then

$$\lim_{n\to\infty} \frac{|x_n - x_{n-1}| + \cdots + |x_n - x_0|}{n} = 0.$$

This can be proven by using the splitting method; we proceed further.

Because $(x_n)_{n\geq 0}$ is convergent, it is also bounded, hence there exists $M > 0$ such that $|x_n| < M$ for all n; it follows that $|x_m - x_n| < 2M$ for all m and n. Further, since $(x_n)_{n\geq 0}$ is convergent, it is also a Cauchy sequence; consequently, for a given $\varepsilon > 0$, there exists a positive integer N such that $|x_m - x_n| < \varepsilon/2$ for all $m, n \geq N$. We then have, for $n > \max\{N, 4NM/\varepsilon\}$,

$$\frac{|x_n - x_{n-1}| + \cdots + |x_n - x_0|}{n} = \frac{|x_n - x_{n-1}| + \cdots + |x_n - x_N|}{n}$$

$$+ \frac{|x_n - x_{N-1}| + \cdots + |x_n - x_0|}{n} < \frac{(n-N)\varepsilon}{2n} + \frac{2NM}{n} < \frac{\varepsilon}{2} + \frac{\varepsilon}{2} = \varepsilon,$$

and the lemma is established.

Now for the problem, we observe that

$$\sum_{k=0}^{n} a_k f\left(\frac{k}{n}\right) - \sum_{k=0}^{n} a_k f(0)$$

$$= \sum_{k=0}^{n-1} (a_n + \cdots + a_{n-k}) \left(f\left(\frac{n-k}{n}\right) - f\left(\frac{n-k-1}{n}\right) \right)$$

by an Abel summation formula again. The triangle inequality and the Lipschitz condition yield

$$\left| \sum_{k=0}^{n} a_k f\left(\frac{k}{n}\right) - \sum_{k=0}^{n} a_k f(0) \right| \le L \cdot \frac{1}{n} \sum_{k=0}^{n-1} |a_n + \cdots + a_{n-k}|$$

$$= L \cdot \frac{1}{n} \sum_{k=0}^{n-1} |S_n - S_{n-k-1}|,$$

where $S_p = a_0 + a_1 + \cdots + a_p$ is the pth partial sum of the series $\sum_{n=0}^{\infty} a_n$. As $(S_n)_{n \ge 0}$ is a convergent sequence, the lemma applies and shows that the right hand side from the previous inequality has limit 0; thus, the left hand side also tends to 0; that is,

$$\lim_{n \to \infty} \left(\sum_{k=0}^{n} a_k f\left(\frac{k}{n}\right) - \sum_{k=0}^{n} a_k f(0) \right) = 0.$$

Now $\lim_{n \to \infty} \sum_{k=0}^{n} a_k f(0) = a f(0)$ completes the proof.

14. **Solution I.** Let $A = \sum_{n=0}^{\infty} |a_n|$; of course, $|a| \le A$, and if $A = 0$, there is nothing to prove, as $a = 0$ and all $a_n = 0$ follow from such an assumption; thus, we may assume that $A > 0$.

Let $\varepsilon > 0$ be given. It is not hard to prove (e.g., by using the Stone–Weierstrass theorem—see the Chapter 7) that the set of Lipschitz functions is dense in the set of continuous functions (all defined on $[0, 1]$, say). Thus, for our continuous function f (and the considered ε) there exists a Lipschitz function $g : [0, 1] \to \mathbb{R}$ such that $|f(x) - g(x)| < \varepsilon/(3A)$ for all $x \in [0, 1]$. Yet, according to the previous problem, for the function g we have

$$\lim_{n \to \infty} \sum_{k=0}^{n} a_k g\left(\frac{k}{n}\right) = a g(0);$$

consequently, there exists a positive integer N such that for all positive integers $n \ge N$, we have

$$\left| \sum_{k=0}^{n} a_k g\left(\frac{k}{n}\right) - a g(0) \right| < \frac{\varepsilon}{3}.$$

Now, by the triangle inequality and all of the above remarks, we have

$$\left| \sum_{k=0}^{n} a_k f\left(\frac{k}{n}\right) - af(0) \right| \leq \sum_{k=0}^{n} |a_k| \left| f\left(\frac{k}{n}\right) - g\left(\frac{k}{n}\right) \right|$$

$$+ \left| \sum_{k=0}^{n} a_k g\left(\frac{k}{n}\right) - ag(0) \right| + |a| |f(0) - g(0)|$$

$$< \frac{\varepsilon}{3A} \sum_{k=0}^{n} |a_k| + \left| \sum_{k=0}^{n} a_k g\left(\frac{k}{n}\right) - ag(0) \right| + \frac{\varepsilon}{3A} |a|$$

$$< \frac{\varepsilon}{3A} \cdot A + \frac{\varepsilon}{3} + \frac{\varepsilon}{3A} \cdot A = \varepsilon$$

for every positive integer $n \geq N$, and the proof is complete.

Solution II. The function f is continuous and therefore bounded on the compact interval $[0, 1]$. Thus there exists $M > 0$ such that $|f(x)| \leq M$ for all $x \in [0, 1]$, implying that $|f(x) - f(y)| \leq 2M$ for all $x, y \in [0, 1]$. Because the sequence of partial sums of the series $\sum_{k=0}^{\infty} |a_k|$ is convergent, it is also a Cauchy sequence, hence, for any positive ε, there exists a positive integer N such that

$$|a_m| + |a_{m+1}| + \cdots + |a_n| < \frac{\varepsilon}{4M},$$

for all positive integers m and n with $N < m \leq n$. Yet, because f is continuous on the compact interval $[0, 1]$, it is also uniformly continuous; thus there exists $\delta > 0$ such that $|f(x) - f(y)| < \varepsilon/(2A)$ whenever $x, y \in [0, 1]$ and $|x - y| < \delta$. And here comes the splitting: namely, for $n > \max\{N, N/\delta\}$, we have

$$\left| \sum_{k=0}^{n} a_k f\left(\frac{k}{n}\right) - \sum_{k=0}^{n} a_k f(0) \right| \leq \sum_{k=0}^{n} |a_k| \left| f\left(\frac{k}{n}\right) - f(0) \right|$$

$$= \sum_{k=0}^{N} |a_k| \left| f\left(\frac{k}{n}\right) - f(0) \right| + \sum_{k=N+1}^{n} |a_k| \left| f\left(\frac{k}{n}\right) - f(0) \right|$$

$$< \frac{\varepsilon}{2A} \sum_{k=0}^{N} |a_k| + 2M \sum_{k=N+1}^{n} |a_k| < \frac{\varepsilon}{2A} \cdot A + 2M \cdot \frac{\varepsilon}{4M} = \epsilon,$$

and we are done.

(Note that for $0 \leq k \leq N$ and $n > N/\delta$, we have $|k/n - 0| = k/n \leq N/n < \delta$.)

This is problem 180, from *Gazeta Matematică—seria A*, 3/2004, solved in the same magazine, 3/2005. The problem, as well as the previous one, was proposed by Dan Ştefan Marinescu and Viorel Cornea.

Chapter 10
The Number e

The basic symbol of mathematical analysis is the well-known number e. It is introduced as the limit of the sequence

$$e_n = \left(1 + \frac{1}{n}\right)^n, \ n \geq 1.$$

We have the following result:

Theorem. *The sequence* $(e_n)_{n \geq 1}$ *is monotonically increasing and bounded. Moreover, the sequence* $(f_n)_{n \geq 1}$ *defined by*

$$f_n = \left(1 + \frac{1}{n}\right)^{n+1}, \ n \geq 1.$$

converges to e and is decreasing.

Proof. Indeed, in order to prove the boundedness, one can easily prove by induction with respect to k the inequality

$$\left(1 + \frac{1}{n}\right)^k \leq 1 + \frac{k}{n} + \frac{k^2}{n^2}, \ 1 \leq k \leq n.$$

Then, by taking $k = n$, we obtain $e_n \leq 3$. On the other hand, Bernoulli's inequality gives

$$\left(1 + \frac{1}{n}\right)^n \geq 1 + n \cdot \frac{1}{n} = 2,$$

© Springer Science+Business Media LLC 2017
T. Andreescu et al., *Mathematical Bridges*, DOI 10.1007/978-0-8176-4629-5_10

so $e_n \geq 2$. We proved that $e_n \in [2,3]$ for all positive integers n. We use again Bernoulli's inequality to establish that $(e_n)_{n \geq 1}$ is monotone; we have:

$$\frac{e_{n+1}}{e_n} = \left(1 - \frac{1}{(n+1)^2}\right)^n \cdot \frac{n+2}{n+1}$$
$$> \frac{n^3 + 3n^2 + 3n + 2}{n^3 + 3n^2 + 3n + 1} > 1,$$

so $(e_n)_{n \geq 1}$ is (strictly) increasing. Clearly, $\lim\limits_{n \to \infty} f_n = e$. Finally, with Bernoulli's inequality,

$$\frac{f_n}{f_{n+1}} = \left(1 + \frac{1}{n(n+2)}\right)^{n+1} \cdot \frac{n+1}{n+2}$$
$$> \left(1 + \frac{n+1}{n(n+2)}\right) \cdot \frac{n+1}{n+2}$$
$$= \frac{n^3 + 4n^2 + 4n + 1}{n^3 + 4n^2 + 4n} > 1,$$

so $(f_n)_{n \geq 1}$ is decreasing. This finishes the proof of the theorem. $\square$

More generally, we can consider the function $e_n : \mathbb{R} \to \mathbb{R}$, given by

$$e_n(x) = \left(1 + \frac{x}{n}\right)^n, \quad x \in \mathbb{R}.$$

As above, the sequence of real numbers $(e_n(x))_{n > -x}$ is increasing, and for all real numbers x,

$$\lim_{n \to \infty} \left(1 + \frac{x}{n}\right)^n = e^x.$$

The following result is particularly important, so we present it as a theorem:

Theorem. *The sequence*

$$x_n = 1 + \frac{1}{1!} + \frac{1}{2!} + \cdots + \frac{1}{n!}$$

converges to e. Moreover, we have the inequality

$$0 < e - x_n < \frac{1}{n \cdot n!}$$

and e is irrational.

Proof. We have, by the binomial formula,

$$\left(1+\frac{1}{n}\right)^n = 1 + \frac{1}{1!} + \frac{1}{2!}\left(1-\frac{1}{n}\right) + \cdots + \frac{1}{n!}\left(1-\frac{1}{n}\right)\left(1-\frac{2}{n}\right)\cdots\left(1-\frac{n-1}{n}\right)$$

$$< 1 + \frac{1}{1!} + \frac{1}{2!} + \cdots + \frac{1}{n!} = x_n.$$

On the other hand, we clearly have

$$x_n < 1 + \left(1 + \frac{1}{2} + \frac{1}{4} + \cdots + \frac{1}{2^{n-1}}\right) = 1 + \frac{1-\frac{1}{2^n}}{1-\frac{1}{2}} < 3.$$

Thus, $(x_n)_{n\geq 1}$ is strictly increasing and bounded from above. From $e_n \leq x_n$, we deduce that $\lim_{n\to\infty} x_n \geq e$. On the other hand, for all p and all $n > p$,

$$e_n > 1 + \frac{1}{1!} + \frac{1}{2!}\left(1-\frac{1}{n}\right) + \cdots + \frac{1}{p!}\left(1-\frac{1}{n}\right)\cdots\left(1-\frac{p-1}{n}\right),$$

and if we take the limit as $n \to \infty$, we obtain

$$e \geq 1 + \frac{1}{1!} + \frac{1}{2!} + \cdots + \frac{1}{p!},$$

for all positive integers p. For $p \to \infty$, $e \geq \lim_{p\to\infty} x_p$. In conclusion,

$$\lim_{n\to\infty}\left(1 + \frac{1}{1!} + \frac{1}{2!} + \cdots + \frac{1}{n!}\right) = e.$$

Now, for any positive integers n and $m \geq 2$,

$$\frac{1}{(n+1)!} + \frac{1}{(n+2)!} + \cdots + \frac{1}{(n+m)!}$$

$$\leq \frac{1}{(n+1)!} + \frac{1}{(n+1)!(n+2)} + \cdots + \frac{1}{(n+1)!(n+2)^{m-1}}$$

$$= \frac{1}{(n+1)!}\cdot\left[1 + \frac{1}{n+2} + \cdots + \frac{1}{(n+2)^{m-1}}\right] = \frac{1}{(n+1)!}\cdot\frac{1-\frac{1}{(n+2)^m}}{1-\frac{1}{n+2}}.$$

Consequently,

$$x_{m+n} - x_n \leq \frac{1}{(n+1)!}\cdot\frac{1-\frac{1}{(n+2)^m}}{1-\frac{1}{n+2}}.$$

For $m \to \infty$, it follows that

$$0 < e - x_n \le \frac{1}{(n+1)!} \cdot \frac{n+2}{n+1},$$

for all positive integers n. Thus

$$0 < e - x_n < \frac{1}{n! \cdot n}.$$

To prove that e is irrational, let us assume, by way of contradiction, that $e = \frac{p}{q}$, with positive integers p, q. Then

$$\frac{p}{q} = 1 + \frac{1}{1!} + \frac{1}{2!} + \cdots + \frac{1}{q!} + \frac{\theta_q}{q! \cdot q}.$$

By multiplying with $q! \cdot q$, we conclude that

$$q! \cdot p = q! \cdot q \left(1 + \frac{1}{1!} + \frac{1}{2!} + \cdots + \frac{1}{q!} \right) + \theta_q,$$

or

$$\theta_q = q! \cdot p - q! \cdot q \left(1 + \frac{1}{1!} + \frac{1}{2!} + \cdots + \frac{1}{q!} \right).$$

This is a contradiction, because $\theta_q \in (0, 1)$ and the right-hand side of the last equality is an integer. In conclusion, e is irrational and the proof is done. $\square$

Problem.   a) Prove that $e^x \ge x + 1$ for all real numbers x.
b) Let $a > 0$ be such that $a^x \ge x + 1$, for all real numbers x. Prove that $a = e$.

Solution.   a) Let us define the function $f : \mathbb{R} \to \mathbb{R}$, given by

$$f(x) = e^x - x - 1.$$

Note that f is differentiable, with $f'(x) = e^x - 1$, so f' is negative on $(-\infty, 0)$ and positive on $(0, \infty)$. Therefore, f decreases on $(-\infty, 0)$ and increases on $(0, \infty)$. It follows that 0 is a point of minimum for f, which finishes the proof.
   Another solution is based on Bernoulli's inequality:

$$\left(1 + \frac{x}{n} \right)^n \ge 1 + \frac{x}{n} \cdot n = 1 + x.$$

Now, by taking the limit as $n \to \infty$ in the previous inequality, we obtain $e^x \ge x + 1$.

b) Let us define the function $f : \mathbb{R} \to \mathbb{R}$, by $f(x) = a^x - x - 1$. The given inequality can be written as $f(x) \geq f(0)$, for all x. Hence $x = 0$ is a point of minimum for f, and according to Fermat's theorem, $f'(0) = 0$. But this is equivalent to $a = e$.

Note that we can use only sequences in order to solve this problem. Indeed, if we take $x = \dfrac{1}{n}$ in the given inequality, we obtain

$$a^{1/n} \geq \frac{1}{n} + 1 \Rightarrow a \geq \left(1 + \frac{1}{n}\right)^n,$$

and for $n \to \infty$, we deduce that $a \geq e$. Similarly, by taking $x = -\dfrac{1}{n+1}$,

$$a^{-1/(n+1)} \geq 1 - \frac{1}{n+1} \Rightarrow a \leq \left(\frac{n+1}{n}\right)^{n+1}$$

or

$$a \leq \left(1 + \frac{1}{n}\right)^{n+1}.$$

For $n \to \infty$, it follows that $a \leq e$. $\square$

If we replace in (10.1) x by $x - 1$, we obtain the inequality

$$e^{x-1} \geq x, \tag{10.1}$$

for all real numbers x. This inequality can be used to deduce the AM-GM inequality

$$\frac{a_1 + a_2 + \cdots + a_n}{n} \geq \sqrt[n]{a_1 a_2 \ldots a_n}, \tag{10.2}$$

for all positive real numbers $a_1, a_2, \ldots, a_n$. The inequality (10.2) is equivalent to

$$a_1 + a_2 + \cdots + a_n \geq n,$$

for all positive real numbers $a_1, a_2, \ldots, a_n$, with $a_1 a_2 \cdots a_n = 1$. Assuming this, note that

$$e^{a_1 - 1} \geq a_1$$
$$e^{a_2 - 1} \geq a_2$$
$$\cdots \cdots \cdots$$
$$e^{a_n - 1} \geq a_n$$

and by multiplication,

$$e^{a_1 + a_2 + \cdots + a_n - n} \geq 1$$

or

$$a_1 + a_2 + \cdots + a_n - n \geq 0 \Leftrightarrow a_1 + a_2 + \cdots + a_n \geq n.$$

This proof is due to G. Pólya and it is still considered to be the most beautiful proof of this famous inequality.

In the framework of this new result, we can give another proof for the monotonicity of the sequence $(e_n)_{n \geq 1}$. Indeed,

$$1 + \frac{1}{n} = \underbrace{\frac{1}{n-1} + \frac{1}{n-1} + \cdots + \frac{1}{n-1}}_{n-1 \text{ times}} + \frac{1}{n}$$

$$> n \sqrt[n]{\frac{1}{n-1} \cdot \frac{1}{n-1} \cdots \frac{1}{n-1} \cdot \frac{1}{n}}$$

$$= \sqrt[n]{\left(\frac{n}{n-1}\right)^{n-1}} = \sqrt[n]{\left(1 + \frac{1}{n-1}\right)^{n-1}}$$

which implies $e_n > e_{n-1}$.

Further, we use in a different way the AM-GM inequality to obtain other interesting results. Indeed,

$$\sqrt[m+n+1]{\left(1 + \frac{1}{n}\right)^n \left(1 - \frac{1}{m+1}\right)^{m+1}} < \frac{n\left(1 + \frac{1}{n}\right) + (m+1)\left(1 - \frac{1}{m+1}\right)}{m+n+1} = 1,$$

thus

$$\left(1 + \frac{1}{n}\right)^n < \left(1 + \frac{1}{m}\right)^{m+1},$$

for all positive integers m, n. With the given notations, this means $e_n < f_m$ for all positive integers m, n. In particular,

$$e_{n(n+2)} < f_{n+1} \Rightarrow \left(1 + \frac{1}{n(n+2)}\right)^{n(n+2)} < \left(1 + \frac{1}{n+1}\right)^{n+2}$$

$$\Rightarrow \frac{(n+1)^{2n}}{n^n(n+2)^n} < \frac{(n+2)^{n+2}}{(n+1)^{n+2}}$$

$$\Rightarrow (n+1)^n(n+1)^{n+1} < n^n(n+2)^{n+1}$$

$$\Rightarrow \left(\frac{n+1}{n}\right)^n < \left(\frac{n+2}{n+1}\right)^{n+1} \Rightarrow e_n < e_{n+1}.$$

This is (yet) another proof of the monotonicity of the sequence $(e_n)_{n \geq 1}$.

Proposed Problems

1. Let $\mathcal{F}$ be the family of all functions $f : \mathbb{R} \to \mathbb{R}$ satisfying the relation

$$f(x + y) = f(x)f(y),$$

   for all real numbers x.

   a) Find all continuous functions from $\mathcal{F}$;
   b) Find all monotone functions from $\mathcal{F}$.

2. Find all functions $f : \mathbb{R} \to \mathbb{R}$ which satisfy the following two conditions:

   a) $f(x + y) \geq f(x)f(y)$, for all real numbers x, y;
   b) $f(x) \geq x + 1$, for all real numbers x.

3. Find all functions $f : (0, \infty) \to \mathbb{R}$ which satisfy the following two conditions:

   a) $f(xy) \leq f(x) + f(y)$, for all positive real numbers x, y;
   b) $f(x) \leq x - 1$, for all positive real numbers x.

4. For all positive integers n, prove that:

   a) $\dfrac{e}{2n + 2} < e - \left(1 + \dfrac{1}{n}\right)^n < \dfrac{e}{2n + 1}$;

   b) $\dfrac{e}{2n + 1} < \left(1 + \dfrac{1}{n}\right)^{n+1} - e < \dfrac{e}{2n}$.

5. Prove that

$$\lim_{n \to \infty} n\left[e - \left(1 + \frac{1}{n}\right)^n\right] = \lim_{n \to \infty} n\left[\left(1 + \frac{1}{n}\right)^{n+1} - e\right] = \frac{e}{2}.$$

6. Find the limit of the sequence $(x_n)_{n \geq 1}$ given by the implicit relation

$$\left(1 + \frac{1}{n}\right)^{n + x_n} = e, \ n \geq 1.$$

7. Find the limit of the sequence $(x_n)_{n \geq 1}$ given by the implicit relation

$$\left(1 + \frac{1}{n}\right)^{n + x_n} = 1 + \frac{1}{1!} + \cdots + \frac{1}{n!}, \ n \geq 1.$$

8. a) Prove that $e^x \geq ex$ for all real numbers x.
   b) Prove that if $a > 0$ has the property $a^x \geq ax$ for all real numbers x, then $a = e$.

9. a) Prove that $e^x \geq x^e$ for all positive numbers x.
   b) Prove that if $a > 0$ has the property $a^x \geq x^a$ for all positive numbers x, then
      $a = e$.

10. Let $a > 0$ be such that $x^x \geq a^a$ for all positive numbers x. Prove that $a = e^{-1}$.

11. Let $x < y$ be two positive real numbers such that $x^x = y^y$. Prove that there exists a positive real r such that

$$x = \left(1 - \frac{1}{r+1}\right)^{r+1}, \quad y = \left(1 - \frac{1}{r+1}\right)^r.$$

12. Let $x < y$ be two positive real numbers such that $x^y = y^x$. Prove that there exists a positive real r such that

$$x = \left(1 + \frac{1}{r}\right)^r, \quad y = \left(1 + \frac{1}{r}\right)^{r+1}.$$

13. Prove that:

   a) $\lim\limits_{n \to \infty} \dfrac{\sqrt[n]{n!}}{n} = \dfrac{1}{e}$;

   b) $\lim\limits_{n \to \infty} \left(\sqrt[n+1]{(n+1)!} - \sqrt[n]{n!} \right) = \dfrac{1}{e}$.

Solutions

1. We have

$$f(x) = f^2\left(\frac{x}{2}\right) \geq 0.$$

If $f(\alpha) = 0$ for some real α, then

$$f(x) = f(\alpha)f(x - \alpha) = 0,$$

for all real x. Otherwise, we can define the function $g : (0, \infty) \to \mathbb{R}$, by $g(x) = \ln f(x)$. We have

$$g(x + y) = \ln f(x + y) = \ln [f(x)f(y)]$$
$$= \ln f(x) + \ln f(y) = g(x) + g(y),$$

so $g(x + y) = g(x) + g(y)$. If f is continuous or monotone, then g has the same properties. Hence $g(x) = ax$, for some real a, and, consequently, $f(x) = e^{ax}$ in both cases f monotone and f continuous.

2. From $f(0) \geq f^2(0)$ and $f(0) \geq 1$, it follows that $f(0) = 1$. By induction and using the observation that $f \geq 0$ (because $f(2x) \geq (f(x))^2 \geq 0$), we deduce that

$$f(x_1 + x_2 + \cdots + x_n) \geq f(x_1)f(x_2)\ldots f(x_n).$$

For $x_1 = x_2 = \cdots = x_n = \dfrac{x}{n}$, we have

$$f(x) \geq f^n\left(\frac{x}{n}\right) \geq \left(1 + \frac{x}{n}\right)^n,$$

so

$$f(x) \geq \left(1 + \frac{x}{n}\right)^n.$$

For $n \to \infty$, we deduce $f(x) \geq e^x$. On the other hand,

$$1 = f(0) \geq f(x)f(-x) \geq e^x e^{-x} = 1,$$

so $f(x) = e^x$.

3. Clearly, $f(1) = 0$. By induction,

$$f(x_1 x_2 \ldots x_n) \leq f(x_1) + f(x_2) + \cdots + f(x_n).$$

For $x_1 = x_2 = \cdots = x_n = \sqrt[n]{x}$, we have

$$f(x) \leq nf\left(\sqrt[n]{x}\right) \leq n\left(\sqrt[n]{x} - 1\right)$$

For $n \to \infty$, we deduce that $f(x) \leq \ln x$. On the other hand,

$$0 = f(1) \leq f(x) + f\left(\frac{1}{x}\right) \leq \ln x + \ln \frac{1}{x} = 0,$$

so $f(x) = \ln x$.

4. a) The inequality is equivalent to

$$\frac{2n}{2n+1}e < \left(1 + \frac{1}{n}\right)^n < \frac{2n+1}{2n+2}e$$

or

$$1 + \ln n - \ln\left(n + \frac{1}{2}\right) < n\ln(n+1) - n\ln n < 1 + \ln\left(n + \frac{1}{2}\right) - \ln(n+1).$$

This double inequality is true for $n = 1, 2, 3$. For the left inequality, let us consider the function $f : [4, \infty) \to \mathbb{R}$ given by

$$f(x) = x \ln(x + 1) - (x + 1) \ln x + \ln\left(x + \frac{1}{2}\right).$$

We have to prove that $f(x) > 1$, for all real numbers $x \geq 1$. We have

$$f'(x) = \frac{x}{x + 1} + \ln(x + 1) - \frac{x + 1}{x} - \ln x + \frac{2}{2x + 1}$$

with

$$f''(x) = \frac{5x^2 + 5x + 1}{x^2(x + 1)^2(2x + 1)^2} > 0.$$

It follows that f' is increasing. Thus for $x > 4$,

$$f'(x) > f'(4) = \ln\frac{5}{4} - \frac{41}{180} > 0.$$

Then f is increasing, so for $x \geq 4$,

$$f(x) \geq f(4) = \ln\left(\frac{5^4}{4^5} \cdot \frac{9}{2}\right) > 0.$$

The right inequality can be proved in a similar way.
b) Let us multiply the inequality

$$\frac{2n}{2n + 1}e < \left(1 + \frac{1}{n}\right)^n < \frac{2n + 1}{2n + 2}e$$

by $1 + \dfrac{1}{n}$, to obtain

$$\frac{2n}{2n + 1} \cdot \frac{n + 1}{n}e < \left(1 + \frac{1}{n}\right)^{n+1} < \frac{2n + 1}{2n + 2} \cdot \frac{n + 1}{n}e$$

or

$$\frac{2n + 2}{2n + 1}e < \left(1 + \frac{1}{n}\right)^{n+1} < \frac{2n + 1}{2n}e.$$

Thus

$$\frac{2n + 2}{2n + 1}e - e < \left(1 + \frac{1}{n}\right)^{n+1} - e < \frac{2n + 1}{2n}e - e.$$

or

$$\frac{e}{2n+1} < \left(1+\frac{1}{n}\right)^{n+1} - e < \frac{e}{2n}.$$

5. By multiplying the inequalities from the previous problem by n, we obtain

$$\frac{ne}{2n+2} < n\left[e - \left(1+\frac{1}{n}\right)^n\right] < \frac{ne}{2n+1}$$

and

$$\frac{ne}{2n+1} < n\left[\left(1+\frac{1}{n}\right)^{n+1} - e\right] < \frac{e}{2}.$$

The conclusion follows.

6. We have

$$\left(1+\frac{1}{n}\right)^{n+x_n} = e \Rightarrow (n+x_n)\ln\left(1+\frac{1}{n}\right) = 1$$

$$\Rightarrow n + x_n = \frac{1}{\ln\left(1+\frac{1}{n}\right)},$$

so

$$x_n = \frac{1}{\ln\left(1+\frac{1}{n}\right)} - n.$$

Using l'Hôpital's rule, we deduce that

$$\lim_{n\to\infty} x_n = \lim_{n\to\infty} \frac{1 - n\ln\left(1+\frac{1}{n}\right)}{\ln\left(1+\frac{1}{n}\right)} = \lim_{x\to 0} \frac{1 - \frac{1}{x}\ln(1+x)}{\ln(1+x)}$$

$$= \lim_{x\to 0} \frac{x - \ln(1+x)}{x^2} = \frac{1}{2}.$$

7. We clearly have

$$x_n = \frac{1}{\ln(1+\frac{1}{n})} - n + \frac{\ln(1+\frac{1}{1!}+\cdots+\frac{1}{n!}) - 1}{\ln(1+\frac{1}{n})}.$$

Using the previous problem, it suffices to show that

$$n\left(\ln\left(1+\frac{1}{1!}+\cdots+\frac{1}{n!}\right) - 1\right)$$

converges to 0. This is clear, because we know that

$$0 > \ln\left(1 + \frac{1}{1!} + \cdots + \frac{1}{n!}\right) - 1 > \ln\left(1 - \frac{1}{en \cdot n!}\right).$$

8. a) This is clear, because we have seen that $e^{x-1} \geq x$ for all x, thus $e^x \geq ex$.
    b) The function $g(x) = a^x - ax$ has minimum at $x = 1$; thus, according to Fermat's theorem, $g'(1) = 0$. This implies $a = e$.

9. a) Let us consider the function

$$f(x) = \frac{\ln x}{x}, \quad x \in (0, \infty).$$

We have

$$f'(x) = \frac{1 - \ln x}{x^2},$$

so f is increasing on $(0, e]$ and decreasing on $[e, \infty)$. Therefore,

$$f(x) \leq f(e) = \frac{1}{e} \Rightarrow \frac{\ln x}{x} \leq \frac{1}{e}$$

$$\Rightarrow e \ln x \leq x \Rightarrow \ln x^e \leq \ln e^x \Rightarrow x^e \leq e^x.$$

   b) If $a^x \geq x^a$, then

$$\ln a^x \geq \ln x^a \Leftrightarrow x \ln a \geq a \ln x \Leftrightarrow \frac{\ln x}{x} \leq \frac{\ln a}{a},$$

hence $a = e$.

10. If we denote

$$f(x) = x \ln x, \quad x \in (0, \infty),$$

then a is an absolute minimum point of f. We have

$$f'(a) = 0 \Leftrightarrow 1 + \ln a = 0 \Leftrightarrow a = e^{-1}.$$

Indeed, $x = e^{-1}$ is an absolute minimum point because f is decreasing on $(0, e^{-1}]$ and increasing on $[e^{-1}, \infty)$.

11. Let r be positive so that $y = \left(1 + \frac{1}{r}\right)x$. Then

$$x^x = y^y \Leftrightarrow x^x = \left(\left(1 + \frac{1}{r}\right)x\right)^{(1+\frac{1}{r})x}$$

$$\Leftrightarrow x = \left(1 + \frac{1}{r}\right)^{\left(1 + \frac{1}{r}\right)} x^{1 + \frac{1}{r}}$$

$$\Leftrightarrow x^{1 - \left(1 + \frac{1}{r}\right)} = \left(1 + \frac{1}{r}\right)^{1 + \frac{1}{r}}$$

$$\Leftrightarrow x^{-\frac{1}{r}} = \left(1 + \frac{1}{r}\right)^{1 + \frac{1}{r}}$$

$$\Leftrightarrow x = \left(1 - \frac{1}{r + 1}\right)^{r + 1}.$$

Finally,

$$y = \left(1 + \frac{1}{r}\right) x = \left(1 - \frac{1}{r + 1}\right)^{r}.$$

12. By taking the $(1/xy)$-th power, we obtain

$$(x^y)^{\frac{1}{xy}} = (y^x)^{\frac{1}{xy}} \Leftrightarrow x^{\frac{1}{x}} = y^{\frac{1}{y}}.$$

Thus

$$\left(\frac{1}{x}\right)^{\frac{1}{x}} = \left(\frac{1}{y}\right)^{\frac{1}{y}}.$$

From the previous problem, there exists a positive real number r such that

$$\frac{1}{x} = \left(1 - \frac{1}{r + 1}\right)^{r} = \left(\frac{r}{r + 1}\right)^{r}$$

and

$$\frac{1}{y} = \left(1 - \frac{1}{r + 1}\right)^{r + 1} = \left(\frac{r}{r + 1}\right)^{r + 1}.$$

Hence

$$x = \left(1 + \frac{1}{r}\right)^{r}, \quad y = \left(1 + \frac{1}{r}\right)^{r + 1}.$$

13. a) As a consequence of the Cesàro-Stolz theorem, we have the root formula,

$$\lim_{n\to\infty} \sqrt[n]{a_n} = \lim_{n\to\infty} \frac{a_{n+1}}{a_n},$$

under the hypothesis that the limit on the right-hand side exists. Here,

$$\lim_{n\to\infty} \frac{\sqrt[n]{n!}}{n} = \lim_{n\to\infty} \sqrt[n]{\frac{n!}{n^n}} = \lim_{n\to\infty} \frac{\frac{(n+1)!}{(n+1)^{n+1}}}{\frac{n!}{n^n}} = \lim_{n\to\infty} \frac{1}{\left(1+\frac{1}{n}\right)^n} = \frac{1}{e}.$$

b) We have

$$\lim_{n\to\infty}\left(\sqrt[n+1]{(n+1)!} - \sqrt[n]{n!}\right) = \lim_{n\to\infty} \sqrt[n]{n!}\left(\frac{\sqrt[n+1]{(n+1)!}}{\sqrt[n]{n!}} - 1\right)$$

$$= \lim_{n\to\infty}\left[n\cdot\frac{\sqrt[n]{n!}}{n}\cdot\frac{\frac{\sqrt[n+1]{(n+1)!}}{\sqrt[n]{n!}} - 1}{\ln\frac{\sqrt[n+1]{(n+1)!}}{\sqrt[n]{n!}}}\cdot\ln\frac{\sqrt[n+1]{(n+1)!}}{\sqrt[n]{n!}}\right]$$

$$= \frac{1}{e}\cdot\lim_{n\to\infty}\left(n\ln\frac{\sqrt[n+1]{(n+1)!}}{\sqrt[n]{n!}}\right) = \frac{1}{e}\cdot\lim_{n\to\infty}\ln\frac{\sqrt[n+1]{[(n+1)!]^n}}{n!}$$

$$= \frac{1}{e}\cdot\lim_{n\to\infty}\ln\sqrt[n+1]{\frac{[(n+1)!]^n}{(n!)^{n+1}}} = \frac{1}{e}\cdot\lim_{n\to\infty}\ln\sqrt[n+1]{\frac{(n+1)^n}{n!}} = \frac{1}{e}.$$

Chapter 11
The Intermediate Value Theorem

Let $I \subseteq \mathbb{R}$ be an interval. We say that a function $f : I \to \mathbb{R}$ has *the intermediate value property* (or IVP, for short) if it takes all intermediate values between any two of its values. More precisely, for every $a, b \in I$ and for any λ between $f(a)$ and $f(b)$, we can find c between a and b such that $f(c) = \lambda$. A direct consequence of this definition is that f has IVP if and only if it transforms any interval into an interval. Equivalently, a function with IVP which takes values of opposite signs must vanish at some point.

Theorem (Intermediate value theorem). The class of continuous functions is strictly included in the class of functions with IVP.

Proof. Let $f : [a, b] \to \mathbb{R}$ be a continuous function. If $f(a)f(b) < 0$, we will prove that there exists $c \in (a, b)$ such that $f(c) = 0$. We assume, without loss of generality, that $f(a) < 0$ and $f(b) > 0$. From the continuity of f, we can find $\varepsilon > 0$ such that f remains negative on $[a, a + \varepsilon) \subset [a, b]$. Let us define the set

$$A = \{x \in (a, b) \mid f \text{ is negative on } [a, x)\}.$$

The set is nonempty, because $a + \varepsilon \in A$. Let $c = \sup A$. We will prove that $f(c) = 0$. Let $(c_n)_{n \geq 1} \subset [a, c)$ be a sequence convergent to c. From $f(c_n) < 0$, we deduce $f(c) \leq 0$ and consequently $c < b$. If $f(c) < 0$, then from the continuity of f at c, we can find $\delta > 0$ such that f remains negative on $[c, c + \delta)$. This means that $c + \delta$ is in A, a contradiction. In conclusion, $f(c) = 0$. $\square$

On the other hand, the first proposed problem gives an example of a function with IVP which is not continuous.

Another important property of IVP-functions which we use here is the following:

Proposition. *Each one-to-one IVP-function $f : I \to \mathbb{R}$ is strictly monotone.*

Proof. Let us assume, by way of contradiction, that there are $a < b < c$ in I for which $f(b)$ is not between $f(a)$ and $f(c)$ (as it would be if f was monotone). One immediately sees that we can assume $f(c) < f(a) < f(b)$, without loss of generality.

© Springer Science+Business Media LLC 2017
T. Andreescu et al., *Mathematical Bridges*, DOI 10.1007/978-0-8176-4629-5_11

Because $f(a)$ is an intermediate value between $f(c)$ and $f(b)$, we can find $d \in (b, c)$ so that $f(d) = f(a)$. But this contradicts the injectivity of f, because $d \neq a$. $\square$

Problem. Prove that there are no functions $f : \mathbb{R} \to \mathbb{R}$ with IVP, such that

$$f(f(x)) = -x, \ \forall \, x \in \mathbb{R}.$$

Solution. If x, y satisfy $f(x) = f(y)$, then $f(f(x)) = f(f(y))$ and further, $x = y$. Thus f is injective. But any injective function with IVP is strictly monotone. Consequently, $f \circ f$ is increasing. This is a contradiction, because $f \circ f = -1_{\mathbb{R}}$ is decreasing. $\square$

Also, the above proposition implies that any increasing function having IVP is continuous. Indeed, such a function has one-sided limits at any point because it is increasing, and because it cannot "jump" values, these one-sided limits are identical. This is the idea that underlies the following problem, proposed for the Romanian National Olympiad in 1998:

Example. Let $f : \mathbb{R} \to \mathbb{R}$ be a differentiable function such that

$$f'(x) \leq f'\left(x + \frac{1}{n}\right)$$

for all $x \in \mathbb{R}$ and all positive integers n. Prove that f' is continuous.

Solution. Consider the following function:

$$f_n(x) = n\left(\left(f(x + \frac{1}{n}\right) - f(x)\right).$$

By the hypothesis, f_n has a nonnegative derivative; therefore it is increasing. Thus, if $x < y$, we have $f_n(x) \leq f_n(y)$ for all n. By making $n \to \infty$, we conclude that $f'(x) \leq f'(y)$. Thus, f' is increasing. Using Darboux's theorem, proved in Chapter 14, as well as the observation preceding the solution, we conclude that f' is continuous. $\square$

We now present a quite challenging problem in which the IVP plays a crucial role. Here is problem E3191 from the American Mathematical Monthly:

Problem. Find all functions $f : \mathbb{R} \to \mathbb{R}$ with IVP such that

$$f(x + y) = f(x + f(y))$$

for all $x, y \in \mathbb{R}$.

Solution. The constant functions are clearly solutions, so we suppose further that f is nonconstant. Let $a = \inf \operatorname{Im} f$, $b = \sup \operatorname{Im} f$, where $\operatorname{Im} f$ denotes the range of f. The assumption we made shows that $a < b$. We have $f(y) = f(f(y))$ for all $y \in \mathbb{R}$ (by setting $x = 0$ in the given relation); thus $f(x) = x$ for $x \in (a, b)$.

Suppose that $a \in \mathbb{R}$. By IVP and $f(x) = x$ for $x \in (a, b)$, we must have $f(a) = a$; thus there is $d > 0$ such that $f(x) = x$ on $[a, a + 2d]$. But then any $0 < t < d$ satisfies $f(a - t) = a + s$ for some $s > d$, because if $s \leq d$ then

$$a = f(t + a - t) = f(t + f(a - t)) = f(a + s + t) = a + s + t,$$

a contradiction. On the other hand, $f(a - t) > a + d$ for all $t \in (0, d)$ and $f(a) = a$ contradict IVP. Thus it is impossible for a to be finite, that is, $a = -\infty$. Similarly, $b = \infty$ and $f(x) = x$, for all $x \in \mathbb{R}$. $\square$

The following problem was the highlight of the Romanian National Olympiad in 2000.

Problem. A function $f : \mathbb{R}^2 \to \mathbb{R}$ is called *olympic* if for any $n \geq 3$ and any $A_1, A_2, \ldots, A_n \in \mathbb{R}^2$ distinct points such that $f(A_1) = \cdots = f(A_n)$, the points $A_1, A_2, \ldots, A_n$ are the vertices of a convex polygon.

Let $P \in \mathbb{C}[X]$ be nonconstant. Prove that $f : \mathbb{R}^2 \to \mathbb{R}, f(x, y) = |P(x + iy)|$ is olympic if and only if all the roots of P are equal.

Solution. If $P(z) = a(z - z_0)^n$ and $f(A_1) = \cdots = f(A_n)$, then all A_i are on a circle of center z_0, thus are the vertices of a convex polygon. Now, suppose that not all the roots of P are equal and consider z_1, z_2 two roots of P such that $|z_1 - z_2| \neq 0$ is minimal.

Let d be the line containing z_1 and z_2, and let $z_3 = \dfrac{z_1 + z_2}{2}$. Denote by s_1, s_2 the half lines determined by z_3. By the minimality of $|z_1 - z_2|$, we must have $f(z_3) > 0$ and because

$$\lim_{\substack{|z| \to \infty \\ z \in s_1}} f(z) = \lim_{\substack{|z| \to \infty \\ z \in s_2}} f(z) = \infty$$

and f has IVP, there exist $z_4 \in s_1$ and $z_5 \in s_2$ with $f(z_3) = f(z_4) = f(z_5)$, a contradiction. $\square$

Example. The continuous function $f : \mathbb{R} \to \mathbb{R}$ has the property that for any real number a, the equation $f(x) = f(a)$ has only a finite number of solutions. Prove that there exist real numbers a, b such that the set of real numbers x such that $f(a) \leq f(x) \leq f(b)$ is bounded.

Solution. First, we will prove that f has limit at ∞ and at $-\infty$. Indeed, if this is not the case, there exist sequences a_n and b_n which tend to ∞ and such that $f(a_n)$, $f(b_n)$ have limit points $a < b$. This implies the existence of a number c such that $f(a_n) < c < f(b_n)$ for sufficiently large n. Using the intermediate value theorem, we deduce the existence of c_n between a_n and b_n such that $f(c_n) = c$ for all sufficiently large n. Clearly, this contradicts the hypothesis. $\square$

Now, let $l_1 = \lim_{x \to \infty} f(x)$ and $l_2 = \lim_{x \to -\infty} f(x)$. We have two cases: the first one is when $l_1 \neq l_2$. Let us suppose, without loss of generality, that $l_1 < l_2$, and let us consider real numbers a, b such that $l_1 < a < b < l_2$. There are numbers A, B such that $f(x) < a$ for $x < A$ and $f(x) > b$ for $x > B$. Using the intermediate value theorem, it follows that $a, b \in \text{Im}(f)$ and clearly the set of those x for which $f(x) \in [a, b]$ is a subset of $[A, B]$, thus bounded. The second case is $l_1 = l_2 = l$.

Clearly, f is not constant, so we can choose $a < b < l$ in the range of f if Im (f) is not a subset of (l, ∞), and we can choose $l < a < b$ in the range of f if Im (f) is not a subset of $(-\infty, l)$. Now, we can argue as in the first case to deduce that the set of those x for which $a \leq x \leq b$ is bounded. This finishes the solution. $\square$

Proposed Problems

1. Prove that the function $f : \mathbb{R} \to \mathbb{R}$, given by

$$f(x) = \begin{cases} \sin \dfrac{1}{x}, & x \in \mathbb{R} \setminus \{0\} \\ 0, & x = 0 \end{cases}$$

   has IVP.
2. Let $f : [0, 1] \to [0, 1]$ be such that $f(0) = 0$ and $f(1) = 1$. Does the surjectivity of f imply that f has IVP?
3. Let $f : [0, \infty) \to \mathbb{R}$ be a function continuous on $(0, \infty)$. Prove that the following assertions are equivalent:

   a) f has IVP;
   b) there exists a sequence $(a_n)_{n \geq 1} \subset (0, \infty)$ for which

$$\lim_{n \to \infty} a_n = 0 \text{ and } \lim_{n \to \infty} f(a_n) = f(0).$$

4. Let $f : [0, \infty) \to [a, b]$ be continuous on $(0, \infty)$ such that

$$\lim_{n \to \infty} f\left(\frac{1}{2n}\right) = a, \quad \lim_{n \to \infty} f\left(\frac{1}{2n + 1}\right) = b.$$

   Prove that f has IVP.
5. Let $f : [a, b] \to \mathbb{R}$ be a continuous function such that $f(a)f(b) < 0$. Prove that for all integers $n \geq 3$, there exists an arithmetic progression $x_1 < x_2 < \cdots < x_n$ such that

$$f(x_1) + f(x_2) + \cdots + f(x_n) = 0.$$

6. Prove that there are no differentiable functions $f : \mathbb{R} \to \mathbb{R}$, such that

$$f'(x) - f(x) = \begin{cases} \sin x, & x \in (-\infty, 0) \\ \cos x, & x \in [0, \infty) \end{cases}.$$

7. Prove that there are no functions $f : \mathbb{R} \to \mathbb{R}$ with IVP such that

$$f(f(x)) = \cos^2 x, \ \forall \, x \in \mathbb{R}.$$

8. Prove that there are no functions $f : \mathbb{R} \to [0, \infty)$ with IVP, such that

$$f(f(x)) = 2^x, \ \forall \, x \in \mathbb{R}.$$

9. Find all functions $f : \mathbb{R} \to \mathbb{R}$ with IVP, such that $f(2) = 8, f(-2) = -8$ and

$$f(f(x)) = f^3(x), \ \forall \, x \in \mathbb{R}.$$

10. Let $f : [0, 1] \to \mathbb{R}$ be integrable. Assume that $0 \le a < c < b < d \le 1$ such that $\displaystyle\int_a^b f(x)\,dx < 0$ and $\displaystyle\int_c^d f(x)\,dx > 0$. Prove that there exist $0 \le \alpha < \beta \le 1$ such that $\displaystyle\int_\alpha^\beta f(x)\,dx = 0$.

11. Find all functions $f : \mathbb{R} \to \mathbb{R}$ having the intermediate value property such that $f^{[0]} + f^{[1]}$ is increasing and there exists such positive integer m that $f^{[0]} + f^{[1]} + \cdots + f^{[m]}$ is decreasing.

    Note that $f^{[n]} = f \circ f \circ \cdots \circ f$ is the nth iterate of f (with $f^{[0]} = 1_{\mathbb{R}}$, the identity function on the reals) and that we call increasing (respectively decreasing) a function h with property that $h(x) \le h(y)$ (respectively $h(x) \ge h(y)$) whenever $x < y$.

12. Let $f : \mathbb{R} \to \mathbb{R}$ be a continuous function and let $a, b \in f(\mathbb{R})$ (i.e., in the range of f) with $a < b$. Prove that there exists an interval I such that $f(I) = [a, b]$.

13. Let $f : [0, 1] \to [0, \infty)$ be a continuous function such that $f(0) = f(1) = 0$ and $f(x) > 0$ for $0 < x < 1$. Show that there exists a square with two vertices in the interval $(0, 1)$ on the x-axis and the other two vertices on the graph of f.

14. Let $a_1, \ldots, a_n$ be real numbers, each greater than 1. If $n \ge 2$, prove that there is exactly one solution in the interval $(0, 1)$ to

$$\prod_{j=1}^n (1 - x^{a_j}) = 1 - x.$$

Solutions

1. We prove that if I is an interval, then $f(I)$ is an interval. If zero is not an accumulating point of I, then f is continuous on I and consequently, $f(I)$ is an interval.

    Obviously, $f(\mathbb{R}) \subseteq [-1, 1]$, so it is sufficient to prove the equality

$$f((0, \varepsilon)) = [-1, 1],$$

for all $\varepsilon > 0$. Indeed, for positive integers $n > \varepsilon^{-1}$, we have

$$\frac{1}{\frac{\pi}{2} + 2n\pi} \; , \;\; \frac{1}{-\frac{\pi}{2} + 2n\pi} \in (0, \varepsilon)$$

and

$$f\left(\frac{1}{\frac{\pi}{2} + 2n\pi}\right) = 1, \quad f\left(\frac{1}{-\frac{\pi}{2} + 2n\pi}\right) = -1.$$

Now, using the continuity of f on the interval $\left[\dfrac{1}{-\frac{\pi}{2} + 2n\pi} \; , \; \dfrac{1}{\frac{\pi}{2} + 2n\pi}\right]$, we deduce that $f((0, \varepsilon)) = [-1, 1]$.

2. The answer is no. The function $f : [0, 1] \to [0, 1]$, given by

$$f(x) = \begin{cases} 2x, & x \in [0, 1/2) \\ 2x - 1, & x \in [1/2, 1] \end{cases}$$

is surjective but it does not have IVP because

$$f\left(\left[\frac{2}{5}, \frac{3}{5}\right]\right) = \left[0, \frac{1}{5}\right] \cup \left[\frac{4}{5}, 1\right].$$

3. First we prove that

$$\forall \, n \in \mathbb{N}^*, \; \exists \, a_n \in \left(0, \frac{1}{n}\right), \quad \text{such that} \quad |f(a_n) - f(0)| < \frac{1}{n},$$

where $a_0 = 1$. Indeed, if there is $n_0 \in \mathbb{N}^*$ for which

$$|f(x) - f(0)| \geq \frac{1}{n_0}, \quad \forall x \in \left(0, \frac{1}{n_0}\right),$$

then

$$f(x) \in \left(-\infty, f(0) - \frac{1}{n_0}\right] \cup \{f(0)\} \cup \left[f(0) + \frac{1}{n_0}, \infty\right),$$

for all $x \in \left[0, \dfrac{1}{n_0}\right)$. This is a contradiction, because $f\left(\left[0, \dfrac{1}{n_0}\right)\right)$ cannot be an interval.

Reciprocally, if $a > 0$ and $\lambda \in (f(0), f(a))$, then from $\lim\limits_{n \to \infty} a_n = 0$, we conclude that there exists $n_1 \in \mathbb{N}$ such that $a_n < a$, $\forall \, n \geq n_1$. Because $\lim\limits_{n \to \infty} f(a_n) = f(0) < \lambda$, there is $n_2 \in \mathbb{N}$ such that $f(a_n) < \lambda$, $\forall \, n \geq n_2$. Now, for $n_0 = \max\{n_1, n_2\}$, we have $a_{n_0} < a$, $f(a_{n_0}) < \lambda$. Further, f is continuous on $[a_{n_0}, a]$ and $f(a_{n_0}) < \lambda, f(a) > \lambda$, so there exists $c \in (a_{n_0}, a) \subset [0, a]$ such that $f(c) = \lambda$.

4. If $I \subseteq [0, \infty)$ is an interval having the origin as an accumulating point, then

$$(a, b) \subseteq f(I) \subseteq [a, b],$$

so $f(I)$ is an interval. To this end, we will use the previous result. Let n_0 be a positive integer such that

$$f\left(\frac{1}{2n}\right) \le f(0) \le f\left(\frac{1}{2n+1}\right), \quad \forall \, n \ge n_0.$$

Note that f is continuous on $\left[\dfrac{1}{2n+1}, \dfrac{1}{2n}\right]$, so we can find $a_n \in \left[\dfrac{1}{2n+1}, \dfrac{1}{2n}\right]$ for which $f(a_n) = f(0)$. In conclusion, $a_n \to 0$ and $f(a_n) \to f(0)$, as $n \to \infty$.

5. Assume that $f(a) < 0$ and $f(b) > 0$, without loss of generality. Let $\varepsilon > 0$ be such that

$$f(x) < 0, \ \forall \, x \in [a, a + \varepsilon], \quad f(x) > 0, \ \forall \, x \in [b - \varepsilon, b]$$

and let us consider the arithmetic progressions

$$a_1 < a_2 < \cdots < a_n \in [a, a + \varepsilon], \quad b_1 < b_2 < \cdots < b_n \in [b - \varepsilon, b].$$

Define $F : [0, 1] \to \mathbb{R}$ by the formula

$$F(t) = \sum_{k=1}^{n} f((1 - t)a_k + tb_k), \ t \in [0, 1].$$

The function F is continuous and

$$F(0) = \sum_{k=1}^{n} f(a_k) < 0, \quad F(1) = \sum_{k=1}^{n} f(b_k) > 0,$$

so we can find $\tau \in (0, 1)$ such that $F(\tau) = 0$, i.e.,

$$\sum_{k=1}^{n} f((1 - \tau)a_k + \tau b_k) = 0.$$

Finally, we can take the arithmetic progression $x_k = (1 - \tau)a_k + \tau b_k$.

6. If we assume by contradiction that such a function exists, then define $g : \mathbb{R} \to \mathbb{R}$, by $g(x) = e^{-x}f(x)$. Its derivative

$$g'(x) = e^{-x}(f'(x) - f(x)) = \begin{cases} e^{-x} \sin x, \ x \in (-\infty, 0) \\ e^{-x} \cos x, \ x \in [0, \infty) \end{cases}$$

has IVP. This is impossible, because g' has at zero a discontinuity of the first kind,

$$\lim_{x \nearrow 0} g'(x) = \lim_{x \to 0} e^{-x} \sin x = 0$$

and

$$\lim_{x \searrow 0} g'(x) = \lim_{x \to 0} e^{-x} \cos x = 1.$$

7. From $f(\cos^2 x) = \cos^2(f(x)) \in [0, 1]$, $x \in \mathbb{R}$, we deduce that $y \in [0, 1] \Rightarrow f(y) \in [0, 1]$. This fact allows us to define the function $g : [0, 1] \to [0, 1]$, by $g(x) = f(x)$, $x \in [0, 1]$. Obviously, $g(g(x)) = \cos^2 x . g$ is injective and g has IVP, so it is strictly monotone. Therefore, $g \circ g$ is increasing, a contradiction.

8. The function f is injective and f has IVP, so f is continuous and strictly monotone. If $f(\alpha) = \alpha$, for some real α, then

$$f(f(\alpha)) = f(\alpha) \Rightarrow 2^\alpha = \alpha,$$

which is impossible. If $f(x) < x$, for all real numbers x, then

$$f(f(x)) < f(x) < x \Rightarrow 2^x < x, \ \forall \, x \in \mathbb{R},$$

a contradiction. Consequently, $f(x) > x$, for all real numbers x and further, f is increasing. The limit $l = \lim_{x \to -\infty} f(x)$ does exist.

Because $f(x) \geq 0$ for all x, we have $l \in \mathbb{R}$ and $f(l) = 0$.

Finally, from the monotonicity, we conclude that $f(l-1) < 0$, a contradiction.

9. Im f is an interval and $f(y) = y^3$, for all real numbers y in Im f. Indeed, if $y = f(x)$, for some real x, then

$$f(y) = f(f(x)) = f^3(x) = y^3.$$

We prove that Im $f = \mathbb{R}$ and so $f(x) = x^3$, for all real numbers x. To this end, define $u_1 = -2$, $v_1 = 2$,

$$u_{n+1} = u_n^3, \quad v_{n+1} = v_n^3, \quad n \geq 1.$$

It is easy to see that $u_n \to -\infty$ and $v_n \to \infty$ as $n \to \infty$. Finally, note that $u_n, v_n \in \text{Im} f$ for all positive integers n. Indeed, if $u_n \in \text{Im} f$, $u_n = f(\alpha)$, then

$$u_{n+1} = u_n^3 = f^3(\alpha) = f(f(\alpha)) \in \text{Im} f.$$

10. The function $F : [0, 1] \to \mathbb{R}$, given by

$$F(t) = \int_{(1-t)a+tc}^{(1-t)b+td} f(x)\, dx, \ t \in [0, 1]$$

is continuous, and consequently it has IVP. According to the hypothesis,

$$F(0) = \int_a^b f(x)\, dx < 0$$

and

$$F(1) = \int_c^d f(x)\, dx > 0,$$

so $F(\tau) = 0$ for some τ in $(0, 1)$. More precisely,

$$f(\tau) = \int_{(1-\tau)a+\tau c}^{(1-\tau)b+\tau d} f(x)\, dx = 0,$$

so we can choose $\alpha = (1 - \tau)a + \tau c, \beta = (1 - \tau)b + \tau d$.

11. Let $f_n = f^{[0]} + f^{[1]} + \cdots + f^{[n]}$ for every $n \geq 0$. Thus, the problem gives us that, for all $x, y \in \mathbb{R}$, with $x \leq y$, we have $f_1(x) \leq f_1(y)$, and $f_m(x) \geq f_m(y)$.

First, f is injective. Indeed, let x, y be real numbers such that $f(x) = f(y)$, and assume, without loss of generality, that $x \leq y$. Clearly we have $f^{[n]}(x) = f^{[n]}(y)$ for all $n \geq 1$; thus $f_m(x) \geq f_m(y)$ becomes $f^{[0]}(x) \geq f^{[0]}(y)$, that is, $x \geq y$, and, consequently, we have $x = y$. Being injective and with the intermediate value property, f must be strictly monotone, which implies that $f^{[2]} = f \circ f$ is strictly increasing; therefore, $f^{[2n]}$ is strictly increasing for all $n \geq 0$. Thus

$$f_{2n} = f_1 + f_1 \circ f^{[2]} + \cdots + f_1 \circ f^{[2n-2]} + f^{[2n]}$$

is strictly increasing, while

$$f_{2n+1} = f_1 + f_1 \circ f^{[2]} + \cdots + f_1 \circ f^{[2n]}$$

is increasing.

Because $f^{[m]}$ is decreasing, it follows that m is odd, and $f^{[m]}$ is constant (being decreasing and increasing at the same time). But, clearly, this can happen only if f_1 is in turn constant. This means that there exists some $c \in \mathbb{R}$ such that $f_1(x) = c \Leftrightarrow f(x) = -x + c$ for all $x \in \mathbb{R}$, and, indeed, the reader can easily check that these functions satisfy all conditions from the problem statement.

This was proposed by Dorel Miheţ for the Romanian National Mathematics Olympiad in 2014.

12. There exist real numbers p and q such that $f(p) = a$ and $f(q) = b$. Of course, $p \neq q$, and we can assume, without loss of generality, that $p < q$. The set of those x in $[p, q]$ for which $f(x) = a$ is bounded and nonempty (it contains p); thus it has a supremum, and let it be α. The set of those $x \in [\alpha, q]$ for which $f(x) = q$ is bounded and nonempty (it contains q); thus it has an infimum, and let it be β. By passing to the limit, and using the continuity of f, we see that $f(\alpha) = a$, and $f(\beta) = b$. We claim that $f([\alpha, \beta]) = [a, b]$.

The inclusion $[a, b] \subseteq f([\alpha, \beta])$ follows immediately from the fact that f has the intermediate value property. Conversely, let us prove that $f([\alpha, \beta]) \subseteq [a, b]$. Let $x \in [\alpha, \beta]$, and suppose that $f(x) < a$, so that we have $f(x) < a < b = f(\beta)$. Again by the intermediate value property, we obtain the existence of some $y \in (x, \beta)$, with $f(y) = a$, but this contradicts the supremum property of α. Similarly, the assumption that $f(x) > b$ (and thus $f(x) > b > a = f(\alpha)$) leads to a contradiction of the infimum property of β. It remains the only possibility that $f(x)$ belongs to $[a, b]$, finishing the proof.

This problem was also proposed for the 11th grade in the Romanian National Mathematics Olympiad, in 2007.

13. The set

$$A = \{x \in [0, 1] : \ x + f(x) \geq 1\}$$

is nonempty (it contains 1) and, of course, bounded; hence it has an infimum, and let it be c. Since $0 + f(0) < 1$ and f is continuous, the inequality $x + f(x) < 1$ holds for all x in a neighborhood of the origin (and in $[0, 1]$, of course); therefore $c > 0$. Yet we have $x + f(x) < 1$ for all $x \in [0, 1]$, $x < c$, and $x + f(x) \geq 1$ for all $x \in A$ (and A contains a sequence with limit c), and these relations immediately yield $c + f(c) = 1$. To summarize: there is a $c \in (0, 1]$ such that $c + f(c) = 1$ and $x + f(x) < 1$ for all $x \in [0, c)$.

Since f is continuous, it is bounded and attains its extrema on the interval $[0, 1]$. Thus there is a $d \in [0, 1]$ such that $f(x) \leq f(d)$ for all $x \in [0, 1]$; actually the conditions from the problem statement imply that $d \in (0, 1)$.

Now let the continuous function $g : [0, 1] \to [0, \infty)$ be defined by $g(x) = x + f(x)$. Since $g(0) = 0 < d < 1 = c + f(c) = g(c)$, there exists $\alpha \in (0, c)$ such that $g(\alpha) = d \Leftrightarrow \alpha + f(\alpha) = d$ (because g has the intermediate value property).

Now consider the function $h : [0, c] \to \mathbb{R}$ defined by $h(x) = f(x + f(x)) - f(x)$ for all $x \in [0, c]$ (for which $0 \leq x + f(x) \leq 1$, thus h is well defined); certainly, h is also a continuous function. We have

$$h(\alpha) = f(\alpha + f(\alpha)) - f(\alpha) = f(d) - f(\alpha) \geq 0,$$

$$h(c) = f(c + f(c)) - f(c) = f(1) - f(c) = -f(c) \leq 0,$$

and also,

$$h(d) = f(d + f(d)) - f(d) \le 0.$$

The last relation holds (at least) if $d \le c$.

Now, as h has the intermediate value property, there exists $\gamma \in [\alpha, c]$ such that $h(\gamma) = 0 \Leftrightarrow f(\gamma + f(\gamma)) = f(\gamma)$, and $\alpha > 0$ assures $\gamma > 0$. However, we cannot be sure of the fact that $\gamma < 1$. If γ is exactly 1, then $c = 1$ follows as well; therefore $d < c$, and we can find another γ in the interval $[\alpha, d]$ with the property that $h(\gamma) = 0$, and in this case, we have $\gamma \le d < 1$. Either way, there exists $\gamma \in (0, 1)$ such that $h(\gamma) = 0$, that is, $f(\gamma + f(\gamma)) = f(\gamma)$.

Finally we see that the quadrilateral with vertices $(\gamma, 0)$, $(\gamma + f(\gamma), 0)$, $(\gamma + f(\gamma), f(\gamma + f(\gamma)))$, and $(\gamma, f(\gamma))$ is a square (the first two points are on the x-axis; the others are on the graph of f).

This is problem 11402 proposed by Cătălin Barboianu in *The American Mathematical Monthly*.

14. We first use (a form of) Bernoulli's inequality, namely, $1 - x^a \le a(1 - x)$ for $a > 1$ and $x \ge 0$. Thus, for $x \in (0, 1)$, we have

$$f(x) = \prod_{j=1}^{n} (1 - x^{a_j}) \le a_1 \cdots a_n (1 - x)^n < 1 - x,$$

the last inequality being satisfied if we choose $x > 1 - 1/\sqrt[n-1]{a_1 \cdots a_n}$.

On the other hand, the inequality $(1 - s_1) \cdots (1 - s_n) > 1 - s_1 - \cdots - s_n$ (somehow also related to Bernoulli's inequality) can be proved by an easy induction for $s_1, \ldots, s_n \in (0, 1)$. Thus, for $x \in (0, 1)$,

$$f(x) = \prod_{j=1}^{n} (1 - x^{a_j}) > 1 - x^{a_1} - \cdots - x^{a_n} > 1 - x,$$

where the last inequality holds if we choose $x < (1/n)^{1/(a-1)}$, with $a = \min_{1 \le j \le n} a_j$.

The intermediate value theorem applied to the function $x \mapsto f(x) - (1-x)$ shows that there is at least one solution in $(0, 1)$ to the equation $f(x) = 1 - x$.

Now suppose that the equation $f(x) = 1 - x$ has two solutions x_1 and x_2, with $0 < x_1 < x_2 < 1$, and let $g(x) = \log f(x) - \log(1 - x)$, and $h(x) = (1 - x)g'(x)$. Of course, 0, x_1 and x_2 would be solutions for $g(x) = 0$, as well, therefore, by Rolle's theorem, $g'(x) = 0$, and consequently, $h(x) = 0$ would have at least two solutions in $(0, 1)$ (one between 0 and x_1, and one in the interval (x_1, x_2)). Finally, this would produce at least one solution in $(0, 1)$ for the equation $h'(x) = 0$ (again by Rolle's theorem). But we have

$$h(x) = -\sum_{j=1}^{n} a_j (x^{a_j - 1} - x^{a_j})(1 - x^{a_j})^{-1} + 1,$$

and

$$h'(x) = -\sum_{j=1}^{n} a_j x^{a_j-2}(x^{a_j} - 1 - a_j(x-1))(1 - x^{a_j})^{-2};$$

hence we see that h' is negative in $(0, 1)$, according to the same inequality of Bernoulli mentioned in the beginning. The contradiction thus obtained finishes the proof.

This is problem 11226, proposed by Franck Beaucoup and Tamás Erdélyi in *The American Mathematical Monthly* 6/2006, with the solution of the Microsoft Research Problem Group in the same *Monthly*, 1/2008. Yet another (a bit more complicated) solution can be found in Chapter 14 about derivatives and their applications.

Chapter 12
The Extreme Value Theorem

The extreme value theorem asserts that any continuous function defined on a compact interval with real values is bounded and it attains its extrema. Indeed, let us consider a continuous function $f : [a, b] \to \mathbb{R}$. If we assume by contradiction that f is unbounded, then for each positive integer n, we can find an element x_n in $[a, b]$ such that $|f(x_n)| > n$. In this way, we define a bounded sequence $(x_n)_{n \geq 1}$. The Bolzano-Weierstrass theorem implies (due to the compactness of $[a, b]$) the existence of a convergent subsequence $(x_{k_n})_{n \geq 1}$. We have $|f(x_{k_n})| > k_n$, for all positive integers n. In particular, the sequence $(f(x_{k_n}))_{n \geq 1}$ is unbounded. This is a contradiction, because the sequence $(f(x_{k_n}))_{n \geq 1}$ is convergent to $f(l)$, where l is the limit of the sequence $(x_{k_n})_{n \geq 1}$. Then suppose that $M = \sup\limits_{x \in [a,b]} f(x)$ is the least upper bound (or supremum) of f (which is finite, as shown above). There exists a sequence $(u_n)_{n \geq 1} \subseteq [a, b]$ such that $\lim\limits_{n \to \infty} f(u_n) = M$. By Bolzano-Weierstrass theorem again, $(u_n)_{n \geq 1}$ has a convergent subsequence; let us call $(u_{k_n})_{n \geq 1}$ this subsequence, and let $c \in [a, b]$ be its limit. We then have $M = \lim\limits_{n \to \infty} f(u_n) = \lim\limits_{n \to \infty} f(u_{k_n}) = f(c)$, and thus f attains its supremum M. In a similar manner, we show that f attains its infimum, thus finishing the proof of the theorem. $\square$

The fundamental reason for which the above proof works lies in the fact that a continuous function carries compact sets into compact sets. Thus if we consider a compact metric space (X, d) and a continuous function f defined on X and having real values, f will also be bounded and will achieve its extrema, because its image will be a compact connected set, thus a compact interval. For the special case discussed in the beginning of this chapter, we present another interesting proof. Using the continuity of f, for every x in $[a, b]$, we can choose $\varepsilon_x > 0$ and an open interval $I_x = (x - \varepsilon_x, x + \varepsilon_x)$ containing x such that $|f(y) - f(x)| < 1$, for all real numbers $y \in I_x \cap [a, b]$. Further, with the notation $M_x = |f(x)| + 1$, we have $|f(y)| < M_x$ for all $y \in I_x$. The family $(I_x)_{x \in [a,b]}$ is an open cover of the compact set $[a, b]$. Consequently, we can find $x_1, x_2, \ldots, x_n \in [a, b]$ such that

$$[a, b] \subset I_{x_1} \cup I_{x_2} \cup \cdots \cup I_{x_n}.$$

© Springer Science+Business Media LLC 2017
T. Andreescu et al., *Mathematical Bridges*, DOI 10.1007/978-0-8176-4629-5_12

Now, one can easily see that f is bounded in absolute value by

$$\max\{M_{x_1}, M_{x_2}, \ldots, M_{x_n}\}.$$

For the second part of the proof, we proceed as above. Alternatively, one can consider the function $x \mapsto g(x) = 1/(M - f(x))$, with $M = \sup\limits_{x \in [a,b]} f(x)$. If we assume $f(x) < M$ for all $x \in [a, b]$, then g is continuous, thus bounded on $[a, b]$, according to the first part of the theorem. However, this cannot happen, as there are values of f as close to M as we want. $\square$

Here is a spectacular application of this theorem in linear algebra:

Theorem. *Any symmetric real matrix is conjugate to a real diagonal matrix.*

Proof. Indeed, take A to be a symmetric matrix of order n and define the quadratic form $q(x) = \langle Ax, x \rangle$, where $\langle \cdot \rangle$ is the standard Euclidean scalar product on $\mathbb{R}^n$. This is clearly a continuous function in $\mathbb{R}^n$ with real values. Consider S^{n-1}, the unit sphere of $\mathbb{R}^n$. This is a compact set, because it is clearly bounded and closed. Therefore, q has a maximum λ on S^{n-1}, attained at a point v. We will prove that λ is an eigenvalue of A. Indeed, by the definition of λ, the quadratic form $q_1(x) = \lambda \|x\|^2 - q(x)$ is positive and vanishes at v. The Cauchy-Schwarz inequality implies $b(u, v)^2 \leq q_1(u)q_1(v)$ for all u, where b is the bilinear form associated to q_1. Thus $b(u, v) = 0$ for all u and this implies $Av = \lambda v$. Thus we have found an eigenvalue of A. Because A is symmetric, the subspace orthogonal to v is invariant by A, and thus applying the previous argument in this new space and repeating this allow us to find a basis consisting of (orthogonal) eigenvectors. Thus A is diagonalizable. $\square$

If you are still not convinced about the power of this result, consider its use in the following proof of the celebrated fundamental theorem of algebra, due to D'Alembert and Gauss:

Theorem. *Any nonconstant polynomial with complex coefficients has at least one complex zero.*

Proof. Consider

$$P(X) = a_n X^n + a_{n-1} X^{n-1} + \cdots + a_1 X + a_0$$

a complex polynomial and suppose that $n \geq 1$ and $a_n \neq 0$. Let $f(z) = |P(z)|$. Because

$$f(z) \geq |a_n||z|^n - |a_{n-1}||z|^{n-1} - \cdots - |a_1||z| - |a_0|,$$

it follows that $\lim\limits_{|z| \to \infty} f(z) = \infty$. In particular, there exists M such that for all $|z| \geq M$, we have $f(z) \geq f(0)$. Because f is continuous on the compact disc of radius M and centered at the origin, its restriction to this compact set has a minimum at a point

z_0 such that $|z_0| \leq M$. Now, if $|z| \geq M$, we have $f(z) \geq f(0) \geq f(z_0)$. Otherwise, we know that $f(z) \geq f(z_0)$. Thus f attains its minimum at z_0. Suppose that P does not vanish on the set of complex numbers. Consider $Q(z) = \dfrac{P(z + z_0)}{P(z_0)}$. It is still a polynomial and $g(z) = |Q(z)|$ attains its minimal value, equal to 1, at 0. Let

$$Q(z) = 1 + b_j z^j + \cdots + b_n z^n,$$

where $b_j \neq 0$. Let $b_j = re^{i\alpha}$ with $r > 0$ and $\alpha \in \mathbb{R}$. The triangle inequality shows that

$$\left| Q\left(\epsilon e^{\frac{i(\pi - \alpha)}{j}} \right) \right| \leq |1 - re^j| + \sum_{k=j+1}^{n} |b_k| \epsilon^{j+1},$$

for all $\epsilon > 0$ such that $\epsilon < \min\left(1, \frac{1}{r}\right)$. Thus, for $\epsilon > 0$ sufficiently small, we have

$$\left| Q\left(\epsilon e^{\frac{i(\pi - \alpha)}{j}} \right) \right| < 1,$$

which is a contradiction. This shows that necessarily P must vanish at some $z \in \mathbb{C}$. $\square$

Let us continue with some concrete examples of application of the extreme value theorem:

Problem. Find all continuous functions $f : [0, 1] \to \mathbb{R}$ such that

$$f(x) = \sum_{n=1}^{\infty} \frac{f(x^n)}{2^n}, \quad \forall\, x \in [0, 1].$$

Solution. Because $\sum_{n=1}^{\infty} \frac{1}{2^n} = 1$, any constant function is solution of the equation. Now let f be a solution, and consider $0 < a < 1$ and $x \in [0, a]$ such that $f(x) = \max_{[0,a]} f$. Because $x^n \in [0, a]$ for all n, we have $f(x^n) \leq f(x)$ for all n and because

$$f(x) = \sum_{n=1}^{\infty} \frac{f(x^n)}{2^n},$$

we must have $f(x^n) = f(x)$ for all n. Take $n \to \infty$ to obtain $f(x) = f(0)$. Therefore $f(0) \geq f(t)$ for all $t \in [0, a]$ and all $a < 1$; thus $f(0) = \max_{[0,1]} f$. But if f is a solution, $-f$ is also a solution; thus $-f(0) = \max_{[0,1]}(-f) = -\min_{[0,1]} f$. Thus, $\min_{[0,1]} f = \max_{[0,1]} f$, and f is constant. $\square$

Here is a problem from the Putnam Competition:

Problem. Let $f, g : [0, 1] \to \mathbb{R}$ and $K : [0, 1] \times [0, 1] \to \mathbb{R}$ be positive continuous functions and suppose that for all $x \in [0, 1]$,

$$f(x) = \int_0^1 K(x, y)g(y)dy, \quad g(x) = \int_0^1 K(x, y)f(y)dy.$$

Prove that $f = g$.

Solution. We may assume that

$$a = \min_{[0,1]} \frac{f}{g} \le b = \min_{[0,1]} \frac{g}{f}$$

and let $x_0 \in [0, 1]$ be such that $a = \dfrac{f(x_0)}{g(x_0)}$. Thus $g(x) \ge af(x)$ for all x. Also,

$$\int_0^1 K(x_0, y)(g(y) - af(y))dy = f(x_0) - ag(x_0) = 0$$

and because $y \to K(x_0, y)(g(y) - af(y))$ is a continuous nonnegative function with average 0, it must be identically 0. Thus, since $K(x_0, y) > 0$ for all y, we have $g = af$. But then

$$f(x) = \int_0^1 K(x, y)g(y)dy = a \int_0^1 K(x, y)f(y)dy = ag(x) = a^2 f(x);$$

thus $a = 1$ and $f = g$. $\square$

We end this theoretical part with a nontrivial problem taken from the American Mathematical Monthly:

Problem. Let $f : \mathbb{R} \to \mathbb{R}$ be a twice continuously differentiable function such that

$$2f(x + 1) = f(x) + f(2x)$$

for all x. Prove that f is constant.

Solution. With $x = 0$ and $x = 1$, we obtain $f(0) = f(1) = f(2)$. Let $F = f''$, and differentiate the given equality twice to obtain

$$2F(x + 1) = F(x) + 4F(2x).$$

Thus

$$F(x) = \frac{1}{2}F\left(\frac{x}{2} + 1\right) - \frac{1}{4}F\left(\frac{x}{2}\right).$$

Now take $a \geq 2$ and let $I = [-a, a]$. Let $M = \max_{x \in I} |f(x)|$. Clearly, if $x \in I$, we have $\dfrac{x}{2} \in I$ and $1 + \dfrac{x}{2} \in I$. Thus

$$|F(x)| \leq \frac{M}{2} + \frac{M}{4} = \frac{3M}{4}$$

for all $x \in I$ and so $M \leq \dfrac{3M}{4}$; hence $M = 0$. Therefore F is identically 0 and f is of the form $ax + b$. Because $f(0) = f(1) = f(2)$, we deduce that $a = 0$, and so f is constant. $\square$

Proposed Problems

1. Let $f, g : [a, b] \to \mathbb{R}$ be continuous such that

$$\sup_{x \in [a,b]} f(x) = \sup_{x \in [a,b]} g(x).$$

Prove that there exists $c \in [a, b]$ such that $f(c) = g(c)$.

2. Let $f, g : [a, b] \to \mathbb{R}$ be continuous such that

$$\inf_{x \in [a,b]} f(x) \leq \inf_{x \in [a,b]} g(x) \leq \sup_{x \in [a,b]} g(x) \leq \sup_{x \in [a,b]} f(x).$$

Prove that there exists $c \in [a, b]$ such that $f(c) = g(c)$.

3. Let $f : [0, \infty) \to \mathbb{R}$ be continuous such that the limit $\lim_{x \to \infty} f(x) = l$ exists and is finite. Prove that f is bounded.

4. Let $f : [0, \infty) \to [0, \infty)$ be continuous such that $\lim_{x \to \infty} f(f(x)) = \infty$. Prove that $\lim_{x \to \infty} f(x) = \infty$.

5. Let $f : \mathbb{R} \to \mathbb{R}$ be a continuous function which transforms every open interval into an open interval. Prove that f is strictly monotone.

6. a) Prove that there are no continuous and surjective functions $f : [0, 1] \to (0, 1)$.
   b) Prove that there are no continuous and bijective functions $f : (0, 1) \to [0, 1]$.

7. Let $f : \mathbb{R} \to \mathbb{R}$ be continuous such that

$$|f(x) - f(y)| \geq |x - y|$$

for all real numbers $x, y \in \mathbb{R}$. Prove that f is surjective.

8. Let $f : \mathbb{R} \to \mathbb{R}$ be continuous and 1-periodic.

   a) Prove that f is bounded above and below, and it attains its minimum and maximum.

b) Prove that there exists a real number x_0 such that $f(x_0) = f(x_0 + \pi)$.

9. Prove that there exists $\varepsilon > 0$ such that

$$|\sin x| + |\sin(x + 1)| > \varepsilon,$$

for all real numbers x. Deduce that the sequence

$$x_n = \frac{|\sin 1|}{1} + \frac{|\sin 2|}{2} + \cdots + \frac{|\sin n|}{n}, \quad n \geq 1,$$

is unbounded.

10. Let $f : [0, 1] \to [0, \infty)$ be a continuous function. Prove that the sequence

$$a_n = \sqrt[n]{f\left(\frac{1}{n}\right)^n + f\left(\frac{2}{n}\right)^n + \cdots + f\left(\frac{n}{n}\right)^n}, \quad n \geq 2$$

is convergent.

11. Let $f : \mathbb{R} \to \mathbb{R}$ be a continuous function. Assume that for any real numbers a, b, with $a < b$, there exist $c_1, c_2 \in [a, b]$, with $c_1 \leq c_2$, such that

$$f(c_1) = \min_{x \in [a,b]} f(x), \quad f(c_2) = \max_{x \in [a,b]} f(x).$$

Prove that f is nondecreasing.

12. Let $f : \mathbb{R} \to \mathbb{R}$ be a continuous function. Assume that for any real numbers a, b, with $a < b$, there exist $c_1, c_2 \in [a, b]$, with $c_1 \neq c_2$, such that

$$f(c_1) = f(c_2) = \max_{x \in [a,b]} f(x).$$

Prove that f is constant.

13. Let $f : \mathbb{R} \to \mathbb{R}$ be a continuous function. Assume that for any real numbers a, b, with $a < b$, there exists $c \in (a, b)$ such that $f(c) \geq f(a)$ and $f(c) \geq f(b)$. Prove that f is constant.

14. Let $f, g : [a, b] \to [a, b]$ be continuous such that $f \circ g = g \circ f$. Prove that there exists $c \in [a, b]$ such that $f(c) = g(c)$.

15. Let (X, d) be a metric space and let $K \subseteq X$ be compact. Prove that every function $f : K \to K$ with the property that

$$d(f(x), f(y)) < d(x, y)$$

for all $x, y \in K, x \neq y$ has exactly one fixed point.

Solutions

1. Let us define the function $h : [a, b] \to \mathbb{R}$ by the formula $h(x) = f(x) - g(x)$. The function h is continuous as a difference of two continuous functions. By the extreme value theorem, we can find $\alpha, \beta \in [a, b]$ such that $f(\alpha) = g(\beta) = M$, where M denotes

$$M = \sup_{x \in [a,b]} f(x) = \sup_{x \in [a,b]} g(x).$$

We have $f(x) \le M$ and $g(x) \le M$ for all real numbers $x \in [a, b]$. Thus

$$h(\alpha) = f(\alpha) - g(\alpha) = M - g(\alpha) \ge 0$$

and

$$h(\beta) = f(\beta) - g(\beta) = f(\beta) - M \le 0,$$

so there exists $c \in [a, b]$ such that $h(c) = 0 \Leftrightarrow f(c) = g(c)$.

2. Let $\alpha, \beta \in [a, b]$ be such that

$$f(\alpha) = \inf_{x \in [a,b]} f(x), \quad f(\beta) = \sup_{x \in [a,b]} f(x).$$

By hypothesis, $f(\alpha) \le g(x) \le f(\beta)$ for every real number $x \in [a, b]$. Define the function $h : [a, b] \to \mathbb{R}$ by $h(x) = f(x) - g(x)$. The function h is continuous as a difference of two continuous functions. Then $h(\alpha) = f(\alpha) - g(\alpha) \le 0$ and $h(\beta) = f(\beta) - g(\beta) \ge 0$, so there exists $c \in [a, b]$ such that $h(c) = 0 \Leftrightarrow f(c) = g(c)$.

3. For $\varepsilon = 1$ in the definition of the limit, there exists $\delta > 0$ for which $|f(x) - l| \le 1$, for all real numbers $x \in [\delta, \infty)$. It follows that $|f(x)| \le 1 + |l|$ for all $x \in [\delta, \infty)$. Also the restriction $f|_{[0,\delta]} : [0, \delta] \to \mathbb{R}$ is a continuous function defined on a compact interval, so it is bounded, say $|f(x)| \le M$, for all $x \in [0, \delta]$. Consequently, $|f(x)| \le \max\{M, 1 + |l|\}$ for all nonnegative real numbers x.

4. Let us assume, by way of contradiction, that there exists a strictly increasing, unbounded sequence $(a_n)_{n \ge 1} \subset [0, \infty)$ such that $f(a_n) \le M$, for all positive integers n and for some real number M. By the Bolzano-Weierstrass theorem, we can assume that the sequence $(f(a_n))_{n \ge 1}$ is convergent (by working eventually with a subsequence of it). If $\lim_{n \to \infty} f(a_n) = L$, then by using the continuity of f at L, we obtain $\lim_{n \to \infty} f(f(a_n)) = f(L)$, which contradicts the fact that $\lim_{x \to \infty} f(f(x)) = \infty$.

5. Because any injective continuous function is strictly monotone, it is sufficient to prove that f is injective. Let us assume, by way of contradiction, that there exist two real numbers $a < b$ such that $f(a) = f(b)$. The function f cannot

be constant on $[a, b]$, because in this case, for every $a < c < d < b$, the set $f((c, d)) = \{f(a)\}$ is not an open interval. Thus the restriction $f|_{[a,b]}$ attains its minimum or maximum on (a, b), say $\min_{x \in [a,b]} f(x) = f(\xi)$, for some real number $\xi \in (a, b)$. In particular, $f(x) \geq f(\xi)$, for all reals $x \in (a, b)$, so $f(a, b)$ cannot be an open interval. This follows from the fact that the set $f((a, b))$ has a minimum and open intervals do not have minima.

6. a) If such a function exists, then $f([0, 1]) = (0, 1)$. This contradicts the fact that $f([0, 1]) = [m, M]$, where

$$m = \inf_{x \in [0,1]} f(x), \quad M = \sup_{x \in [0,1]} f(x),$$

which is a consequence of the extreme value theorem.

   b) Suppose that such a function exists. The function f is continuous and injective, so it will be strictly monotone, say increasing. By surjectivity, there exists $c \in (0, 1)$ such that $f(c) = 0$. Then for every real number $0 < x < c$, by monotony, it follows that $f(x) < f(c) = 0$, which is impossible.

   Alternatively, if a function f has the given properties, then its inverse $f^{-1} : [0, 1] \to (0, 1)$ is continuous and surjective, which contradicts a).

7. If $x_1, x_2 \in \mathbb{R}$ are so that $f(x_1) = f(x_2)$, then from the relation

$$|f(x_1) - f(x_2)| \geq |x_1 - x_2|, \Leftrightarrow 0 \geq |x_1 - x_2|,$$

it follows that $x_1 = x_2$. Hence f is injective. But f is continuous, so it is strictly monotone. For $y = 0$, we obtain $|f(x)| \geq -|f(0)| + |x|$, so $\lim_{|x| \to \infty} |f(x)| = \infty$. For example, if f is strictly increasing, then

$$\lim_{x \to -\infty} f(x) = -\infty, \quad \lim_{x \to \infty} f(x) = \infty$$

and by continuity, f is surjective.

8. a) Observe that because of the periodicity of f, we have $f(x) = f(x - [x])$ for all real numbers x. This shows that the range of f is exactly $f([0, 1])$. This is a compact interval by the extreme value theorem, so f is bounded and attains its extrema.

   b) Let us define the function $\phi : \mathbb{R} \to \mathbb{R}$

$$\phi(x) = f(x + \pi) - f(x),$$

for all real numbers x. The function f is continuous, so ϕ is continuous. If we show that the function ϕ takes values of opposite signs, then the problem is solved. Let α be such that $f(\alpha) \leq f(x)$ for all real numbers x. The existence of such a number follows from a). Then

$$\phi(\alpha) = f(\alpha + \pi) - f(\alpha) \geq 0$$

and

$$\phi(\alpha - \pi) = f(\alpha) - f(\alpha - \pi) \le 0.$$

9. Let us consider the continuous function $f : [0, 2\pi] \to \mathbb{R}$, given by

$$f(x) = |\sin x| + |\sin(x + 1)|.$$

Let $x_0 \in [0, 2\pi]$ be such that $f(x_0) = \min\limits_{x \in [0,2\pi]} f(x)$. We prove that $\varepsilon = f(x_0)$ is positive. Indeed, $\varepsilon \ge 0$, and if $\varepsilon = 0$, then

$$|\sin x_0| + |\sin(x_0 + 1)| = 0$$

$$\Rightarrow |\sin x_0| = 0 \quad \text{and} \quad |\sin(x_0 + 1)| = 0.$$

We can find integers m, n such that $x_0 = m\pi$, $x_0 + 1 = n\pi$, so

$$m\pi = n\pi - 1 \Rightarrow \pi = \frac{1}{n - m},$$

which is impossible because the above fraction on the right is smaller than 1 in absolute value. Now, for this ε, we have

$$|\sin x| + |\sin(x + 1)| \ge \varepsilon, \ \forall \, x \in [0, 2\pi].$$

Moreover, this inequality holds for all real numbers x, because sin is 2π-periodic. For the second part, note that

$$x_{2n} = \left(\frac{|\sin 1|}{1} + \frac{|\sin 2|}{2} \right) + \left(\frac{|\sin 3|}{3} + \frac{|\sin 4|}{4} \right) + \cdots$$

$$+ \left(\frac{|\sin(2n - 1)|}{2n - 1} + \frac{|\sin 2n|}{2n} \right)$$

$$> \frac{|\sin 1| + |\sin 2|}{2} + \frac{|\sin 3| + |\sin 4|}{4} + \cdots + \frac{|\sin(2n - 1)| + |\sin 2n|}{2n}$$

$$> \frac{\varepsilon}{2} + \frac{\varepsilon}{4} + \cdots + \frac{\varepsilon}{2n} = \frac{\varepsilon}{2} \left(1 + \frac{1}{2} + \cdots + \frac{1}{n} \right),$$

and the conclusion follows from the fact that

$$\lim_{n \to \infty} \left(1 + \frac{1}{2} + \cdots + \frac{1}{n} \right) = \infty.$$

One can prove (but the proof is much more involved) that if f is a continuous periodic function such that the series $\dfrac{|f(1)|}{1} + \dfrac{|f(2)|}{2} + \cdots$ converges, then $f(n) = 0$ for all positive integers n.

10. We will prove that $\lim\limits_{n\to\infty} a_n = M$, where $M = \max\limits_{x\in[0,1]} f(x)$. Assume that $f(x_0) = M$. Let $\varepsilon > 0$. There exists $\delta > 0$ for which

$$f(x) > M - \varepsilon, \ \forall \, x \in (x_0 - \delta, x_0 + \delta) \cap [0, 1].$$

For each $n > 1/\delta$, we can find $1 \leq k_0 \leq n$ such that

$$\frac{k_0}{n} \in (x_0 - \delta, x_0 + \delta).$$

Then

$$a_n = \sqrt[n]{f\left(\frac{1}{n}\right)^n + f\left(\frac{2}{n}\right)^n + \cdots + f\left(\frac{n}{n}\right)^n}$$

$$\geq \sqrt[n]{f\left(\frac{k_0}{n}\right)^n} = f\left(\frac{k_0}{n}\right) \geq M - \varepsilon.$$

We derive

$$M - \varepsilon \leq a_n \leq \sqrt[n]{n}M$$

for all integers $n \geq 1/\delta$; thus $(a_n)_{n\geq 1}$ converges to M.

11. Let us assume that $f(a) > f(b)$ for some $a < b$. Define

$$A = \{x \in [a, b] \mid f(x) \geq f(a)\}.$$

A is nonempty, $a \in A$, so let $c = \sup A$. If $(a_n)_{n\geq 1} \subset A$ is a sequence that converges to c, then $f(a_n) \geq f(a)$. By taking $n \to \infty$, we obtain $f(c) \geq f(a)$; thus $c < b$.

If $(r_n)_{n\geq 1} \subset (c, b)$, then $f(r_n) < f(a)$, and moreover, if $(r_n)_{n\geq 1}$ converges to c, we deduce that $f(c) \leq f(a)$. In conclusion, $f(c) = f(a)$. Now, let $c \leq c_1 \leq c_2 \leq b$ be such that

$$f(c_1) = \min_{x\in[c,b]} f(x), \quad f(c_2) = \max_{x\in[c,b]} f(x).$$

Because $f(c) > f(x)$, for all x in $(c, b]$, we must have $c_2 = c$, which is impossible.

12. Let $M = \max\limits_{x\in[a,b]} f(x)$ and define the set

$$A = \{x \in [a,b] \mid f(x) = M\}.$$

Let $c = \inf A$; then $f(c) = M$. We will prove that $f(a) = M$. Indeed, if $f(a) < M$, then $a < c$, and the assumption of the problem fails on the interval $[a,c]$.

We proved that for all $a < b, f(a) = \max\limits_{x\in[a,b]} f(x)$ and analogously, $f(b) = \max\limits_{x\in[a,b]} f(x)$. In conclusion, f is constant.

13. Let $a < b$ and let $c_1 \in [a,b]$ be such that $f(c_1) = \max\limits_{x\in[a,b]} f(x)$. According to the hypothesis, we can find $c_2 \in (c_1, b)$ such that $f(c_2) \geq f(c_1)$ and $f(c_2) \geq f(b)$. Thus $f(c_1) = f(c_2) = \max\limits_{x\in[a,b]} f(x)$. The conclusion follows from the previous problem.

14. We have to prove that the function $h : [a,b] \to \mathbb{R}$ given by $h(x) = f(x) - g(x)$ has a zero in $[a,b]$. If we suppose that this is not the case, then, because the function h is continuous, it keeps a constant sign on $[a,b]$, say $h > 0$. According to the extreme value theorem, there exists $x_0 \in [a,b]$ for which $h(x_0) = \inf\limits_{x\in[a,b]} h(x) > 0$. It follows that $f(x) - g(x) > h(x_0)$ for all real numbers x in $[a,b]$. Thus

$$f(f(x)) - g(g(x)) = [f(f(x)) - g(f(x))] + [f(g(x)) - g(g(x))]$$
$$> h(x_0) + h(x_0).$$

By induction,

$$f^{[n]}(x) - g^{[n]}(x) > nh(x_0),$$

for all real numbers $x \in [a,b]$ and all positive integers n. The last inequality cannot be true, if we take into account that f and g are bounded (they are continuous on a compact interval).

15. If $c, c' \in K, c \neq c'$ are so that $f(c) = c$, and $f(c') = c'$, then for $x = c$ and $y = c'$, we obtain

$$d(f(c), f(c')) < d(c, c') \Leftrightarrow d(c, c') < d(c, c'),$$

a contradiction that proves the uniqueness of a fixed point of f (if any). Let us define the application $\phi : K \to \mathbb{R}$ by the formula $\phi(x) = d(x, f(x))$ for all $x \in K$. The function ϕ is continuous, because f is continuous. According to the extreme value theorem, there exists $x_0 \in K$ such that $\phi(x_0) = \min\limits_{x\in K} \phi(x)$. Hence $d(x_0, f(x_0)) \leq d(x, f(x))$ for all $x \in K$, and we will prove that $f(x_0) = x_0$. If $f(x_0) \neq x_0$, then by hypothesis,

$$d(f(x_0), f(f(x_0))) < d(x_0, f(x_0)) \Leftrightarrow \phi(f(x_0)) < \phi(x_0),$$

which contradicts the minimality of $\phi(x_0)$.

Chapter 13
Uniform Continuity

We say that a function $f : D \subseteq \mathbb{R} \to \mathbb{R}$ is *uniformly continuous* if for all $\varepsilon > 0$, there exists $v(\varepsilon) > 0$ such that:

$$x, x' \in D, \ \left|x - x'\right| \leq v(\varepsilon) \ \Rightarrow \ \left|f(x) - f(x')\right| \leq \varepsilon.$$

Note that this is equivalent to the fact that for all $\varepsilon > 0$, there exists $v(\varepsilon) > 0$ such that:

$$x, x' \in D, \ \left|x - x'\right| < v(\varepsilon) \ \Rightarrow \ \left|f(x) - f(x')\right| < \varepsilon.$$

Also, it is easy to deduce that every uniformly continuous function is continuous, but the converse is not always true, as we can see from the following example.

Example. The function $f : \mathbb{R} \to \mathbb{R}, \ f(x) = x^2$ is continuous, but it is not uniformly continuous.

Proof. If f is uniformly continuous, then $|f(x) - f(y)| \leq \varepsilon$, for all real numbers x, y, with $|x - y| \leq v(\varepsilon)$. In particular, for $\varepsilon = 1$, we can find $v > 0$ such that $|f(x) - f(y)| \leq 1$, for all real numbers x, y, with $|x - y| \leq v$. Consequently, $|f(y + v) - f(y)| \leq 1$, for all real numbers y. Therefore

$$\left|(y + v)^2 - y^2\right| \leq 1 \Leftrightarrow |2vy + v^2| \leq 1$$

for all y, which is clearly impossible. $\square$

The following theoretical result is a criterion for uniformly continuous functions.

Proposition. *Let $f : D \subseteq \mathbb{R} \to \mathbb{R}$. The following assertions are equivalent:*

a) *f is uniformly continuous.*
b) *for all sequences $(x_n)_{n \geq 1}, (y_n)_{n \geq 1} \subset D$ with $\lim\limits_{n \to \infty} (x_n - y_n) = 0$, we have*

© Springer Science+Business Media LLC 2017

T. Andreescu et al., *Mathematical Bridges*, DOI 10.1007/978-0-8176-4629-5_13

$$\lim_{n\to\infty} (f(x_n) - f(y_n)) = 0.$$

Proof. a)$\Rightarrow$b). Let $\varepsilon > 0$. The uniform continuity of f implies the existence of $v(\varepsilon) > 0$ such that

$$|f(x) - f(y)| \le \varepsilon, \tag{13.1}$$

for all real numbers $x, y \in D$, with $|x - y| \le v(\varepsilon)$. Because $\lim_{n\to\infty} (x_n - y_n) = 0$, there exists a positive integer $n(\varepsilon)$ such that $|x_n - y_n| \le v(\varepsilon)$, for all integers $n \ge n(\varepsilon)$. Using (13.1), we derive $|f(x_n) - f(y_n)| \le \varepsilon$, for all $n \ge n(\varepsilon)$, so

$$\lim_{n\to\infty} (f(x_n) - f(y_n)) = 0.$$

b)$\Rightarrow$a). Assuming by contradiction that f is not uniformly continuous, we deduce the existence of a real number $\varepsilon_0 > 0$ so that

$$\forall v > 0, \ \exists x_v, y_v, \ |x_v - y_v| \le v, \quad |f(x_v) - f(y_v)| \ge \varepsilon_0.$$

If $v = \dfrac{1}{n}, n \ge 1$, then we can find sequences $(x_n)_{n\ge1}, (y_n)_{n\ge1}$ for which

$$|x_n - y_n| \le \frac{1}{n} \quad \text{and} \quad |f(x_n) - f(y_n)| \ge \varepsilon_0,$$

for all positive integers n. This contradicts $\lim_{n\to\infty} (f(x_n) - f(y_n)) = 0$. $\square$

We recall that f is a *Lipschitz function* if

$$|f(x) - f(y)| \le L|x - y|,$$

for all x, y in D and some positive real L. It is immediate that every Lipschitz function is uniformly continuous: for each $\varepsilon > 0$, it is enough to take $v(\varepsilon) := \frac{\varepsilon}{L}$.

One method to recognize a Lipschitz function is to verify that it has a bounded derivative. In this case, if $\sup_{x \in D} |f'(x)| = L < \infty$, then, using Lagrange's mean value theorem,

$$|f(x) - f(y)| = |f'(c)| \cdot |x - y| \le L \cdot |x - y|.$$

As a direct consequence, a differentiable function with a bounded derivative is uniformly continuous.

We have seen that uniform continuity implies continuity, but the converse is not true. However, we have the following useful result.

Theorem. *If $f : D \to \mathbb{R}$ is continuous and D is compact, then f is uniformly continuous.*

Proof. Assuming the falsity of the conclusion, we can find $\varepsilon_0 > 0$ and sequences $(x_n)_{n\geq 1}, (y_n)_{n\geq 1}$ for which $\lim_{n\to\infty} (x_n - y_n) = 0$ and $|f(x_n) - f(y_n)| \geq \varepsilon_0$, for all integers $n \geq 1$.

Because of compactness of D, the sequence $(x_n)_{n\geq 1}$ has a convergent subsequence denoted $(x_{k_n})_{n\geq 1}$. Because $\lim_{n\to\infty} (x_n - y_n) = 0$, the sequence $(y_{k_n})_{n\geq 1}$ is also convergent to the same limit, say a. The inequality $|f(x_{k_n}) - f(y_{k_n})| \geq \varepsilon_0$, therefore, contradicts the continuity of f at a. $\square$

This result is extremely useful in a variety of problems concerning continuous functions defined on compact spaces. Here is a nontrivial application:

Problem. Let U be the set of complex numbers of absolute value 1 and let $f : [0, 1] \to U$ be a continuous function. Prove the existence of a real continuous function g such that $f(x) = e^{ig(x)}$ for all $x \in [0, 1]$.

Solution. We will begin with a definition: we will say that f has a *lifting* if there exists such a function g. First of all, we will prove the following.

Lemma. *Any continuous function f which is not onto has a lifting.*

Indeed, we may assume that 1 is not in the range of f. Then, for all x, there exists some $g(x) \in (0, 2\pi)$ such that $f(x) = e^{ig(x)}$. We claim that this function g is a lifting of f; that is, g is continuous. Indeed, fix a number x_0 and suppose that g is not continuous at x_0. Thus, there exists a sequence a_n converging to x_0 and an $\epsilon > 0$ such that $|g(a_n) - g(x_0)| \geq \epsilon$. Because the sequence $(g(a_n))$ is bounded, it has a convergent subsequence, and by replacing eventually a_n with the corresponding subsequence, we may assume that $g(a_n)$ converges to some $l \in [0, 2\pi]$. But then $f(a_n)$ converges to e^{il}, and by continuity of f, we must have $e^{il} = e^{ig(x_0)}$, which means that $l - g(x_0)$ is a multiple of 2π. But since $g(x_0) \in (0, 2\pi)$, this forces $l = g(x_0)$, and thus we contradict the inequality $|g(a_n) - g(x_0)| \geq \varepsilon$ for sufficiently large n. Thus g is continuous and f has a lifting. $\square$

Now, we will employ the uniform continuity of f. Take N sufficiently large such that

$$\left| \frac{f\left(\frac{kx}{N}\right)}{f\left(\frac{(k-1)x}{N}\right)} - 1 \right| < 1$$

for all x and all $1 \leq k \leq N$. This is possible, since the maximal distance between $\frac{(k-1)x}{N}$ and $\frac{kx}{N}$ (taken over all x) tends to 0 as $N \to \infty$ and since f is uniformly continuous. But then, if we define

$$h_k(x) = \frac{f\left(\frac{kx}{N}\right)}{f\left(\frac{(k-1)x}{N}\right)},$$

these functions take their values in U and are not onto, since -1 is not in their range. Thus, by the lemma, they have liftings $g_1, g_2, \ldots, g_N$. By taking some a such that $f(0) = e^{ia}$, we observe immediately that $a + g_1 + \cdots + g_N$ is a lifting for f. $\square$

We continue with an easier problem. However, it requires some geometric interpretation:

Problem. Let $f : [0, \infty) \to [0, \infty)$ be a continuous function such that

$$f\left(\frac{x+y}{2}\right) \geq \frac{f(x) + f(y)}{2}$$

for all x, y. Prove that f is uniformly continuous.

Solution. First, we are going to prove that f is concave. Indeed, an immediate induction shows that

$$f\left(\frac{x_1 + x_2 + \cdots + x_{2^n}}{2^n}\right) \geq \frac{f(x_1) + f(x_2) + \cdots + f(x_{2^n})}{2^n}$$

for all nonnegative numbers $x_1, x_2, \ldots, x_{2^n}$ and all positive integers n. Thus, by fixing some $0 \leq k \leq 2^n$ and two nonnegative integers x, y, we deduce that

$$f\left(\frac{k}{2^n}x + \left(1 - \frac{k}{2^n}\right)y\right) \geq \frac{k}{2^n}f(x) + \left(1 - \frac{k}{2^n}\right)f(y).$$

But, as we have seen in the chapter concerning density, the set of numbers of the form $\frac{k}{2^n}$ with $0 \leq k \leq 2^n - 1$ is dense in $[0, 1]$ and the continuity of f implies the inequality

$$f(ax + (1-a)y) \geq af(x) + (1-a)f(y)$$

for all $x, y \geq 0$ and all $a \in [0, 1]$. That is, f is concave.

Now, take some sequences a_n and b_n such that $a_n - b_n$ converges to 0 and $a_n < b_n$ for all n and suppose that $f(a_n) - f(b_n)$ does not converge to 0. Thus, $a_n \to \infty$ (otherwise, it would have a convergent subsequence, and the same subsequence of b_n would converge to the same limit, thus contradicting the continuity of f). Take some $c > d > 1$ and observe that the concavity of f implies the inequality

$$f(b_n) - f(a_n) \leq (b_n - a_n) \cdot \frac{f(d) - f(c)}{d - c}.$$

Now, let us prove that f is increasing. This is quite clear, because for any $a \geq 0$, the function

$$h_a(x) = \frac{f(x) - f(a)}{x - a}$$

is decreasing for $x > a$ (since f is concave) and $h_a(x) \geq -\frac{f(a)}{x-a}$, the last quantity converging to 0 as $x \to \infty$. Thus h_a must be nonnegative on (a, ∞), and so f is increasing. Finally, this argument combined with the previous inequality gives the estimation

$$0 \leq f(b_n) - f(a_n) \leq (b_n - a_n) \cdot \frac{f(d) - f(c)}{d - c},$$

which shows that $f(b_n) - f(a_n)$ converges to 0. Using the theoretical results, we finally deduce that f is uniformly continuous. $\square$

The following problem is much more challenging:

Problem. Let $I \subset \mathbb{R}$ be an interval and let $f : I \to \mathbb{R}$ be uniformly continuous. Define

$$\delta(\varepsilon) = \sup\{\delta > 0 \mid \forall\, x_1, x_2 \in I,\ |x_1 - x_2| \leq \delta \Rightarrow |f(x_1) - f(x_2)| \leq \varepsilon\}.$$

Prove that:

a) For all $\varepsilon > 0$ and all $x, y \in I$,

$$|f(x) - f(y)| \leq \frac{\varepsilon}{\delta(\varepsilon)}|x - y| + \varepsilon.$$

b) $\lim\limits_{\varepsilon \to 0} \dfrac{\delta(\varepsilon)}{\varepsilon} = 0$ if and only if f is not Lipschitz on I.

Solution. a) Let $n = \left[\dfrac{|x - y|}{\delta(\varepsilon)}\right]$. Everything is clear if $n = 0$, by definition of $\delta(\varepsilon)$, so assume that $n \geq 1$. Also, suppose that $x < y$. Let

$$x_1 = x + \delta(\varepsilon),\ x_2 = x + 2\delta(\varepsilon), \ldots,\ x_n = x + n\delta(\varepsilon).$$

Then

$$|f(x) - f(x_1)| \leq \varepsilon,\ |f(x_1) - f(x_2)| \leq \varepsilon, \ldots, |f(x_n) - f(y)| \leq \varepsilon,$$

thus

$$|f(x) - f(y)| \leq (n + 1)\varepsilon \leq \frac{|x - y|}{\delta(\varepsilon)}\varepsilon + \varepsilon.$$

b) Let us suppose that f is not Lipschitz and assume that $\lim\limits_{\varepsilon \to 0} \dfrac{\delta(\varepsilon)}{\varepsilon}$ is not 0 (it may not exist); thus $\lim\limits_{\varepsilon \to 0} \dfrac{\varepsilon}{\delta(\varepsilon)}$ is not ∞, so there is a decreasing sequence $\varepsilon_n \to 0$ and a real number γ such that $\dfrac{\varepsilon_n}{\delta(\varepsilon_n)} \to \gamma$. Let $C = \sup\limits_{n} \dfrac{\varepsilon_n}{\delta(\varepsilon_n)} < \infty$. Then by a), we have

$$|f(x) - f(y)| \leq C|x - y| + \varepsilon_n, \ \forall\, n, \ \forall\, x, y \in I.$$

For $x, y \in I$ fixed, let $n \to \infty$. We obtain $|f(x) - f(y)| \leq C|x - y|$; thus f is C-Lipschitz, a contradiction.

Now, suppose that $\lim\limits_{\varepsilon \to 0} \dfrac{\delta(\varepsilon)}{\varepsilon} = 0$, but f is Lipschitz. Let $C > 0$ be such that $|f(x) - f(y)| \leq C|x - y|$ for all $x, y \in I$. Take $\varepsilon > 0$. For $x_1, x_2 \in I$ with $|x_1 - x_2| \leq \dfrac{\varepsilon}{C}$, we have $|f(x_1) - f(x_2)| \leq \varepsilon$, thus $\delta(\varepsilon) \geq \dfrac{\varepsilon}{C} \Rightarrow \dfrac{\delta(\varepsilon)}{\varepsilon} \geq \dfrac{1}{C}$. This contradicts the fact that $\lim\limits_{\varepsilon \to 0} \dfrac{\delta(\varepsilon)}{\varepsilon} = 0$. $\square$

We continue with another quite difficult exercise:

Problem. Let $f, g : \mathbb{R} \to \mathbb{R}$ be continuous functions. Suppose that f is periodic and nonconstant, and $\lim\limits_{x \to \infty} \dfrac{g(x)}{x} = \infty$. Prove that $f \circ g$ is not periodic.

Solution. Suppose the contrary. Then by proposed problem 13, $f \circ g$ is uniformly continuous. Let $T > 0$ be a period for f and $u, v \in [0, T]$ be such that $f(u) = \min\limits_{[0,T]} f$, $f(v) = \max\limits_{[0,T]} f$. Let $f(u) = m, f(v) = M$, and $\varepsilon = \dfrac{M - m}{2}$. By uniform continuity, there is δ such that

$$|x - y| \leq \delta \Rightarrow |f(g(x)) - f(g(y))| \leq \frac{M - m}{2}. \tag{13.1}$$

We claim that there is an interval of length δ such that the variation of g on this interval is greater than T. Indeed, otherwise

$$\left| \frac{g(n\delta)}{n\delta} \right| \leq \frac{|g(0)| + |g(\delta) - g(0)| + \cdots + |g(n\delta) - g((n-1)\delta)|}{n\delta} \leq \frac{|g(0)|}{n\delta} + \frac{T}{\delta}$$

and this contradicts the hypothesis made on g.

Thus, there are $x_1, x_2 \in \mathbb{R}$ with $|x_1 - x_2| \leq \delta$ such that $|g(x_1) - g(x_2)| > T$. Thus there are a, b between $g(x_1)$ and $g(x_2)$ such that $f(a) = m, f(b) = M$. By continuity of g, there are x, y between x_1, x_2 such that $a = g(x), b = g(y)$. Thus $|x - y| \leq \delta$ and

$$|f(g(x)) - f(g(y))| = |f(a) - f(b)| = M - m > \frac{M - m}{2},$$

which contradicts (13.1). This finishes the solution. $\square$

Proposed Problems

1. Prove that the function $f : (-1, 1) \to \mathbb{R}, f(x) = x^2$ is uniformly continuous.

2. Prove that the function $f : (0, \infty) \to \mathbb{R}, f(x) = \sin \dfrac{1}{x}$ is not uniformly continuous.

3. Prove that the function $f : (0, \infty) \to \mathbb{R}, f(x) = \sin(x^2)$ is not uniformly continuous.

4. Let $f, g : \mathbb{R} \to \mathbb{R}, f(x) = x, g(x) = \sin x$. Prove that f and g are uniformly continuous, but their product fg is not uniformly continuous.

5. Let $f : D \subseteq \mathbb{R} \to \mathbb{R}$ be a function with the property that

$$|f(x) - f(y)| \leq \sqrt{|x - y|},$$

   for all $x, y \in D$. Prove that f is uniformly continuous. Is this condition necessary for uniform continuity?

6. Let $f : D \subseteq \mathbb{R} \to \mathbb{R}$ be uniformly continuous and bounded. Prove that f^2 is uniformly continuous.

7. For what α is the function $f : [0, \infty) \to \mathbb{R}, f(x) = x^\alpha$ uniformly continuous?

8. Prove that the function $f : (a, \infty) \to \mathbb{R}, f(x) = \ln x$ is uniformly continuous if and only if $a > 0$.

9. Does there exist a uniformly continuous function that is not a Lipschitz function?

10. Let $f : [0, 1) \to \mathbb{R}$ be uniformly continuous. Prove that the function f has finite limit at 1.

11. Let $f : \mathbb{R} \to \mathbb{R}$ be continuous, having horizontal asymptotes at $\pm\infty$. Prove that f is uniformly continuous.

12. Let $f, g : [0, \infty) \to \mathbb{R}$ be continuous such that

$$\lim_{x \to \infty} [f(x) - g(x)] = 0.$$

   Prove that f is uniformly continuous if and only if g is uniformly continuous.

13. Let $f : \mathbb{R} \to \mathbb{R}$ be continuous and periodic. Prove that f is uniformly continuous.

14. Let $f : \mathbb{R} \to \mathbb{R}$ be uniformly continuous. Prove that

$$|f(x)| \leq a|x| + b,$$

   for all x and some a, b.

15. Let $f : \mathbb{R} \to \mathbb{R}$ be a uniformly continuous function. Prove that the function $g : \mathbb{R}^2 \to \mathbb{R}$, given by

$$g(x, y) = \begin{cases} xf\left(\dfrac{y}{x}\right), & x \neq 0 \\ 0, & x = 0 \end{cases}$$

   is continuous.

Solutions

1. Let $(x_n)_{n \geq 1}$, $(y_n)_{n \geq 1}$ be sequences with elements in $(-1, 1)$ such that

$$\lim_{n \to \infty} (x_n - y_n) = 0.$$

Then

$$|f(x_n) - f(y_n)| = |x_n^2 - y_n^2| = |x_n - y_n| \cdot (x_n + y_n) \leq 2 |x_n - y_n|.$$

We used the inequality $|x_n + y_n| \leq 2$, which follows from $x_n, y_n \in (-1, 1)$. Now, from the inequalities

$$0 \leq |f(x_n) - f(y_n)| \leq 2 |x_n - y_n|,$$

we derive $\lim_{n \to \infty} (f(x_n) - f(y_n)) = 0$. Hence f is uniformly continuous.

In fact, the uniform continuity of f can be established by using the fact that f is differentiable with bounded derivative, $|f'(x)| \leq 2 |x| < 2$, for all x in $(-1, 1)$.

On other hand, we can easily see that if a function is uniformly continuous on D, then every restriction $f|_{D_0}$, with $D_0 \subset D$ remains uniformly continuous. In our case, the given function f can be considered as a restriction of the function $F : [-2, 2] \to \mathbb{R}$, given by the law $F(x) = x^2$. Finally, F is uniformly continuous on the compact set $[-2, 2]$.

2. We can take the sequences

$$x_n = \frac{1}{2n\pi}, \quad y_n = \frac{1}{2n\pi - \pi/2}, \quad n \geq 1.$$

Easily, $\lim_{n \to \infty} (x_n - y_n) = 0$ and

$$\lim_{n \to \infty} (f(x_n) - f(y_n)) = \lim_{n \to \infty} \left(\sin 2n\pi - \sin \left(2n\pi - \frac{\pi}{2} \right) \right) = 1,$$

so f is not uniformly continuous.

This proof uses sequences in a neighborhood of the origin, so a natural question is if the restriction $g = f|_{[a,\infty)}$, with $a > 0$, is uniformly continuous or not. In this case, we have

$$|g'(x)| = \frac{|\cos \frac{1}{x}|}{x^2} \leq \frac{1}{x^2} \leq \frac{1}{a^2},$$

for all x in $[a, \infty)$. Hence $f|_{[a,\infty)}$ is uniformly continuous, because it has bounded derivative.

3. First, note that the derivative $f'(x) = 2x \cos x^2$ is bounded on intervals of the form $(0, a]$, $a > 0$, so $f|_{(0,a]}$ is uniformly continuous. Indeed,

$$|f'(x)| = 2x |\cos x^2| \le 2x \le 2a,$$

for all x in $(0, a]$. On $(0, \infty)$, we can consider the sequences

$$x_n = \sqrt{2n\pi}, \quad y_n = \sqrt{2n\pi - \pi/2}, \quad n \ge 1.$$

We have

$$\lim_{n \to \infty} (x_n - y_n) = \lim_{n \to \infty} \left(\sqrt{2n\pi} - \sqrt{2n\pi - \pi/2} \right)$$

$$= \frac{\pi}{2} \lim_{n \to \infty} \frac{1}{\sqrt{2n\pi} + \sqrt{2n\pi - \pi/2}} = 0.$$

However,

$$\lim_{n \to \infty} (f(x_n) - f(y_n)) = \lim_{n \to \infty} \left(\sin 2n\pi - \sin \left(2n\pi - \frac{\pi}{2} \right) \right) = 1,$$

with $\lim_{n \to \infty} (x_n - y_n) = 0$. Thus, f is not uniformly continuous.

4. The functions f and g are uniformly continuous, because they have bounded derivatives on $\mathbb{R}$,

$$|f'(x)| = 1, \quad |g'(x)| = |\cos x| \le 1, \quad x \in \mathbb{R}.$$

The product function $h(x) = x \sin x$, $x \in \mathbb{R}$, is not uniformly continuous. Indeed, let us consider the sequences

$$x_n = 2n\pi, \quad y_n = 2n\pi - \frac{1}{n}, \quad n \ge 1.$$

We have $\lim_{n \to \infty} (x_n - y_n) = 0$ and

$$h(x_n) - h(y_n) = \left(2n\pi - \frac{1}{n} \right) \cdot \sin \left(\frac{1}{n} \right)$$

converges to 2π. Hence h is not uniformly continuous.

5. Let $(x_n)_{n \ge 1}$, $(y_n)_{n \ge 1} \subset D$ be sequences such that $\lim_{n \to \infty} (x_n - y_n) = 0$. From the inequality

$$|f(x_n) - f(y_n)| \le \sqrt{|x_n - y_n|},$$

it follows that

$$\lim_{n \to \infty} |f(x_n) - f(y_n)| = 0,$$

so f is uniformly continuous. Clearly, there are uniformly continuous functions which do not satisfy this condition, for example, $f(x) = x$.

6. Let us assume that for some $M > 0$, we have $|f(x)| \le M$, for all $x \in D$. We then deduce that

$$\left| f^2(x) - f^2(y) \right| = |f(x) + f(y)| \cdot |f(x) - f(y)|$$

so

$$\left| f^2(x) - f^2(y) \right| \le 2M \cdot |f(x) - f(y)| .$$

Now let $(x_n)_{n \ge 1}$, $(y_n)_{n \ge 1} \subset D$ be any sequences such that

$$\lim_{n \to \infty} |x_n - y_n| = 0.$$

The function f is uniformly continuous, so $\lim_{n \to \infty} |f(x_n) - f(y_n)| = 0$. Hence

$$\lim_{n \to \infty} \left| f^2(x_n) - f^2(y_n) \right| = 0,$$

and the function f^2 is uniformly continuous.

7. First of all, let $\alpha > 1$. If f is uniformly continuous, then there is $\nu > 0$ so that

$$x, y \in [0, \infty), \ |x - y| \le \nu \Rightarrow |x^\alpha - y^\alpha| \le 1.$$

In particular,

$$|(x + \nu)^\alpha - x^\alpha| \le 1,$$

for all nonnegative reals x. This contradicts the fact that

$$\lim_{x \to \infty} [(x + \nu)^\alpha - x^\alpha] = \infty,$$

as follows immediately from Lagrange's mean value theorem.

For $0 < \alpha \le 1$, we can see that f is uniformly continuous on $[0, 1]$ (continuous on a compact set) and f' is bounded on $[1, \infty)$. It follows that f is uniformly continuous on $[0, \infty)$.

8. First assume that $a > 0$. The function f is uniformly continuous because it has bounded derivative,

$$f'(x) = \frac{1}{x} < \frac{1}{a},$$

for all $x \in (a, \infty)$. In the case $a = 0$, consider the sequences

$$x_n = \frac{1}{n}, \quad y_n = \frac{1}{n^2}, \quad n \geq 1.$$

We have

$$\lim_{n \to \infty} (x_n - y_n) = \lim_{n \to \infty} \left(\frac{1}{n} - \frac{1}{n^2} \right) = 0$$

and

$$\lim_{n \to \infty} (f(x_n) - f(y_n)) = \lim_{n \to \infty} \left(\ln \frac{1}{n} - \ln \frac{1}{n^2} \right) = \lim_{n \to \infty} \ln n = \infty,$$

so f is not uniformly continuous.

9. Yes. An example is $f(x) = \sqrt{x}$, $x \in [0, \infty)$. As we have already proved, f is uniformly continuous. We have

$$\sup_{x \neq y} \left| \frac{f(x) - f(y)}{x - y} \right| = \sup_{x \neq y} \left| \frac{\sqrt{x} - \sqrt{y}}{x - y} \right| = \sup_{x \neq y} \left| \frac{1}{\sqrt{x} + \sqrt{y}} \right| = \infty,$$

so f is not Lipschitz.

10. First of all, it is clear that if $(x_n)_{n \geq 1}$ converges to 1, then $(f(x_n))_{n \geq 1}$ is a Cauchy sequence. Indeed, for $\epsilon > 0$, take ν such that $|x - y| \leq \nu$ implies $|f(x) - f(y)| \leq \epsilon$. Then (because $(x_n)_{n \geq 1}$ is a Cauchy sequence) for sufficiently large m, n, we have $|x_m - x_n| \leq \nu$, thus $|f(x_n) - f(x_m)| \leq \epsilon$ for all sufficiently large m, n. Thus $(f(x_n))_{n \geq 1}$ is a Cauchy sequence, thus convergent. We claim that all such sequences have the same limit. This is easily seen by considering the sequence $x_1, y_1, x_2, y_2, \ldots$. Call its terms $z_1, z_2, \ldots$. Then we know that $(f(z_n))_{n \geq 1}$ converges. Since $(f(x_n))_{n \geq 1}$ and $(f(y_n))_{n \geq 1}$ are subsequences of $(f(z_n))_{n \geq 1}$, they must have the same limit. This shows that $(f(x_n))_{n \geq 1}$ converges to a certain l for all $(x_n)_{n \geq 1}$ convergent to 1. Thus f has a finite limit at 1.

11. Let us denote

$$a = \lim_{x \to -\infty} f(x), \quad b = \lim_{x \to \infty} f(x).$$

Suppose that f is not uniformly continuous. By the theoretical part of the text, we know that there exist two sequences $(a_n)_{n \geq 1}$ and $(b_n)_{n \geq 1}$ such that $\lim_{n \to \infty} (a_n - b_n) = 0$ and $\epsilon > 0$ such that $|f(a_n) - f(b_n)| > \epsilon$ for all n. Next, $(a_n)_{n \geq 1}$ has a limit point in $\mathbb{R} \cup \{\pm\infty\}$; that is, for some increasing sequence $(k_n)_{n\geq 1}$, we have $\lim_{n\to\infty} a_{k_n} = l$. Clearly, $\lim_{n\to\infty} b_{k_n} = l$.

Now, if l is real, the inequality $|f(a_{k_n}) - f(b_{k_n})| > \epsilon$ cannot hold for all n because f is continuous at l. Suppose for instance that $l = \infty$. Then

$$\lim_{n\to\infty} f(a_{k_n}) = b = \lim_{n\to\infty} f(b_{k_n}),$$

and again we reach a contradiction.

As a consequence, a continuous function having oblique asymptotes at $\pm\infty$ is in fact uniformly continuous. Indeed, if

$$\lim_{x\to\infty} (f(x) - mx - n) = 0,$$

with $m, n \in \mathbb{R}$, then $g(x) = f(x) - mx - n$ defined on a neighborhood of $+\infty$ is uniformly continuous, because it has horizontal asymptote $y = 0$. Finally,

$$f(x) = g(x) + (mx + n)$$

is uniformly continuous, as a sum of two uniformly continuous functions.

12. If we assume that f is uniformly continuous, then

$$\forall \varepsilon > 0, \ \exists \nu_1 > 0, \ |x - y| < \nu_1 \Rightarrow |f(x) - f(y)| < \frac{\varepsilon}{6}.$$

Pick $M > 0$ with the following property:

$$x \in [M, \infty) \Rightarrow |f(x) - g(x)| < \frac{\varepsilon}{6}.$$

The function g is uniformly continuous on $[0, M]$, so we can find $\nu_2 > 0$ for which

$$x, y \in [0, M], \ |x - y| < \nu_2 \Rightarrow |g(x) - g(y)| < \frac{\varepsilon}{2}.$$

Now, with $\nu = \min\{\nu_1, \nu_2\}$, we can prove the implication

$$x, y \in [0, \infty), \ |x - y| < \nu \Rightarrow |g(x) - g(y)| < \varepsilon,$$

which solves the problem. The last implication is true in case $x, y \in [0, M]$. If $x, y \in [M, \infty)$, then

$$|g(x) - g(y)| \leq |g(x) - f(x)| + |f(x) - f(y)| + |f(y) - g(y)|$$
$$\leq \frac{\varepsilon}{6} + \frac{\varepsilon}{6} + \frac{\varepsilon}{6} = \frac{\varepsilon}{2}.$$

If $x \in [0, M], y \in [M, \infty)$, then

$$|g(x) - g(y)| \le |g(x) - g(M)| + |g(y) - g(M)|$$

$$\le \frac{\varepsilon}{2} + \frac{\varepsilon}{2} = \varepsilon.$$

This result can be easily extended to continuous functions $f, g : \mathbb{R} \to \mathbb{R}$ for which

$$\lim_{x \to \pm\infty} [f(x) - g(x)] = 0.$$

As a direct consequence, continuous functions with oblique asymptotes are uniformly continuous. Indeed, if

$$\lim_{x \to \pm\infty} [f(x) - mx - n] = 0,$$

then $g(x) = mx + n$ is uniformly continuous, and so f is uniformly continuous. More generally, the result remains true even if f has different asymptotes at $-\infty$ and $+\infty$.

13. Let $T > 0$ be a period of f. By the uniform continuity of f on the compact set $[0, T]$, we can state the following:

$$\forall \varepsilon > 0, \exists v > 0, \ x, y \in [0, T], \ |x - y| < v \Rightarrow |f(x) - f(y)| \le \frac{\varepsilon}{2}.$$

The value v can be chosen less than T. We prove now the implication

$$x, y \in \mathbb{R}, \ |x - y| \le v \Rightarrow |f(x) - f(y)| \le \varepsilon.$$

If $0 \le y - x \le v$, then $x, y \in [(k - 1)T, kT]$ or

$$x \in [(k - 1)T, kT], \quad y \in [kT, (k + 1)T].$$

In the first case,

$$|f(x) - f(y)| \le |f(x - (k - 1)T) - f(y - (k - 1)T)| \le \frac{\varepsilon}{2} < \varepsilon,$$

because

$$x - (k - 1)T, \ y - (k - 1)T \in [0, T].$$

Otherwise,

$$|f(x) - f(y)| \le |f(x) - f(kT)| + |f(y) - f(kT)|$$

$$= |f(x - (k - 1)T) - f(0)| + |f(y - kT) - f(0)|$$

$$\le \frac{\varepsilon}{2} + \frac{\varepsilon}{2} = \varepsilon.$$

In conclusion, f is uniformly continuous.

14. Let $v > 0$ for which we have the implication

$$x, y \in \mathbb{R}, \ |x - y| < v \Rightarrow |f(x) - f(y)| < 1.$$

Let $x \in \mathbb{R}, x \neq 0$ and let $n = \left[\frac{|x|}{v}\right] + 1$, which is a positive integer. Then

$$\left|\frac{kx}{n} - \frac{(k-1)x}{n}\right| < v,$$

and it follows that

$$|f(x)| \le |f(0)| + \sum_{k=0}^{n-1}\left|f\left(\frac{kx}{n}\right) - f\left(\frac{k+1}{n}x\right)\right|$$

$$\le |f(0)| + n \le \frac{|x|}{v} + |f(0)| + 1.$$

This condition is not sufficient for uniform continuity. For example, the function $f(x) = x \sin x, \ x \in \mathbb{R}$, is not uniformly continuous, as we have proved, but $|f(x)| \le |x|$, for all x.

15. Let a, b be real numbers satisfying $|f(x)| \le a|x| + b$, for all real numbers x. Then

$$0 \le |g(x, y)| = |x| \cdot \left|f\left(\frac{y}{x}\right)\right| \le |x| \cdot \left(a \cdot \left|\frac{y}{x}\right| + b\right) = a \cdot |y| + b \cdot |x|.$$

Obviously, g is continuous at the origin, and consequently, it is continuous on $\mathbb{R} \times \mathbb{R}$.

Chapter 14
Derivatives and Functions' Variation

The derivatives of a differentiable function $f : [a, b] \to \mathbb{R}$ give us basic information about the variation of the function. For instance, it is well-known that if $f' \geq 0$, then the function f is increasing and if $f' \leq 0$, then f is decreasing. Also, a function defined on an interval having the derivative equal to zero is in fact constant. All of these are consequences of some very useful theorems due to Fermat, Cauchy, and Lagrange. Fermat's theorem states that the derivative of a function vanishes at each interior extremum point of f. The proof is not difficult: suppose that x_0 is a local extremum, let us say a local minimum. Then $f(x_0 + h) - f(x_0) \geq 0$ for all h in an open interval $(-\delta, \delta)$. By dividing by h and passing to the limit when h approaches 0, we deduce that $f'(x_0) \geq 0$ (for $h > 0$) and $f'(x_0) \leq 0$ (for $h < 0$); thus $f'(x_0) = 0$. Using this result, we can now easily prove the following useful theorem:

Theorem (Rolle). *Let $f : [a, b] \to \mathbb{R}$ be continuous on $[a, b]$ and differentiable on (a, b). If $f(a) = f(b)$, then there exists c in (a, b) satisfying $f'(c) = 0$.*

As we said, the proof is not difficult. Because f is continuous on the compact set $[a, b]$, it is bounded and attains its extrema. We may of course assume that f is not constant, so at least one extremum is not attained at a or b (here we use the fact that $f(a) = f(b)$). Thus there exists $c \in (a, b)$ a point of global maximum or minimum for f. By Fermat's theorem, f' vanishes at this point. $\square$

As a consequence, we obtain the following:

Theorem (Lagrange's mean value theorem). *Let $f : [a, b] \to \mathbb{R}$ be continuous on $[a, b]$ and differentiable on (a, b). Then there exists c in (a, b) satisfying the equality*

$$f(b) - f(a) = (b - a)f'(c).$$

The proof is immediate using Rolle's theorem: the function

$$h(x) = (b - a)(f(x) - f(a)) - (f(b) - f(a))(x - a)$$

© Springer Science+Business Media LLC 2017
T. Andreescu et al., *Mathematical Bridges*, DOI 10.1007/978-0-8176-4629-5_14

satisfies the conditions of Rolle's theorem; thus its derivative vanishes at least at one point, which is equivalent to the statement of the theorem. □
In exactly the same way, by considering the auxiliary function

$$h(x) = (g(b) - g(a))(f(x) - f(a)) - (f(b) - f(a))(g(x) - g(a)),$$

one can prove Cauchy's theorem as an immediate consequence of Rolle's theorem.

Theorem (Cauchy's mean value theorem). *Let* $f, g : [a, b] \to \mathbb{R}$ *be both continuous on* $[a, b]$ *and differentiable on* (a, b). *Assume also that the derivative of* g *is never zero on* (a, b). *Then* $g(b) - g(a) \neq 0$ *and there exists* c *in* (a, b) *such that*

$$\frac{f(b) - f(a)}{g(b) - g(a)} = \frac{f'(c)}{g'(c)}.$$

Note that Lagrange's theorem appears now to be a particular case of this last theorem, and also note that it proves the assertions from the beginning of the chapter. Yet another important result is the following.

Theorem (Taylor's formula). *Let* $f : I \to \mathbb{R}$ *be an* $n + 1$ *times differentiable function on the interval* I. *Then, for every* x_0 *and* x *in* I, *there exists* $c \in I$ *(depending on* x_0 *and* x *and between* x_0 *and* x*) such that*

$$f(x) = \sum_{k=0}^{n} \frac{f^{(k)}(x_0)}{k!}(x - x_0)^k + \frac{f^{(n+1)}(c)}{(n+1)!}(x - x_0)^{n+1}.$$

Proof. Let F and G be defined by

$$F(t) = f(t) - \sum_{k=0}^{n} \frac{f^{(k)}(x_0)}{k!}(t - x_0)^k \quad \text{and} \quad G(t) = \frac{(t - x_0)^{n+1}}{(n+1)!}$$

for all $t \in I$. One immediately sees that F and G are $n + 1$ times differentiable and that $F^{(k)}(x_0) = G^{(k)}(x_0) = 0$ for $0 \leq k \leq n$, while $F^{(n+1)}(t) = f^{(n+1)}(t)$ and $G^{(n+1)}(t) = 1$ for all $t \in I$. Cauchy's theorem applies to F and G on the interval from x_0 to x and yields the existence of some $c_1 \in I$ with property

$$\frac{F(x)}{G(x)} = \frac{F(x) - F(x_0)}{G(x) - G(x_0)} = \frac{F'(c_1)}{G'(c_1)}.$$

By Cauchy's theorem again (for F' and G' on the interval from x_0 to c_1), we obtain some $c_2 \in I$ such that

$$\frac{F'(c_1)}{G'(c_1)} = \frac{F'(c_1) - F'(x_0)}{G'(c_1) - G'(x_0)} = \frac{F''(c_2)}{G''(c_2)}.$$

We continue this procedure based on Cauchy's theorem until we get

$$\frac{F(x)}{G(x)} = \frac{F'(c_1)}{G'(c_1)} = \frac{F''(c_2)}{G''(c_2)} = \cdots = \frac{F^{(n+1)}(c_{n+1})}{G^{(n+1)}(c_{n+1})}$$

for some $c_1, c_2, \ldots, c_{n+1} \in I$ (actually between x_0 and x). Now, with $c = c_{n+1}$, the equality

$$\frac{F(x)}{G(x)} = \frac{F^{(n+1)}(c)}{G^{(n+1)}(c)} = f^{(n+1)}(c)$$

is precisely what we wanted to prove. $\Box$

So, we can deduce from Lagrange's theorem the relationship between the derivative's sign and the monotonicity of f. Moreover, the second derivative f'' is closely related to the notion of convexity. Indeed, f is convex on intervals where f'' is nonnegative and is concave on the intervals where f'' is negative. Recall that f is called *convex* if for all $t \in [0, 1]$ and $x, y \in [a, b]$, the following inequality holds

$$f((1-t)x + ty) \leq (1-t)f(x) + tf(y),$$

and it is *concave* if the reversed inequality holds for all $t \in [0, 1]$ and $x, y \in [a, b]$. Any convex function satisfies Jensen's inequality: for each positive integer n, for all $\alpha_1, \ldots, \alpha_n \in [0, 1]$ with sum 1, and for every $x_1, \ldots, x_n \in [a, b]$,

$$f(\alpha_1 x_1 + \cdots + \alpha_n x_n) \leq \alpha_1 f(x_1) + \cdots + \alpha_n f(x_n).$$

In particular,

$$f\left(\frac{x_1 + \cdots + x_n}{n}\right) \leq \frac{f(x_1) + \cdots + f(x_n)}{n},$$

for every $x_1, \ldots, x_n \in [a, b]$. When the function is concave, each of the above inequalities is reversed.

Let us see now how we can use the previous results in a few concrete problems.

Problem. Let $a_1, a_2, \ldots, a_n$ be positive real numbers. Prove that the inequality

$$a_1^x + a_2^x + \cdots + a_n^x \geq n,$$

holds for all real numbers x if and only if $a_1 a_2 \cdots a_n = 1$.

Solution. If the product of all the a_i is 1, we have, using the AM-GM inequality,

$$a_1^x + a_2^x + \cdots + a_n^x \geq n\sqrt[n]{a_1^x a_2^x \cdots a_n^x} = n\sqrt[n]{(a_1 a_2 \cdots a_n)^x} = n.$$

For the converse, we give two approaches.

Method I. Let us define the function $f : \mathbb{R} \to \mathbb{R}$, given by

$$f(x) = a_1^x + a_2^x + \cdots + a_n^x - n.$$

The given inequality can be written as $f(x) \geq f(0)$, so 0 is a point of minimum. According to Fermat's theorem, $f'(0) = 0$. But

$$f'(x) = a_1^x \ln a_1 + a_2^x \ln a_2 + \cdots + a_n^x \ln a_n,$$

thus $\ln a_1 + \ln a_2 + \cdots + \ln a_n = 0$, which is exactly the desired condition.

Method II. One can easily establish that

$$\lim_{x \to 0} \left(\frac{a_1^x + a_2^x + \cdots + a_n^x}{n} \right)^{1/x} = \sqrt[n]{a_1 a_2 \cdots a_n}.$$

For $x > 0$, we have

$$\left(\frac{a_1^x + a_2^x + \cdots + a_n^x}{n} \right)^{1/x} \geq 1 \tag{14.1}$$

and for $x < 0$,

$$\left(\frac{a_1^x + a_2^x + \cdots + a_n^x}{n} \right)^{1/x} \leq 1. \tag{14.2}$$

Now, if $x \searrow 0$ in (14.1) and $x \nearrow 0$ in (14.1), then we deduce that $\sqrt[n]{a_1 a_2 \cdots a_n} \geq 1$ and $\sqrt[n]{a_1 a_2 \cdots a_n} \leq 1$, *respectively*. In conclusion, $a_1 a_2 \cdots a_n = 1$. $\square$

Problem. Solve in $\mathbb{R}$ the equation $4^x - 3^x = x$.

Solution. For each fixed real number x, define the function $f : [3, 4] \to \mathbb{R}$, by $f(y) = y^x$, with $f'(y) = xy^{x-1}$. According to Lagrange's mean value theorem,

$$f(4) - f(3) = (4 - 3)f'(c) \Leftrightarrow 4^x - 3^x = xc^{x-1},$$

for some $c \in (3, 4)$. The given equation can be written as $xc^{x-1} = x$. One solution is $x = 0$. For $x \neq 0$, we derive $c^{x-1} = 1 \Leftrightarrow x = 1$. In conclusion, there are two solutions: $x = 0$ and $x = 1$. $\square$

Lagrange's theorem also has important applications in transcendental number theory. We present here Liouville's famous theorem, which shows that algebraic irrational numbers are badly approximable with rational numbers. This is the very first nontrivial result in this active field of research:

Theorem (Liouville). *Let P be a nonzero polynomial with integer coefficients and degree n. Suppose that a is an irrational zero of P. Then there exists a constant C > 0 such that for all rational numbers $\frac{p}{q}$ with $p, q \in \mathbb{Z}$, and $q \neq 0$, we have*

$$\left| a - \frac{p}{q} \right| \geq \frac{C}{|q|^n}.$$

Solution. It is clear that there exists $\delta > 0$ such that the only zero of f in $I = [a - \delta, a + \delta]$ is a. Because P' is continuous on I, it is bounded, and we can pick M such that $M \geq |P'(x)|$ for all $x \in I$. Using Lagrange's mean value theorem, we deduce that $\left| a - \frac{p}{q} \right| \geq \frac{1}{M} \left| P\left(\frac{p}{q}\right) \right|$ for all $x = \frac{p}{q} \in I$. Because P has integer coefficients, the number $|q|^n \left| P\left(\frac{p}{q}\right) \right|$ is a nonzero integer; thus it is at least equal to 1. Thus for all $\frac{p}{q} \in I$, we have $|a - x| \geq \frac{1}{M|q|^n}$. If x does not belong to I, then $|x - a| \geq \delta$. Thus by taking $C = \min\left(\delta, \frac{1}{M}\right)$, we always have $\left| a - \frac{p}{q} \right| \geq \frac{C}{|q|^n}$. □

The following result is often useful for studying the behavior of a function when we know some of its zeros and the behavior of its derivatives. One can also see it as a factorization result, similar to the well-known result for polynomials.

Problem. Let f be a n times continuously differentiable function defined on $[0, 1]$, with real values, and let $a_1, a_2, \ldots, a_n$ be n distinct real numbers in $[0, 1]$ such that $f(a_1) = f(a_2) = \cdots = f(a_n) = 0$. Then for all $x \in (0, 1)$, there exists $c \in (0, 1)$ such that

$$f(x) = \frac{(x - a_1)(x - a_2) \cdots (x - a_n)}{n!} f^{(n)}(c).$$

Solution. The case when x is one of $a_1, a_2, \ldots, a_n$ being immediate, let us assume that $x, a_1, a_2, \ldots, a_n$ are pairwise distinct. Consider the function

$$g(t) = (t - a_1)(t - a_2) \cdots (t - a_n) f(x) - (x - a_1)(x - a_2) \cdots (x - a_n) f(t),$$

which is n times differentiable and satisfies $g(x) = g(a_1) = g(a_2) = \cdots = g(a_n) = 0$. Let $b_0 < b_1 < \cdots < b_n$ be the numbers $x, a_1, \ldots, a_n$ when arranged in increasing order. By Rolle's theorem, g' has at least n zeros, one in each (b_i, b_{i+1}). Applying again Rolle's theorem and repeating the argument, we deduce that $g^{(n)}$ vanishes at a point $c \in (0, 1)$. Thus

$$n! f(x) = (x - a_1)(x - a_2) \cdots (x - a_n) f^{(n)}(c),$$

which is exactly what we needed. □

We continue with a particularly nice and surprising property of derivatives.

Theorem (Darboux). *The derivative of a differentiable function has the intermediate value property.*

Proof I. The proof is not difficult at all. Take f a differentiable function and suppose that $a < b$ and $f'(a) < f'(b)$, for instance. Consider $y \in (f'(a), f'(b))$. We need to prove that there exists $x \in (a, b)$ such that $y = f'(x)$. Let $g(t) = f(t) - yt$. Then $g'(a) < 0$; thus there exists $a' > a$ such that $g(a') < g(a)$. Similarly, there exists $b' < b$ such that $g(b') < g(b)$. Because g is continuous on $[a, b]$, it attains its minimum at a point x. The previous remarks show that $x \in (a, b)$. Thus by Fermat's theorem, we have $g'(x) = 0$, that is, $y = f'(x)$. $\square$

Proof II. (See the note *A New Proof of Darboux's Theorem* by Lars Olsen in *The American Mathematical Monthly*, 8/2004). Let (for $c \in [a, b]$) h_c be the function defined for all $x \in [a, b]$ by

$$h_c(x) = \begin{cases} \frac{f(x)-f(c)}{x-c}, & \text{for } x \neq c \\ f'(c), & \text{for } x = c. \end{cases}$$

Clearly, any h_c is continuous on the interval $[a, b]$, and we have $h_a(a) = f'(a)$, $h_b(b) = f'(b)$, and $h_a(b) = h_b(a)$. Thus, if y is between $f'(a)$ and $f'(b)$, it is also either between $h_a(a)$ and $h_a(b)$ or between $h_b(a)$ and $h_b(b)$. Suppose we are in the first case. By the continuity of h_a, we can find c between a and b such that $y = h_a(c) = (f(c) - f(a))/(c - a)$. But Lagrange's mean value theorem ensures the existence of some x between a and c such that $(f(c) - f(a))/(c - a) = f'(x)$; thus $y = f'(x)$, and we are done.

When y is between $h_b(a)$ and $h_b(b)$, we proceed analogously, by using the fact that h_b, being continuous, has the intermediate value property. $\square$

We now present a very useful criterion for proving that a function is continuously differentiable k times. It is also a consequence of Lagrange's theorem:

Proposition. *Let I be a nontrivial interval of the set of real numbers, $x_0 \in I$ and $f : I \to \mathbb{R}$ continuous on I, differentiable at all points of I except x_0 and such that $f'(x)$ has a finite limit when x tends to x_0. Then $f'(x_0)$ exists, and it is equal to this limit.*

Proof. Let $\delta, \epsilon > 0$ be such that $|f'(x) - l| < \epsilon$ for all $x \in V = (x_0 - \delta, x_0 + \delta) \cap I$, $x \neq x_0$. From Lagrange's theorem, for all $x \in V$, $x > x_0$, we have the existence of some $c(x)$ between x and x_0 such that $\frac{f(x)-f(x_0)}{x-x_0} = f'(c(x))$. Therefore

$$\left| l - \frac{f(x) - f(x_0)}{x - x_0} \right| < \epsilon.$$

A similar argument shows that the last relation actually holds for all $x \in V$, $x \neq x_0$. This shows that $f'(x_0) = l$. $\square$

Problem. Let $x_0 \in \mathbb{R}$ and $\epsilon > 0$. Prove that there exists an infinitely many times differentiable function $f : \mathbb{R} \to \mathbb{R}$, not identically zero and which vanishes outside $(x_0 - \epsilon, x_0 + \epsilon)$.

Solution. First of all, consider the function $g(x) = e^{-\frac{1}{x}}$ for $x > 0$ and 0 otherwise. We claim that this function is infinitely many times differentiable on $\mathbb{R}$. This is clear at all points except 0. Using the above proposition, it is enough to prove that $g^{(k)}$ has a limit equal to 0 at 0 for all k. An inductive argument combined with the proposition would finish the proof. But it is very easy to prove by induction that $g^{(k)}(x)$ equals 0 for $x < 0$ and $P_k(\frac{1}{x})e^{-\frac{1}{x}}$ if $x > 0$, where P_k is a polynomial with real coefficients and degree $2k$. This shows that $g^{(k)}(x)$ tends to 0 as x tends to 0 (from the right), and so g is indeed infinitely many times differentiable at 0 and thus everywhere differentiable. Now, consider the function

$$f(x) = g(\epsilon^2 - (x - x_0)^2).$$

This is also an infinitely many times differentiable function and, clearly, it is not identically zero. However, $f(x) > 0$ implies $\epsilon^2 > (x-x_0)^2$, that is, $x \in (x_0-\epsilon, x_0+\epsilon)$. This shows that this function is a solution of the problem. $\square$

We continue with some problems taken from the Putnam Mathematical Competition.

Problem. Let $f, g : \mathbb{R} \to \mathbb{R}$ with f twice differentiable $g(x) \geq 0$ for all x and

$$f''(x) + f(x) = -xg(x)f'(x).$$

Prove that f is bounded.

Solution. Observe that $f'(x)f''(x) + f(x)f'(x) = -xg(x)(f'(x))^2$ is negative for positive x and positive for negative x. On the other hand, $f'f'' + ff' = \frac{1}{2}(f^2 + f'^2)'$, so $f^2 + f'^2$ has a maximum at 0. Thus $|f(x)| \leq \sqrt{f^2(0) + f'^2(0)}$ for all x and the conclusion follows. $\square$

Problem. Let $f : \mathbb{R} \to [-1, 1]$ twice differentiable such that $f(0)^2 + f'(0)^2 = 4$. Prove that there exists x_0 such that $f(x_0) + f''(x_0) = 0$.

Solution. Let $g(x) = f(x)^2 + f'(x)^2$. By Lagrange's theorem there is $a \in (0, 2)$ with $f'(a) = \dfrac{f(2) - f(0)}{2}$. Clearly, $|f'(a)| \leq 1$, thus $g(a) \leq 2$. Working in the same way with $(-2, 0)$, we find $b \in (-2, 0)$ with $g(b) \leq 2$. Since $g(0) = 4$, the maximum of g on $[-2, 2]$ appears at some $c \in (-2, 2)$. Because $g(c) \geq g(0) = 4, f'(c) \neq 0$. But $g'(c) = 0$, thus $f(c) + f''(c) = 0$. $\square$

Problem. How many real solutions does the equation $2^x = 1 + x^2$ have?

Solution. Any such root is nonnegative since $2^x \geq 1$. If $f(x) = 2^x - x^2 - 1$ then $f(0) = 0, f(1) = 0, f(4) < 0$ and $f(5) > 0$, so f has at least three zeros. If f has at least four zeros, by Rolle's theorem f'' would have at least two zeros, which is clearly wrong because $f''(x) = (\ln 2)^2 \cdot 2^x - 2$ is injective. Hence, the equation has three real solutions. $\square$

Proposed Problems

1. Solve in $[1, \infty)$ the equation $9^x - 8^x = x^2$.
2. Solve in $\mathbb{R}$ the system $\begin{cases} 2^x + 2^y = 16 \\ 3^x + 3^y = 54 \end{cases}$.
3. Find all positive real numbers a, such that the inequality

$$a^x + 10^x \geq 5^x + 6^x$$

   holds for every real number x.
4. Let $f : \mathbb{R} \to \mathbb{R}$ be a twice differentiable function whose graph meets the line $y = x$ three times. Prove that $f''(\xi) = 0$, for at least one real value ξ.
5. Let a, b, c, m, n, p be real numbers. Prove that

$$a \sin x + b \sin 2x + c \sin 3x + m \cos x + n \cos 2x + p \cos 3x = 0$$

   for at least one real value x.
6. Prove that the function $f : \mathbb{R} \to \mathbb{R}$, given by $f(x) = \sin x + \sin(x^2)$ is not periodic.
7. Let $f, g : [a, b] \to \mathbb{R}$ be differentiable, with continuous, positive, increasing derivatives. Prove that there exists c such that

$$\frac{f(b) - f(a)}{b - a} \cdot \frac{g(b) - g(a)}{b - a} = f'(c)g'(c).$$

8. Let $f : \mathbb{R} \to \mathbb{R}$ be a differentiable, bounded function. Prove that $f(c)f'(c) = c$, for some real number c.
9. Let $f : \mathbb{R} \to \mathbb{R}$ be a convex function so that

$$\lim_{x \to \infty} (f(x) - x - 1) = 0.$$

   Prove that $f(x) \geq x + 1$, for all real numbers x.
10. Let $f : \mathbb{R} \to \mathbb{R}$ be a concave, strictly monotone function. Prove that f is unbounded from below.
11. Let $f : \mathbb{R} \to \mathbb{R}$ be a three times differentiable function. Prove that there exists a real number c such that

$$f(c)f'(c)f''(c)f'''(c) \geq 0.$$

12. Let $f : \mathbb{R} \to \mathbb{R}$ be a function with fourth derivative such that $f(x) \geq 0$, $f^{(4)}(x) \leq 0$, for all real numbers x. Prove that there exist real numbers a, b, c such that $f(x) = ax^2 + bx + c$, for all real x.

13. Let $f : \mathbb{R} \to \mathbb{R}$ be a function with continuous third derivative such that $f(x), f'(x), f''(x), f'''(x)$ are positive and $f'''(x) \leq f(x)$, for all real numbers x. Prove that $(f''(0))^2 < 2f(0)f'(0)$.

14. Prove the following equality: $2 \arctan \pi + \arcsin \dfrac{2\pi}{1 + \pi^2} = \pi$.

15. Let f be a differentiable function defined on the set of real numbers, with values in the same set. If f has infinitely many zeros, prove that its derivative also has infinitely many zeros.

16. Determine all differentiable functions $f : \mathbb{R} \to \mathbb{R}$ with the following two properties:

   (i) $f'(x) = 0$ for every integer x.
   (ii) If $x \in \mathbb{R}$ and $f'(x) = 0$, then $f(x) = 0$.

17. Let $a_1, \ldots, a_n$ be real numbers, each greater than 1. If $n \geq 2$, prove that there is exactly one solution in the interval $(0, 1)$ to

$$\prod_{j=1}^{n}(1 - x^{a_j}) = 1 - x.$$

18. Let a, b, c, d be positive real numbers. Prove that

$$\frac{a}{b+c} + \frac{b}{c+d} + \frac{c}{d+a} + \frac{d}{a+b} \geq 2.$$

19. Prove that for all real t, and all $\alpha \geq 2$,

$$e^{\alpha t} + e^{-\alpha t} - 2 \leq (e^t + e^{-t})^\alpha - 2^\alpha.$$

20. Let $a > 1$ be a real number. Prove that

$$\sum_{k=1}^{\infty} \frac{1}{(n+k)^a} > \frac{1}{(n+1)\left((n+1)^{a-1} - n^{a-1}\right)},$$

   for any positive integer n.

21. Prove that the inequality

$$\frac{a}{a^3 + 1} + \frac{b}{b^3 + 1} + \frac{c}{c^3 + 1} \leq \frac{3}{2}$$

holds for all positive real numbers a, b, and c for which $abc = 1$.

Solutions

1. The solution is $x = 1$. Let us define the function $f : [8, 9] \to \mathbb{R}$, by $f(y) = y^x$, where $x > 0$ is arbitrary and fixed. There exists $c \in (8, 9)$ such that

$$f(9) - f(8) = f'(c)(9 - 8) \Leftrightarrow 9^x - 8^x = xc^{x-1}.$$

Now, the given equation can be written as $xc^{x-1} = x^2$. We have $x \neq 0$; thus $c^{x-1} = x$. For $x > 1$, we have

$$c^{x-1} \geq e^{x-1} > x.$$

2. The solution is $x = y = 3$. Let us define

$$f(x) = 2^x, \quad g(x) = 3^x, \quad x \in \mathbb{R}.$$

If $x < 3$, then $y > 3$, and the system is equivalent to

$$\begin{cases} f(x) + f(y) = 2f(3) \\ g(x) + g(y) = 2g(3) \end{cases} \Leftrightarrow \begin{cases} f(y) - f(3) = f(3) - f(x) \\ g(y) - g(3) = g(3) - g(x) \end{cases},$$

so

$$\frac{f(y) - f(3)}{g(y) - g(3)} = \frac{f(3) - f(x)}{g(3) - g(x)},$$

which is impossible. Indeed, by Cauchy's theorem,

$$\frac{f(y) - f(3)}{g(y) - g(3)} = \frac{f'(c_1)}{g'(c_1)}, \quad \text{and} \quad \frac{f(3) - f(x)}{g(3) - g(x)} = \frac{f'(c_2)}{g'(c_2)}$$

for some $c_1 \in (3, y)$ and $c_2 \in (x, 3)$. The function $x \mapsto \dfrac{f'(x)}{g'(x)}$ is injective, so the equality

$$\frac{f'(c_1)}{g'(c_1)} = \frac{f'(c_2)}{g'(c_2)}$$

cannot be true.

3. The answer is $a = 3$. Let $f : \mathbb{R} \to \mathbb{R}$, given by

$$f(x) = a^x + 10^x - 5^x - 6^x.$$

According to the hypothesis, f is nonnegative, which is equivalent to $f(x) \geq f(0)$. Hence $f'(0) = 0$. We have

$$f'(x) = a^x \ln a + 10^x \ln 10 - 5^x \ln 5 - 6^x \ln 6,$$

so $f'(0) = \ln\dfrac{a}{3} = 0 \Rightarrow a = 3$. If $a = 3$, the given inequality is

$$(2^x - 1)(5^x - 3^x) \geq 0,$$

which is true.

4. Assume that the function $g(x) = f(x) - x$, $x \in \mathbb{R}$, vanishes at $a, b, c, a < b < c$. According to Rolle's theorem,

$$g(a) = g(b) = 0 \Rightarrow g'(\xi_1) = 0 \Rightarrow f'(\xi_1) = 1$$

$$g(b) = g(c) = 0 \Rightarrow g'(\xi_2) = 0 \Rightarrow f'(\xi_2) = 1,$$

with $\xi_1 \in (a, b)$, $\xi_2 \in (b, c)$. Further,

$$f'(\xi_1) = f'(\xi_2) \Rightarrow f''(\xi) = 0,$$

for some $\xi \in (\xi_1, \xi_2)$.

5. Let us define the function $f : \mathbb{R} \to \mathbb{R}$, given by

$$f(x) = -a\cos x - \frac{b}{2}\cos 2x - \frac{c}{3}\cos 3x + m\sin x + \frac{n}{2}\sin 2x + \frac{p}{3}\sin 3x.$$

Obviously,

$$f(0) = f(2\pi) = -a - \frac{b}{2} - \frac{c}{3},$$

so we can find $\xi \in (0, 2\pi)$ so that

$$f'(\xi) = 0 \Leftrightarrow a\sin\xi + b\sin 2\xi + c\sin 3\xi + m\cos\xi + n\cos 2\xi + p\cos 3\xi = 0.$$

6. If we assume by contradiction that $f(x) = \sin x + \sin(x^2)$ is periodic, then its derivative must be also periodic. But then

$$f'(x) = \cos x + 2x\cos(x^2)$$

would be continuous and periodic; hence f' must be bounded. This is a contradiction, because for all positive integers n, we have

$$f'\left(\sqrt{2n\pi}\right) = \cos\sqrt{2n\pi} + 2\sqrt{2n\pi} \to \infty \quad \text{as} \quad n \to \infty.$$

7. Let $c_1, c_2 \in (a, b)$, $c_1 \leq c_2$ be such that

$$\frac{f(b) - f(a)}{b - a} = f'(c_1), \qquad \frac{g(b) - g(a)}{b - a} = g'(c_2).$$

Consider the function $F : [0, 1] \rightarrow \mathbb{R}$ given by

$$F(t) = f'(t)g'(t) - f'(c_1)g'(c_2).$$

Then $F(c_1) \leq 0$, $F(c_2) \geq 0$, so there exists $c \in [0, 1]$ such that $F(c) = 0$. That is,

$$f'(c)g'(c) = f'(c_1)g'(c_2) = \frac{f(b) - f(a)}{b - a} \cdot \frac{g(b) - g(a)}{b - a}.$$

8. Define the function $g(x) = f^2(x) - x^2$, $x \in \mathbb{R}$. Because of boundedness of f, we have

$$\lim_{x \to -\infty} g(x) = \lim_{x \to \infty} g(x) = -\infty. \tag{14.1}$$

If $g'(x) \neq 0$, for all reals x, then g is strictly monotone, which contradicts (14.1). Thus there is c such that

$$g'(c) = 0 \Leftrightarrow 2f(c)f'(c) - 2c = 0 \Leftrightarrow f(c)f'(c) = c.$$

9. Let us suppose that $f(a) < a + 1$, for some a. Because

$$\lim_{x \to \infty} (f(x) - x - 1) = 0,$$

we can find $b > a$ such that

$$m = \frac{f(b) - f(a)}{b - a} > 1.$$

From the convexity of f, we deduce that the function $\phi : [b, \infty) \rightarrow \mathbb{R}$,

$$\phi(x) = \frac{f(x) - f(a)}{x - a}$$

is increasing, so

$$\phi(x) \geq \phi(b) \Rightarrow \frac{f(x) - f(a)}{x - a} \geq m$$

$$\Rightarrow f(x) \geq m(x - a) + f(a)$$

$$\Rightarrow f(x) - x - 1 \geq (m - 1)x + f(a) - ma - 1.$$

Now, if $x \to \infty$, we obtain $0 \geq \infty$, which is false.

10. Suppose, by way of contradiction, that $f > 0$ and that f is decreasing. If $a, b \in \mathbb{R}$, $a < b$, let $M(x_0, 0)$ be the intersection point of the line joining $(a, f(a))$ and $(b, f(b))$ with the x-axis. Then, from the concavity of f, it follows that $f(x) \leq 0$, for all $x \geq x_0$, a contradiction.

11. Let us suppose, by way of contradiction, that

$$f(x)f'(x)f''(x)f'''(x) < 0, \ \forall \, x \in \mathbb{R}.$$

In particular, f, f', f'', f''' do not vanish on $\mathbb{R}$, so they have constant sign. Assume that $f > 0$, so

$$f'(x)f''(x)f'''(x) < 0, \ \forall \, x \in \mathbb{R}.$$

If $f' < 0, f'' < 0$ and $f''' < 0$, then f is strictly decreasing, concave, and bounded from below, which is a contradiction. If $f' < 0, f'' > 0, f''' > 0$, then $g = -f'$ is strictly decreasing, concave, and $g > 0$, which is impossible. The case $f' > 0$ is proved analogously to be impossible.

Alternatively, one can use Taylor's theorem (with the remainder in Lagrange's form); thus we have that, for all real x and t, there exist $c_{x,t}$ such that

$$f(x) = f(t) + \frac{f'(t)}{1!}(x - t) + \frac{f''(c_{x,t})}{2!}(x - t)^2,$$

therefore, $f(x) > 0$ and $f''(x) < 0$ together lead to the contradiction

$$0 < f(t) + \frac{f'(t)}{1!}(x - t)$$

for all x and t. Indeed, for fixed (but arbitrary) t, this is possible for all x only if $f'(t) = 0$. (See also the following problem for such a reasoning.) But if f' is identically zero, the same holds for f'', which is not the case. Similarly, $f(x) < 0$ and $f''(x) > 0$ for all x are incompatible. Thus, if f has the same sign on $\mathbb{R}$, and f' also has constant sign on $\mathbb{R}$, then necessarily $f(x)f''(x) > 0$ for all $x \in \mathbb{R}$. Of course, this also holds for f', that is, our assumption leads to $f'(x)f'''(x) > 0$ for all $x \in \mathbb{R}$ and the contradiction follows.

This is problem A3 from the 59th William Lowell Putnam Mathematical Competition, 1998. A generalization (for a $4k + 3$ times differentiable function) appeared as problem 11472, proposed by Mahdi Makhul in *The American Mathematical Monthly*.

12. By Taylor's theorem, for all real numbers x and t, there exists $c_{x,t}$ between x and t such that

$$f(x) = f(t) + \frac{f'(t)}{1!}(x - t) + \frac{f''(t)}{2!}(x - t)^2 + \frac{f^{(3)}(t)}{3!}(x - t)^3 + \frac{f^{(4)}(c_{x,t})}{4!}(x - t)^4.$$

Now, using the hypothesis, we have

$$0 \le f(x) \le f(t) + \frac{f'(t)}{1!}(x-t) + \frac{f''(t)}{2!}(x-t)^2 + \frac{f^{(3)}(t)}{3!}(x-t)^3$$

for all x and t. Considering t fixed, the right-hand side is an at most third-degree polynomial in x that must have only nonnegative values. Clearly, this can happen only if the coefficient of x^3 is 0 (i.e., the polynomial's degree must be even). Thus we must have $f^{(3)}(t) = 0$ for any real number t—and this evidently means that f is either a second-degree polynomial or a nonnegative constant function.

Of course, a similar statement (with an identical proof) holds for a real function that has a $2n$-th order derivative (with n a positive integer), namely, if $f(x) \ge 0$ and $f^{(2n)}(x) \le 0$ for all $x \in \mathbb{R}$, then f is a polynomial function of even degree which is at most $2n - 2$.

13. Let us define the function $\phi : (-\infty, 0] \to \mathbb{R}$ by the formula

$$\phi(x) = \frac{f(0)}{2} \cdot x^2 + xf''(0) + f(0) - f'(x),$$

for every real number x. We have

$$\phi'(x) = xf(0) + f''(0) - f''(x)$$

and $\phi''(x) = f(0) - f'''(x)$. The function f is increasing, because $f' > 0$. Then for every $x < 0$, using the hypothesis again, we obtain $f'''(x) \le f(x) < f(0)$ so $\phi'' > 0$. Hence ϕ' is increasing and for $x < 0, \phi'(x) < \phi'(0) = 0$. Further, ϕ is decreasing, so for all $x < 0$,

$$\phi(x) > \phi(0) = f(0) - f'(0).$$

Thus

$$g(x) = \frac{f(0)}{2} \cdot x^2 + xf''(0) + f'(0) > f'(x) > 0,$$

for all $x \in (-\infty, 0)$. Because $f(0), f'(0), f''(0)$ are positive, it follows that for all nonnegative x,

$$g(x) = \frac{f(0)}{2} \cdot x^2 + xf''(0) + f'(0) > 0.$$

Hence this inequality holds for all real x, so the discriminant of $g(x)$ is negative:

$$(f''(0))^2 - 2f(0)f'(0) < 0,$$

which is what we wanted to prove.

14. Define the function

$$f(x) = 2\arctan x + \arcsin \frac{2x}{1+x^2}, \quad x > 1,$$

with

$$f'(x) = \frac{2}{1+x^2} + \frac{1}{\sqrt{1 - \left(\frac{2x}{1+x^2}\right)^2}} \cdot \left(\frac{2x}{1+x^2}\right)' = 0;$$

hence f is constant on $(1, \infty)$ (try showing this as an exercise!). Thus

$$f(\pi) = \lim_{x \to \infty} f(x) = \pi.$$

15. Let $(a_n)_{n \geq 1}$ be a sequence of distinct zeros of f. One may say that the problem is obvious, because by Rolle's theorem, f' vanishes in all intervals (a_n, a_{n+1}). However, we must be careful, because there is no reason for these zeros to be increasingly ordered. The idea is that any infinite sequence has a monotone subsequence. So, by working with a subsequence of $(a_n)_{n \geq 1}$, we may assume that the sequence is monotone, say increasing. Then Rolle's theorem says that f' has a zero inside each of the intervals (a_n, a_{n+1}); thus it has infinitely many zeros.

16. Clearly the identically 0 function satisfies both conditions from the problem statement. We prove that this is the only solution of the problem.

First note that, by (i) and (ii), we need to have $f(x) = f'(x) = 0$ for all $x \in \mathbb{Z}$. Suppose there is a $a \in \mathbb{R}$ with $f(a) \neq 0$. Of course, a is not an integer, thus $a \in (n, n+1)$, where $n = [a]$ (the integral part of a). Being continuous, f is bounded and attains its extrema on $[n, n+1]$; hence we can find $b, c \in [n, n+1]$ with $f(b) \leq f(x) \leq f(c)$ for all $x \in [n, n+1]$. If we have $f(a) > 0$, this implies $f(c) > 0$; thus c can be neither n nor $n+1$ (since f is 0 at these points), that is, $c \in (n, n+1)$. But in this case, Fermat's theorem tells us that $f'(c) = 0$, while property (ii) of f implies $f(c) = 0$, which contradicts $f(c) > 0$.

Similarly, $f(a) < 0$ leads to $f(b) < 0$, $b \in (n, n+1)$, $f'(b) = 0$ and finally to the contradiction $f(b) = 0$. Thus $f(a) \neq 0$ is impossible (for any a), and the only function that satisfies both conditions (i) and (ii) remains $f : \mathbb{R} \to \mathbb{R}$ given by $f(x) = 0$ for all x, as claimed.

This problem was proposed for the 11th grade in the Romanian National Mathematics Olympiad in 2015.

17. First we solve the problem in the case when all the numbers a_j are equal. We begin by proving the following:

Lemma. *Let $a > 1$ and $0 < b < 1$ be given real numbers. The equation*

$$x^a + (1-x)^b = 1$$

has exactly one solution in the interval $(0, 1)$.

An immediate consequence is that the equation

$$(1 - x^a)^n = 1 - x$$

has exactly one solution in the interval $(0, 1)$ for any real number a greater than 1 and any positive integer $n \geq 2$.

Proof. Consider the function $f : [0, 1] \to (0, \infty)$ defined by

$$f(x) = x^a + (1 - x)^b, \forall x \in [0, 1].$$

We have $f(0) = f(1) = 1$ and

$$f'(x) = ax^{a-1} - \frac{b}{(1 - x)^{1-b}}, \forall x \in [0, 1).$$

Now we see that the graphs of the functions

$$g(x) = ax^{a-1} \text{ and } h(x) = b/(1 - x)^{1-b}, \ x \in [0, 1)$$

either intersect at two points or they do not intersect at all, since g increases on this interval from 0 to a (its limit in 1) and it is convex or concave (as $a > 2$ and $a < 2$, respectively), while h increases from b to ∞ and it is convex (the convexity of the functions is easy to check by using their second derivatives). But if g and h do not meet, f' has the same sign (it is negative) on $[0, 1)$, as one can easily see, which means that f decreases on $[0, 1)$. Since f is continuous on the entire interval $[0, 1]$, this would lead to the conclusion that $f(0) > \lim_{x \to 1} f(x) = f(1)$, which is false; it remains the only possibility that the graphs of g and h intersect at two points $0 < \alpha < \beta < 1$ in the interval $(0, 1)$ and that f' is negative on $(0, \alpha)$ and $(\beta, 1)$ and positive on (α, β); therefore, f decreases from $f(0) = 1$ to $f(\alpha) < 1$, then it increases from $f(\alpha)$ to $f(\beta)$, and, finally, it decreases again, from $f(\beta)$ to $f(1) = 1$ (hence $f(\beta)$ must be greater than 1). Clearly, f takes the value 1 exactly once, in the interval (α, β) (apart from the endpoints 0 and 1 of the interval), which proves our first claim.

As for the second claim, this is, as we said above, just a simple consequence of the first (proved above) part of the lemma; indeed, the equation $(1 - x^a)^n = 1 - x$ is equivalent to $x^a + (1 - x)^{1/n} = 1$, and $1/n$ is positive and less than 1.

Thus the lemma solves the case when the numbers $a_1, \ldots, a_n$ from the problem statement are all equal; with its help, we can also prove the existence of a solution of the equation

$$\prod_{j=1}^{n}(1 - x^{a_j}) = 1 - x$$

in the interval $(0, 1)$. To do this, consider the function $F : (0, 1) \to \mathbb{R}$,

$$F(x) = \left(\prod_{j=1}^{n} (1 - x^{a_j}) \right) + x - 1, \forall x \in (0, 1).$$

and suppose that $a = \min_{1 \le j \le n} a_j$, $b = \max_{1 \le j \le n} a_j$. We may assume that $a < b$, since the case $a = b$ (that is when all the numbers are equal) has already been treated. Of course, we have $a > 1$ and $b > 1$. Because the above product has at least two factors (and all the factors are positive), we get the inequalities

$$(1 - x^a)^n + x - 1 < \left(\prod_{j=1}^{n} (1 - x^{a_j}) \right) + x - 1 < (1 - x^b)^n + x - 1, \forall x \in (0, 1).$$

According to the lemma, each of the equations $(1 - x^a)^n + x = 1$ and $(1 - x^b)^n + x = 1$ has precisely one solution in the interval $(0, 1)$: there are c and d in this interval such that $(1 - c^a)^n + c = 1$ and $(1 - d^b)^n + d = 1$. From the above inequalities (replacing x with c and d, respectively), it follows that $F(c) > 0$ and $F(d) < 0$; now a typical continuity argument allows the conclusion that F takes the value 0 between c and d, that is, in the interval $(0, 1)$; in other words, the equation from the problem statement has at least one solution in $(0, 1)$.

In order to show the uniqueness of this solution, we rewrite the equation as

$$\frac{1}{\prod\limits_{j=1}^{n} (1 - x^{a_j})} = \frac{1}{1 - x}$$

and observe that both functions $u, v : [0, 1) \to (0, \infty)$, defined as

$$u(x) = \frac{1}{\prod\limits_{j=1}^{n} (1 - x^{a_j})}$$

and

$$v(x) = \frac{1}{1 - x},$$

for all $x \in [0, 1)$, are strictly increasing, are strictly convex, and have the line $x = 1$ as a vertical asymptote. The graphs of both u and v start from the point with coordinates $(0, 1)$; therefore they must have at most one other common point, and this is exactly what we intended to prove.

We still have to prove the claimed properties of the functions u and v. There is no problem with their monotony, and, also, the convexity of v follows easily since its second derivative is

$$v''(x) = \frac{2}{(1-x)^3}, \forall x \in [0, 1).$$

The convexity of u is more complicated to establish.
We may write $u(x) = 1/w(x)$ for

$$w(x) = \prod_{j=1}^{n}(1 - x^{a_j});$$

hence

$$u''(x) = \frac{2(w'(x))^2 - w(x)w''(x)}{(w(x))^3}.$$

On the other hand, we have

$$\log w(x) = \sum_{j=1}^{n} \log(1 - x^{a_j}),$$

and, by differentiation, we get

$$\frac{w'(x)}{w(x)} = -\sum_{j=1}^{n} \frac{a_j x^{a_j-1}}{1 - x^{a_j}}.$$

Differentiating one more time yields

$$\frac{w(x)w''(x) - (w'(x))^2}{(w(x))^2} = -\sum_{j=1}^{n} \frac{a_j[(a_j - 1)x^{a_j-2} + x^{2a_j-2}]}{(1 - x^{a_j})^2},$$

and the inequality

$$w(x)w''(x) - (w'(x))^2 < 0$$

follows for any $x \in (0, 1)$. From this inequality, the positivity of the numerator of $u''(x)$ is a direct consequence; hence (since the denominator is also positive) $u''(x) > 0$ for all $x \in (0, 1)$—and the convexity of u—is established, completing the proof.

This is problem 11226, proposed by Franck Beaucoup and Tamás Erdélyi in *The American Mathematical Monthly* 6/2006. A simpler solution can be found in Chapter 11.

18. The function $f(x) = \dfrac{1}{x}$, $x > 0$, is convex and we can use Jensen's inequality. Assume, without loss of generality, that $a + b + c + d = 1$. Then

$$\frac{a}{b+c} + \frac{b}{c+d} + \frac{c}{d+a} + \frac{d}{a+b}$$

$$= af(b+c) + bf(c+d) + cf(d+a) + df(a+b)$$

$$\geq f(a(b+c) + b(c+d) + c(d+a) + d(a+b))$$

$$= \frac{1}{ab + 2ac + ad + bc + 2bd + cd} \geq 2,$$

because

$$2(ab + 2ac + ad + bc + 2bd + cd) \leq (a+b+c+d)^2 = 1$$

$$\Leftrightarrow (a-c)^2 + (b-d)^2 \geq 0,$$

which is true.

19. **Claim 1.** For all real numbers $\beta \geq 1$ and $y \geq 1$, the inequality

$$(y-1)(y+1)^\beta - y^{\beta+1} + 1 \geq 0$$

holds. For $0 < y \leq 1$, the reverse inequality is true.

Proof. The function $y \mapsto y^\beta$ is strictly convex on the interval $(0, \infty)$ for $\beta > 1$, and if $y \geq 1$, the numbers $1/y$ and $(y-1)/y$ are both nonnegative and sum to 1. Then Jensen's inequality yields

$$\frac{1}{y} \cdot 1^\beta + \frac{y-1}{y}(y+1)^\beta \geq \left(\frac{1}{y} \cdot 1 + \frac{y-1}{y}(y+1)\right)^\beta = y^\beta,$$

which can be rearranged to get the desired inequality. For $0 < y \leq 1$, we obtain, again with Jensen's inequality,

$$y \cdot y^\beta + (1-y)(y+1)^\beta \geq (y \cdot y + (1-y)(y+1))^\beta = 1.$$

In both cases, the inequality is strict except for $y = 1$.

Claim 2. For $\alpha \geq 2$, the function $f : (0, \infty) \to \mathbb{R}$, defined by

$$f(x) = \left(x + \frac{1}{x}\right)^\alpha - x^\alpha - \frac{1}{x^\alpha}, \quad \forall x \in (0, \infty),$$

attains its minimum value at $x = 1$.

Proof. The function f is differentiable and has the derivative

$$f'(x) = \frac{\alpha}{x^{\alpha+1}}[(x^2 - 1)(x^2 + 1)^{\alpha-1} - x^{2\alpha} + 1], \quad \forall x > 0.$$

For $y = x^2$ and $\beta = \alpha - 1 \geq 1$ in claim 1, one sees that the expression in the square brackets is positive for $x > 1$ and negative for $0 < x < 1$ (and 0 only for $x = 1$), respectively. It follows that f decreases in the interval $(0, 1]$ and increases on $[1, \infty)$, having thus (as claimed) an absolute minimum at 1.

This implies, of course, the inequality $f(x) \geq f(1)$ for all $x > 0$ or

$$\left(x + \frac{1}{x}\right)^\alpha - x^\alpha - \frac{1}{x^\alpha} \geq 2^\alpha - 2, \quad \forall x > 0.$$

Just put here $x = e^t$ (for any real t) and rearrange a bit to get precisely the inequality stated in the problem.

This is problem 11369 proposed by Donald Knuth in *The American Mathematical Monthly*, 6/2008. The interested reader will find another solution, a generalization, and other related inequalities in Grahame Bennett's paper *p-Free l^p Inequalities*, in *The American Mathematical Monthly*, 4/2010.

20. We want to prove that

$$\zeta(a) - \zeta_n(a) > \frac{1}{(n+1)\left((n+1)^{a-1} - n^{a-1}\right)}, \quad \forall n \geq 1,$$

where

$$\zeta_n(a) = \sum_{j=1}^{n} \frac{1}{j^a}$$

and $\zeta(a) = \lim_{n\to\infty} \zeta_n(a)$ is the Riemann zeta function. We can rewrite the desired inequality as

$$\zeta(a) > \zeta_n(a) + \frac{1}{(n+1)\left((n+1)^{a-1} - n^{a-1}\right)}, \quad \forall n \geq 1$$

and we can prove this if we show that the sequence on the right hand side tends to $\zeta(a)$ and is increasing. So denote

$$x_n(a) = \zeta_n(a) + \frac{1}{(n+1)\left((n+1)^{a-1} - n^{a-1}\right)}, \forall n \geq 1,$$

and notice first that $(x_n(a))_{n \geq 1}$ has the limit $\zeta(a)$, since

$$\lim_{n \to \infty} x_n(a) = \zeta(a) + \lim_{x \to 0} \frac{x^a}{(x+1)\left((x+1)^{a-1} - 1\right)}$$

$$= \zeta(a) + \lim_{x \to 0} \frac{x^{a-1}}{x+1} \frac{x}{(x+1)^{a-1} - 1} = \zeta(a).$$

Next

$$x_{n+1}(a) - x_n(a) = \frac{1}{(n+1)^a} + \frac{1}{(n+2)\left((n+2)^{a-1} - (n+1)^{a-1}\right)}$$

$$- \frac{1}{(n+1)\left((n+1)^{a-1} - n^{a-1}\right)},$$

for any $n \geq 1$; thus the inequality $x_{n+1}(a) > x_n(a)$ becomes

$$\frac{1}{(n+2)((n+2)^{a-1} - (n+1)^{a-1})} > \frac{1}{(n+1)((n+1)^{a-1} - n^{a-1})} - \frac{1}{(n+1)^a},$$

or

$$\frac{1}{(n+2)((n+2)^{a-1} - (n+1)^{a-1})} > \frac{n^{a-1}}{(n+1)^a((n+1)^{a-1} - n^{a-1})}, \forall n \geq 1.$$

After some small calculation, we find this is equivalent to

$$(n+1)^{2a-1} + n^{a-1}(n+1)^{a-1} > n^{a-1}(n+2)^a, \forall n \geq 1$$

and (after multiplication by $n+1$) to

$$(n+1)^{2a} + n^{a-1}(n+1)^a > n^a(n+2)^a + n^{a-1}(n+2)^a.$$

Finally, we need to prove that

$$\left(n^2 + 2n + 1\right)^a - \left(n^2 + 2n\right)^a > n^{a-1}\left((n+2)^a - (n+1)^a\right), \forall n \geq 1.$$

Lagrange's theorem for the function $x \mapsto x^a$, on the intervals $(n^2 + 2n, n^2 + 2n + 1)$ and $(n + 1, n + 2)$ yields the existence of some

$$c \in \left(n^2 + 2n, n^2 + 2n + 1\right), \quad \text{and} \quad d \in (n + 1, n + 2),$$

such that

$$\left(n^2 + 2n + 1\right)^a - \left(n^2 + 2n\right)^a = ac^{a-1},$$

and

$$(n + 2)^a - (n + 1)^a = ad^{a-1}.$$

Since $a - 1 > 0$, we have

$$\left(n^2 + 2n + 1\right)^a - \left(n^2 + 2n\right)^a > a \left(n^2 + 2n\right)^{a-1},$$

and

$$(n + 2)^a - (n + 1)^a < a \left(n + 2\right)^{a-1}.$$

Thus our inequality follows like this:

$$\left(n^2 + 2n + 1\right)^a - \left(n^2 + 2n\right)^a > a \left(n^2 + 2n\right)^{a-1}$$

$$= an^{a-1} \left(n + 2\right)^{a-1} > n^{a-1} \left((n + 2)^a - (n + 1)^a\right),$$

and the proof is now complete.

21. **Lemma 1.** We have

$$\frac{x}{x^3 + 1} + \frac{y}{y^3 + 1} \leq \frac{2\alpha}{\alpha^3 + 1}$$

for any $\alpha \in \left[1/\sqrt[3]{2}, 1\right]$ and any positive x, y such that $xy = \alpha^2$.

Proof. By replacing y with α^2/x, we get the equivalent form

$$\frac{x}{x^3 + 1} + \frac{\alpha^2 x^2}{x^3 + \alpha^6} \leq \frac{2\alpha}{\alpha^3 + 1}$$

of the inequality that we have to prove. After clearing the denominators and some more simple (but prosaic) calculations, this becomes:

$$2\alpha x^6 - (\alpha^5 + \alpha^2)x^5 - (\alpha^3 + 1)x^4 + 2(\alpha^7 + \alpha)x^3$$
$$- (\alpha^5 + \alpha^2)x^2 - (\alpha^9 + \alpha^6)x + 2\alpha^7 \geq 0.$$

As expected, the sixth-degree polynomial on the left-hand side is divisible by $(x - \alpha)^2$; thus we can rewrite this as

$$(x - \alpha)^2 (2\alpha x^4 + (3\alpha^2 - \alpha^5)x^3 - (2\alpha^6 - 3\alpha^3 + 1)x^2 + (3\alpha^4 - \alpha^7)x + 2\alpha^5) \geq 0.$$

Now, the hypothesis $\alpha \in \left[1/\sqrt[3]{2}, 1\right]$ implies $\alpha^3 \in [1/2, 1]$; thus

$$-(2\alpha^6 - 3\alpha^3 + 1) = (2\alpha^3 - 1)(1 - \alpha^3) \geq 0$$

and, also,

$$3\alpha^2 - \alpha^5 = 2\alpha^2 + \alpha^2(1 - \alpha^3) > 0 \text{ and } 3\alpha^4 - \alpha^7 = 2\alpha^4 + \alpha^4(1 - \alpha^3) > 0;$$

therefore

$$2\alpha x^4 + (3\alpha^2 - \alpha^5)x^3 - (2\alpha^6 - 3\alpha^3 + 1)x^2 + (3\alpha^4 - \alpha^7)x + 2\alpha^5$$

is positive, as a sum of nonnegative terms (some of them being even strictly positive), and the first lemma is proved.

Lemma 2. The inequality

$$\frac{2t^2}{t^3 + 1} + \frac{t^2}{t^6 + 1} \leq \frac{3}{2}$$

is true for any positive t.

Proof. After clearing the denominators and some further calculations, the inequality is equivalent to

$$3t^9 - 4t^8 + 3t^6 - 2t^5 + 3t^3 - 6t^2 + 3 \geq 0$$

$$\Leftrightarrow (t - 1)^2(3t^7 + 2t^6 + t^5 + 3t^4 + 3t^3 + 3t^2 + 6t + 3) \geq 0,$$

which is evident for $t > 0$.

Now let us solve our problem. Because we have $abc = 1$ (and $a, b, c > 0$), one of the three numbers has to be greater than (or equal to) 1. Suppose (without loss of generality, due to the symmetry) that this is c and consider two cases.

(i) First, assume that $c \in \left[1, \sqrt[3]{4}\right]$; then

$$\frac{1}{\sqrt{c}} \in \left[\frac{1}{\sqrt[3]{2}}, 1\right]$$

and $ab = \frac{1}{c} = \left(\frac{1}{\sqrt{c}}\right)^2$. According to the first lemma, we have

$$\frac{a}{a^3 + 1} + \frac{b}{b^3 + 1} \leq \frac{2\frac{1}{\sqrt{c}}}{\left(\frac{1}{\sqrt{c}}\right)^3 + 1} = \frac{2c}{c\sqrt{c} + 1}$$

and this yields

$$\frac{a}{a^3 + 1} + \frac{b}{b^3 + 1} + \frac{c}{c^3 + 1} \leq \frac{2c}{c\sqrt{c} + 1} + \frac{c}{c^3 + 1}.$$

But lemma 2 (for $t = \sqrt{c}$) tells us that

$$\frac{2c}{c\sqrt{c} + 1} + \frac{c}{c^3 + 1} \leq \frac{3}{2},$$

so we get the desired inequality:

$$\frac{a}{a^3 + 1} + \frac{b}{b^3 + 1} + \frac{c}{c^3 + 1} \leq \frac{2c}{c\sqrt{c} + 1} + \frac{c}{c^3 + 1} \leq \frac{3}{2}.$$

ii) Suppose now that c is greater than $\sqrt[3]{4}$. One can easily observe that the function

$$f : (0, \infty) \to (0, \infty), f(t) = \frac{t}{t^3 + 1}, \forall t > 0$$

has the derivative

$$f'(t) = \frac{1 - 2t^3}{(t^3 + 1)^2}, \forall t > 0;$$

therefore the function increases from 0 to $\frac{1}{\sqrt[3]{2}}$ and decreases from $\frac{1}{\sqrt[3]{2}}$ to ∞, having an absolute maximum at $\frac{1}{\sqrt[3]{2}}$:

$$f(t) \leq f\left(\frac{1}{\sqrt[3]{2}}\right) = \frac{1}{3}\sqrt[3]{4}, \forall t > 0.$$

In particular, $f(a)$ and $f(b)$ are less than (or equal to) $\frac{1}{3}\sqrt[3]{4}$. Regarding $f(c)$, we have

$$c \geq \sqrt[3]{4} \Rightarrow f(c) \leq f(\sqrt[3]{4}) = \frac{1}{5}\sqrt[3]{4},$$

because of the monotony of the function f. Finally

$$\frac{a}{a^3+1}+\frac{b}{b^3+1}+\frac{c}{c^3+1}=f(a)+f(b)+f(c)$$

$$\leq\left(2\cdot\frac{1}{3}+\frac{1}{5}\right)\sqrt[3]{4}=\frac{13}{15}\sqrt[3]{4}<\frac{13}{15}\cdot\frac{8}{5}=\frac{104}{75}<\frac{3}{2}$$

(the inequality $\sqrt[3]{4}<1.6$ is equivalent to $4<1.6^3=4.096$; thus it is true) and the proof ends here.

This is problem 245 (rephrased) from *Gazeta Matematică, seria A*, 3/2007. In the same magazine, number 3/2008, one can find two more solutions (due to Marius Olteanu and Ilie Bulacu) and a generalization from Ilie Bulacu.

Remarks. 1) One can also prove lemma 2 by using the derivative of $t\mapsto$ $2t^2/(t^3+1)+t^2/(t^6+1)$. (It is just a matter of taste; the computations are of the same difficulty.) Indeed, the derivative is

$$\frac{-2t\left(t^3(t^{12}-1)+3t^6(t^3-1)+3(t^9-1)\right)}{(t^3+1)^2(t^6+1)^2}$$

and the function has a maximum at 1 (equal to $3/2$).

2) The reader is invited to prove, in the same vein, that

$$\frac{a}{a^4+1}+\frac{b}{b^4+1}+\frac{c}{c^4+1}\leq\frac{3}{2},$$

for all positive a, b, and c with $abc=1$.

Chapter 15
Riemann and Darboux Sums

Let $f : [a, b] \to \mathbb{R}$ be continuous and positive. By the *subgraph* of f, we mean the region from the *xy*-plane delimited by the *x*-axis, the lines $x = a, x = b$, and the curve $y = f(x)$. More precisely, the subgraph is the set

$$\{(x, y) \in \mathbb{R}^2 \mid a \leq x \leq b,\ 0 \leq y \leq f(x)\}.$$

We study the problem of estimation of the area of the subgraph. A method is to consider a *division* (or *partition*) of the interval $[a, b]$, i.e.,

$$\Delta = (a = x_0 < x_1 < \cdots < x_{n-1} < x_n = b).$$

The real number denoted

$$\|\Delta\| = \max_{1 \leq k \leq n} (x_k - x_{k-1})$$

is called *the norm of* Δ. Next, choose arbitrary points

$$\xi_k \in [x_{k-1}, x_k],\ 1 \leq k \leq n,$$

which we call a *system of intermediate points*. Then the area of the subgraph is approximated by the following sum denoted

$$\sigma_\Delta(f, \xi_k) = \sum_{k=1}^{n} f(\xi_k)(x_k - x_{k-1})$$

and called a *Riemann sum*. Our intuition says that this estimation becomes better if the norm of the division is smaller.

© Springer Science+Business Media LLC 2017
T. Andreescu et al., *Mathematical Bridges*, DOI 10.1007/978-0-8176-4629-5_15

This result remains true for a more general class of functions, namely, *Riemann integrable functions.*

Definition. A function $f : [a, b] \to \mathbb{R}$ is called *Riemann integrable* (or just *integrable*) if there exists a real number I having the following property: for every $\varepsilon > 0$, there exists $\delta(\varepsilon) > 0$ such that for every division Δ of $[a, b]$ with $\|\Delta\| < \delta(\varepsilon)$ and for every system of intermediate points $(\xi_k)_{1 \le k \le n}$, we have $|\sigma_\Delta(f, \xi_k) - I| < \varepsilon$.

If it exists, I with this property is unique. (We encourage the reader to prove this uniqueness property.) Thus we may use a special notation for I, namely,

$$I = \int_a^b f(x)\, dx$$

and we call I the *Riemann integral*, or simply the (definite) integral of f on the interval $[a, b]$—or from a to b. Continuous functions and, also, monotone functions are Riemann integrable—the reader can find the proofs of these properties in any calculus book.

Clearly, the definition tells us that

$$\int_a^b f(x)\, dx = \lim_{n \to \infty} \sigma_{\Delta_n}\left(f, \xi_k^{(n)}\right)$$

for every sequence of partitions $(\Delta_n)_{n \ge 1}$, with $\lim\limits_{n \to \infty} \|\Delta_n\| = 0$, and for any choice of the intermediate points $\xi_k^{(n)}$ in the intervals of Δ_n, $n \ge 1$. For instance, consider the partitions Δ_n determined by the points $x_k^{(n)} = a + k\dfrac{b-a}{n}$, $0 \le k \le n$,

$$a < a + \frac{b-a}{n} < a + 2\frac{b-a}{n} < \cdots < a + (n-1)\frac{b-a}{n} < a + n\frac{b-a}{n} = b,$$

and the system of intermediate points

$$\xi_k^{(n)} = a + k\frac{b-a}{n} \in \left[a + (k-1)\frac{b-a}{n}, a + k\frac{b-a}{n}\right], \quad 1 \le k \le n.$$

Since $x_k^{(n)} - x_{k-1}^{(n)} = (b-a)/n$, this Δ_n is called an equidistant partition (all its intervals have the same length), and the corresponding Riemann sum is

$$\sigma_{\Delta_n}\left(f, \xi_k^{(n)}\right) = \frac{b-a}{n} \sum_{k=1}^{n} f\left(a + k\frac{b-a}{n}\right).$$

Thus we have a method to compute the limit of a class of convergent sequences:

Proposition. *For every Riemann integrable function* $f : [a, b] \to \mathbb{R}$, *we have*

$$\lim_{n \to \infty} \frac{b - a}{n} \sum_{k=1}^{n} f\left(a + k\frac{b - a}{n}\right) = \int_a^b f(x)\, dx.$$

In particular, if $f : [0, 1] \to \mathbb{R}$ *is Riemann integrable, then*

$$\lim_{n \to \infty} \frac{1}{n} \sum_{k=1}^{n} f\left(\frac{k}{n}\right) = \int_0^1 f(x)\, dx.$$

One of the most important results comes now. Because it connects the operations of differentiation and integration, the theorem that follows is called *the fundamental theorem of calculus*:

Theorem. a) Let $f : [a, b] \to \mathbb{R}$ be any continuous function on $[a, b]$, and let

$$F(x) = \int_a^x f(t)\, dt,$$

for all $x \in [a, b]$. Then F is differentiable on $[a, b]$, and

$$F'(x) = f(x)$$

for all $x \in [a, b]$.

b) Let $f : [a, b] \to \mathbb{R}$ be an integrable function on $[a, b]$. Moreover, assume that f has an antiderivative $F : [a, b] \to \mathbb{R}$. That is, F is differentiable on $[a, b]$ and $F'(x) = f(x)$ for every $x \in [a, b]$. Then

$$\int_a^b f(x)dx = F(b) - F(a).$$

Proof. a) We have, for (fixed, but arbitrary) $x_0 \in [a, b]$,

$$F(x) - F(x_0) = \int_{x_0}^x f(t)\, dt = (x - x_0)f(c_x)$$

for some c_x between x_0 and x (the last equality follows from the mean value theorem for the definite integral—see the first proposed problem). Thus

$$\frac{F(x) - F(x_0)}{x - x_0} = f(c_x)$$

has limit $f(x_0)$ for x tending to x_0 (which makes c_x tend to x_0 as well; remember the continuity of f). This means that $F'(x_0)$ exists and equals $f(x_0)$ at any $x_0 \in [a, b]$, which we intended to prove.

b) Let Δ_n be a sequence of divisions with the sequence of their norms convergent to 0 (e.g., we can take the equidistant divisions of $[a, b]$). If

$$a = x_0^{(n)} < x_1^{(n)} < \cdots < x_{p_n}^{(n)} = b$$

are the points that define Δ_n, we can apply to F Lagrange's mean value theorem, thus getting some $\xi_k^{(n)}$ in each interval $[x_{k-1}^{(n)}, x_k^{(n)}]$ such that

$$F\left(x_k^{(n)}\right) - F\left(x_{k-1}^{(n)}\right) = \left(x_k^{(n)} - x_{k-1}^{(n)}\right) f\left(\xi_k^{(n)}\right)$$

(do not forget that the derivative of F is f). Consequently,

$$F(b) - F(a) = \sum_{k=1}^{p_n} \left(F\left(x_k^{(n)}\right) - F\left(x_{k-1}^{(n)}\right)\right) = \sum_{k=1}^{p_n} \left(x_k^{(n)} - x_{k-1}^{(n)}\right) f\left(\xi_k^{(n)}\right),$$

that is,

$$F(b) - F(a) = \sigma_{\Delta_n}\left(f, \xi_k^{(n)}\right),$$

for all $n \geq 1$. Now we only have to take the limit as $n \to \infty$, in order to obtain the desired result:

$$F(b) - F(a) = \lim_{n \to \infty} \sigma_{\Delta_n}\left(f, \xi_k^{(n)}\right) = \int_a^b f(x)\, dx. \ \square$$

The second part of the fundamental theorem of calculus is usually named *the Newton-Leibniz formula* and serves to evaluate definite integrals with the help of derivatives. The notation

$$F(x)\Big|_a^b$$

is often used for $F(b) - F(a)$. Thus

$$\int_a^b f(x)dx = F(x)\Big|_a^b = F(b) - F(a).$$

Also note that if f is a continuous function defined on any interval I, and a is any point in I, then F defined by

$$F(x) = \int_a^x f(t)\, dt$$

for all $x \in I$ is the antiderivative of f on I.

Example. Compute $\lim_{n\to\infty} a_n$, where $a_n = \dfrac{1}{n+1} + \dfrac{1}{n+2} + \cdots + \dfrac{1}{n+n}$.

Solution. We have

$$a_n = \sum_{k=1}^{n} \frac{1}{n+k} = \frac{1}{n} \sum_{k=1}^{n} \frac{1}{1 + \frac{k}{n}} = \frac{1}{n} \sum_{k=1}^{n} f\left(\frac{k}{n}\right),$$

where $f(x) = \dfrac{1}{1+x}, x \in [0,1]$. According to the above theoretical results,

$$\lim_{n\to\infty} a_n = \int_0^1 \frac{1}{1+x}\,dx = \ln(1+x)\Big|_0^1 = \ln 2. \quad \square$$

Another method to introduce the notion of integrability is due to Darboux. Let $f : [a,b] \to \mathbb{R}$ be a bounded function. For a division

$$a = x_0 < x_1 < \cdots < x_{n-1} < x_n = b$$

define

$$m_k = \inf_{x \in [x_{k-1}, x_k]} f(x), \quad M_k = \sup_{x \in [x_{k-1}, x_k]} f(x), \quad 1 \le k \le n.$$

The sums denoted by

$$s_\Delta(f) = \sum_{k=1}^{n} m_k(x_k - x_{k-1}), \quad S_\Delta(f) = \sum_{k=1}^{n} M_k(x_k - x_{k-1})$$

are called *the inferior Darboux sum,* and *the superior Darboux sum* of f, respectively. We have the following results (the interested reader can find their proofs in any calculus book—see, for instance, [12]):

Theorem. *Let $f : [a,b] \to \mathbb{R}$ be bounded. The following assertions are equivalent:*

a) *f is Riemann integrable.*
b) *for all $\varepsilon > 0$, there exists $\delta(\varepsilon) > 0$ such that for all divisions Δ with $\|\Delta\| < \delta(\varepsilon)$, we have*

$$S_\Delta(f) - s_\Delta(f) < \varepsilon.$$

This also holds in the following slightly different form:

Theorem. The function f is Riemann integrable on $[a,b]$ if and only if for every $\epsilon > 0$, there exists a partition Δ of $[a,b]$ for which $S_\Delta(f) - s_\Delta(f) < \epsilon$.
We also mention a powerful result of Lebesgue.

Theorem (Lebesgue). *A bounded function* $f : [a, b] \rightarrow \mathbb{R}$ *is integrable if and only if it is continuous almost everywhere in* $[a, b]$, *that is, for any* $\varepsilon > 0$, *there is a family of intervals with sum of lengths at most* ε *such that any point of discontinuity of* f *lies in an interval of the family.* (*We also say that the set of discontinuities of* f *has Lebesgue measure zero or that it is a* null set.)

The proof of this result is highly nontrivial and will not be presented here. We show however some consequences of this theorem: first of all, the product of two Riemann integrable functions is Riemann integrable. Trying to prove this using the definition is not an easy task, but noting that

$$fg = \frac{(f + g)^2 - f^2 - g^2}{2}$$

reduces the problem to proving that the square of an integrable function is integrable. This follows immediately from Lebesgue's criterion, because the discontinuities of f^2 are among those of f. A more general result (with essentially the same proof) is the following:

Problem. Let $f : [a, b] \rightarrow [\alpha, \beta]$ be a Riemann integrable function, and let $g : [\alpha, \beta] \rightarrow \mathbb{R}$ be continuous. Prove that $g \circ f$ is Riemann integrable on $[a, b]$.

Solution I. Let ϵ be an arbitrary positive number. Because g is continuous on the compact interval $[\alpha, \beta]$, it is also uniformly continuous; therefore we can find a positive δ' such that

$$|g(x) - g(y)| < \frac{\epsilon}{2(b - a)} \quad \text{whenever} \ \ x, y \in [\alpha, \beta] \ \ \text{and} \ \ |x - y| < \delta'.$$

Yet, g is bounded, so there exists $M > 0$ such that $|g(x)| \le M$ for all $x \in [\alpha, \beta]$.

Now let δ be a positive number, less than both δ' and $\epsilon/(4M)$. Because f is Riemann integrable on $[a, b]$, there exists $\eta > 0$ such that whenever Δ is a partition of $[a, b]$ with $||\Delta|| < \eta$, we have

$$S_\Delta(f) - s_\Delta(f) < \delta^2.$$

Let us consider such a partition $\Delta = (a = x_0 < x_1 < \cdots < x_n = b)$ of $[a, b]$, with $||\Delta|| < \eta$. We denote by $m_i(f)$ and $M_i(f)$ the lower and upper bound of f, respectively, in the interval $[x_{i-1}, x_i]$; $m_i(g \circ f)$ and $M_i(g \circ f)$ have similar significance for $g \circ f$. We have

$$S_\Delta(g \circ f) - s_\Delta(g \circ f) = \sum_{i=1}^{n}(x_i - x_{i-1})(M_i(g \circ f) - m_i(g \circ f))$$

$$= \overset{'}{\sum}(x_i - x_{i-1})(M_i(g \circ f) - m_i(g \circ f)) + \overset{''}{\sum}(x_i - x_{i-1})(M_i(g \circ f) - m_i(g \circ f)).$$

In the first sum (i.e., in $\sum'$), we collect the terms corresponding to indices $1 \leq i \leq n$ for which $M_i(f) - m_i(f) < \delta$ (and, therefore, $M_i(f) - m_i(f) < \delta'$, too), while the second sum is over those i for which $M_i(f) - m_i(f) \geq \delta$. Consequently,

$$M_i(g \circ f) - m_i(g \circ f) < \frac{\epsilon}{2(b-a)}$$

for every i in the first sum (as $|x - y| \leq M_i(f) - m_i(f) < \delta'$ for all $x, y \in [x_{i-1}, x_i]$).

We further have

$$\delta^2 > \sum{}''(x_i - x_{i-1})(M_i(f) - m_i(f)) \geq \sum{}''(x_i - x_{i-1})\delta,$$

hence

$$\sum{}''(x_i - x_{i-1}) < \delta.$$

Putting all these together, we finally get

$$S_\Delta(g \circ f) - s_\Delta(g \circ f) = \sum_{i=1}^{n}(x_i - x_{i-1})(M_i(g \circ f) - m_i(g \circ f))$$

$$= \sum{}'(x_i - x_{i-1})(M_i(g \circ f) - m_i(g \circ f)) + \sum{}''(x_i - x_{i-1})(M_i(g \circ f) - m_i(g \circ f))$$

$$< \frac{\epsilon}{2(b-a)}\sum{}'(x_i - x_{i-1}) + 2M\sum{}''(x_i - x_{i-1})$$

$$\leq \frac{\epsilon}{2(b-a)}(b-a) + 2M\delta < \frac{\epsilon}{2} + \frac{\epsilon}{2} = \epsilon,$$

and this happens for every partition Δ of $[a, b]$, with $\|\Delta\| < \eta$. Of course, this means that $g \circ f$ is integrable on $[a, b]$, which we intended to prove. $\square$

Solution II. Lebesgue's integrability criterion allows an almost one-line proof of this result. Indeed, we clearly have $g \circ f$ bounded (for g is continuous on the compact $[\alpha, \beta]$). On the other hand, if D_f and $D_{g \circ f}$ are the sets of discontinuities of f and $g \circ f$, respectively, on $[a, b]$, we obviously have $D_{g \circ f} \subseteq D_f$ (for continuous g). (Equivalently we can say that if f is continuous at some point $t \in [a, b]$, then $g \circ f$ is also continuous at t.) So, if f is Riemann integrable, then D_f has null Lebesgue measure, implying that $D_{g \circ f}$ has null Lebesgue measure as well. Thus, $g \circ f$ is also Riemann integrable, finishing the proof. $\square$

A slightly more involved result is the following:

Problem. If $f : [a, b] \to \mathbb{R}$ has the property that $f + \sin(f)$ is Riemann integrable, then so is f.

Solution. Using Lebesgue's theorem, this is not difficult. Indeed, first of all note that $h = f + \sin(f)$ is bounded, so f is also bounded, because $|f(x)| \leq 1 + |h(x)|$. Now, note that the function $g(x) = x + \sin x$ is continuous and bijective. Indeed, it is increasing and has limits $\pm\infty$ at $\pm\infty$. Suppose that x_0 is a point of discontinuity of f, without being a point of discontinuity of h. Then there exists a sequence $(y_n)_{n\geq 1}$ that converges to x_0 and such that $(f(y_n))_{n\geq 1}$ does not converge to $f(x_0)$. However, we know that $(g(f(y_n)))_{n\geq 1}$ converges to $g(f(x_0))$. Let l be any limit point of the sequence $(f(y_n))_{n\geq 1}$ (i.e., l is the limit of some subsequence of $(f(y_n))_{n\geq 1}$). Then $g(l) = g(f(x_0))$ and so $l = f(x_0)$. Thus the bounded sequence $(f(y_n))_{n\geq 1}$ has the property that all its convergent subsequences have the same limit, $f(x_0)$. It means that it converges to $f(x_0)$, a contradiction. Thus the discontinuity points of f are among the discontinuity points of h, and thus the conditions of Lebesgue's criterion are satisfied for f, which shows that f is Riemann integrable. $\square$

An important consequence of Lebesgue's theorem is the following fact:

Problem. Any function $f : [a, b] \rightarrow \mathbb{R}$ which has finite one-sided limits at any point is Riemann integrable.

Solution. Indeed, let us prove first of all that f is bounded. Assuming the contrary, we can find a sequence $x_n \in [a, b]$ such that $|f(x_n)| > n$ for all n. This sequence has a convergent subsequence $(x_{k_n})_{n\geq 1}$, whose limit is $l \in [a, b]$. The inequality $|f(x_{k_n})| > k_n$ shows that one of the one-sided limits of f at l is infinite, which is a contradiction. Thus f is bounded on $[a, b]$.

Next, let us prove that f is continuous almost everywhere. It is enough to prove that the set of discontinuities of f is at most countable. For any x, let

$$f(x+) = \lim_{t \searrow x} f(t)$$

and

$$f(x-) = \lim_{t \nearrow x} f(t).$$

Clearly, it is enough to prove that the set of points x where $f(x+) > f(x-)$ is at most countable. But this set is the union of $A_{p,q}$, where $A_{p,q}$ is the set of real numbers $x \in [a, b]$ such that $f(x-) < p < q < f(x+)$ and the union is taken over all pairs (p, q) of rational numbers such that $p < q$. Clearly, any point of $A_{p,q}$ is isolated. So, because the union is taken over a countable set of pairs (p, q), it is enough to prove that each $A_{p,q}$ is countable. This will follow if we manage to prove that a set X of real numbers all of whose points are isolated is at most countable. Let $I_1, I_2, \ldots$ be all open intervals whose extremities are rational numbers. For all $x \in X$, we know that there exists $n(x)$ such that x is the only common point of X and $I_{n(x)}$. The function n defined on X is clearly injective, and because its values are positive integers, X is at most countable. Thus f is continuous almost everywhere (we leave as an easy exercise for the reader to prove that a countable set can be covered with a family of intervals whose sum of lengths is at most ε, for any $\varepsilon > 0$), and so f is Riemann integrable. $\square$

The following result is weaker than Lebesgue's theorem, but still nontrivial:

Problem. A Riemann integrable function $f : [0, 1] \to \mathbb{R}$ is continuous on a dense set of $[0, 1]$.

Solution. Let $0 \le a < b \le 1$. We will prove that f is continuous at a point in $[a, b]$. Using Darboux's criterion, we can construct by induction a sequence of nested intervals $I_n = [a_n, b_n]$ such that I_{n+1} is a subset of (a_n, b_n), $I_0 = [a, b]$, $b_n - a_n$ converges to 0, and

$$\sup_{x \in I_n} f(x) - \inf_{x \in I_n} f(x) < \frac{1}{n}$$

for all n. Indeed, suppose we constructed I_n. There exists $\delta > 0$ such that for any division Δ with $\|\Delta\| < \delta$ we have

$$S_\Delta(f) - s_\Delta(f) < \frac{b_n - a_n}{2(n+1)}.$$

Let $r = \min(\frac{1}{n+1}, \frac{b_n - a_n}{4}, \delta)$. For any division $\Delta = (x_0, x_1, \dots, x_k)$ such that $\|\Delta\| < r$, let a'_{n+1} be the smallest x_i among those which belong to (a_n, b_n) and let b'_{n+1} be the greatest such x_i. Clearly, $[a'_{n+1}, b'_{n+1}]$ is a subset of (a_n, b_n) and $b'_{n+1} - a'_{n+1} \ge \frac{b_n - a_n}{2}$. Also, let

$$x_{i_0} = a'_{n+1} < x_{i_0+1} < \dots < x_{i_0+p} = b'_{n+1}$$

be the points of Δ that belong to $[a'_{n+1}, b'_{n+1}]$. Finally, let

$$S_i = \sup_{x \in [x_i, x_{i+1}]} f(x) - \inf_{x \in [x_i, x_{i+1}]} f(x).$$

Then

$$\frac{b_n - a_n}{2(n+1)} > \sum_{j=0}^{p-1} S_{i_0+j}(x_{i_0+j+1} - x_{i_0+j}).$$

If S_{i_0+l} is the smallest among the S_{i_0+j}, we deduce that

$$S_{i_0+l} < \frac{b_n - a_n}{2(b'_{n+1} - a'_{n+1})(n+1)} \le \frac{1}{n+1}.$$

Therefore we can choose $a_{n+1} = x_{i_0+l}$ and $b_{n+1} = x_{i_0+l+1}$. Now, using the lemma of nested intervals, there exists x_0 belonging to the intersection of all I_n. Clearly, f is continuous at x_0. $\square$

Finally, a beautiful application of Riemann sums and subtle estimations appears in the solution to the following problem, taken from a Romanian Olympiad.

Problem. Let $f : [0, \infty) \to \mathbb{R}$ be a 1-periodic and Riemann integrable function on $[0, 1]$. For a strictly increasing, unbounded sequence $(x_n)_{n \geq 0}$, with $x_0 = 0$ and $\lim_{n \to \infty} (x_{n+1} - x_n) = 0$, denote

$$r(n) = \max \{ k \in \mathbb{N} \mid x_k \leq n \}.$$

a) Prove that

$$\lim_{n \to \infty} \frac{1}{n} \sum_{k=1}^{r(n)} (x_{k+1} - x_k) f(x_k) = \int_0^1 f(x) \, dx.$$

b) Prove that

$$\lim_{n \to \infty} \frac{1}{\ln n} \sum_{k=1}^{n} \frac{f(\ln k)}{k} = \int_0^1 f(x) \, dx.$$

Solution. a) Let us denote

$$s_p = \sum_{p-1 < x_k \leq p} (x_{k+1} - x_k) f(x_k), \quad p \geq 1.$$

Then, if

$$a_n = \frac{1}{n} \sum_{k=1}^{r(n)} (x_{k+1} - x_k) f(x_k),$$

we have $a_n = \frac{1}{n} \sum_{p=1}^{n} s_p$ and according to the Cesàro-Stolz theorem, $\lim_{n \to \infty} a_n = \lim_{n \to \infty} s_n$, if the limit on the right-hand side exists. However, note that

$$s_n = \sum_{n-1 < x_k \leq n} (x_{k+1} - x_k) f(x_k) = \sum_{0 < x_k - (n-1) \leq 1} (y_{k+1} - y_k) f(y_k)$$

with $y_k = x_k - (n-1)$, represents the Riemann sum related to the function f and the division $(y_k)_{r(n-1) < k \leq r(n)}$ which tends to zero in norm, as $n \to \infty$. Also note that $f(x_k) = f(y_k)$ because f is 1-periodic. Thus,

$$\lim_{n \to \infty} s_n = \int_0^1 f(x)\, dx.$$

b) For $x_n = \ln n$, we have

$$s_n = \frac{1}{n} \sum_{k=1}^{[e^n]} \ln \frac{k+1}{k} f(\ln k) \to \int_0^1 f(x)\, dx.$$

We also have

$$\lim_{n \to \infty} s_{[\ln n]} = \int_0^1 f(x)\, dx,$$

so

$$\frac{1}{[\ln n]} \sum_{k=1}^{\left[e^{[\ln n]}\right]} \ln \frac{k+1}{k} f(\ln k) \to \int_0^1 f(x)\, dx, \quad \text{as} \quad n \to \infty.$$

Now,

$$\frac{1}{\ln n} \sum_{k=1}^{n} \ln \frac{k+1}{k} f(\ln k) = \frac{1}{\ln n} \sum_{k=1}^{\left[e^{[\ln n]}\right]} \ln \frac{k+1}{k} f(\ln k) + \frac{1}{\ln n} \sum_{k=\left[e^{[\ln n]}\right]+1}^{n} \ln \frac{k+1}{k} f(\ln k).$$

Let us prove that

$$\lim_{n \to \infty} \frac{1}{\ln n} \sum_{k=\left[e^{[\ln n]}\right]+1}^{n} \ln \frac{k+1}{k} f(\ln k) = 0.$$

First, with $M = \sup_{x \in [0,1]} |f(x)|$,

$$\left| \sum_{k=\left[e^{[\ln n]}\right]+1}^{n} \ln \frac{k+1}{k} f(\ln k) \right| \leq M \cdot \sum_{k=\left[e^{[\ln n]}\right]+1}^{n} \ln \frac{k+1}{k} = M \ln \frac{n+1}{\left[e^{[\ln n]}\right]+1},$$

which tends to zero, as $n \to \infty$. Finally,

$$\left| \frac{1}{\ln n} \sum_{k=1}^{n} \left(\frac{1}{k} - \ln \frac{k+1}{k} \right) f(\ln k) \right| \leq M \cdot \frac{1}{\ln n} \sum_{k=1}^{n} \left(\frac{1}{k} - \ln \frac{k+1}{k} \right)$$

$$= M \cdot \frac{1 + \frac{1}{2} + \cdots + \frac{1}{n} - \ln(n+1)}{\ln n} \to 0, \quad \text{as} \quad n \to \infty. \quad \square$$

Proposed Problems

1. a) Prove the monotonicity property of the Riemann integral, namely, that if $f, g :$ $[a, b] \to \mathbb{R}$ are two Riemann integrable functions such that $f \leq g$ on $[a, b]$, then

$$\int_a^b f(x) \, dx \leq \int_a^b g(x) \, dx.$$

   b) Prove the mean value property for the Riemann integral. Namely, show that if $f : [a, b]$ is a continuous function, then there exists $c \in [a, b]$ such that

$$\int_a^b f(x) \, dx = (b - a)f(c).$$

2. Find the limit of the sequence $a_n = \displaystyle\sum_{k=1}^n \frac{\sin \frac{k\pi}{n}}{\sqrt{n^2 + k}}$.

3. Find the limit of the sequence $a_n = \displaystyle\sum_{k=1}^n \frac{1}{k + \sqrt{n^2 + kn + k}}, n \geq 1$.

4. Does there exist a Riemann integrable function $f : [0, 1] \to \mathbb{R}$ such that for every $p, q \in (0, 1), p < q$ there exist $c, d \in (p, q)$ for which $f(c) = c^2$ and $f(d) = d^3$?

5. Let $f : [0, 1] \to \mathbb{R}$ be a Riemann integrable function such that

$$0 \leq f\left(\frac{m}{n}\right) \leq \frac{1}{n - m}$$

   for all positive integers $m < n$, $\gcd(m, n) = 1$. Compute $\displaystyle\int_0^1 f(x) \, dx$.

6. Let $f : [0, 1] \to \mathbb{R}$ be integrable such that

$$\frac{f(x) + f(y)}{2} \leq f(\sqrt{xy}),$$

   for every x, y in $[0, 1]$. Prove that if f is continuous at e^{-1}, then

$$\int_0^1 f(x) \, dx \leq f(e^{-1}).$$

7. Let $f : [a, b] \to \mathbb{R}$ be with the property that for all $\varepsilon > 0$, the set

$$\{x \in [a, b] \mid |f(x)| > \varepsilon\}$$

   is finite or empty. Prove that f is integrable and $\displaystyle\int_0^1 f(x) \, dx = 0$.

8. Let $f : [0, 1] \rightarrow \mathbb{R}$ be an integrable function such that for every $p, q \in (0, 1), p < q$, there exists $\xi \in (p, q)$ for which $f(\xi) = 0$. Prove that
$$\int_0^1 f(x)\, dx = 0.$$

9. Let $f : [0, 1] \rightarrow \mathbb{R}$ be an integrable function such that for every $p, q \in (0, 1), p < q$, there exist $c, d \in (p, q)$ for which $f(c) + f(d) = 2$. Prove that
$$\int_0^1 f(x)\, dx = 1.$$

10. Let $f, g : [0, 1] \rightarrow \mathbb{R}$ be increasing, with $g(0) > 0$ and let $(a_n)_{n\geq 1} \subset (0, \infty)$ be such that $(na_n)_{n\geq 1}$ converges to 1 and is monotonically increasing. Compute
$$\lim_{n\to\infty} \frac{f(a_n) + f(2a_n) + \cdots + f(na_n)}{g(a_n) + g(2a_n) + \cdots + g(na_n)}.$$

11. Prove that the function $f : [0, 1] \rightarrow \mathbb{R}$, given by
$$f(x) = \begin{cases} \dfrac{1}{2^n}, & x = \dfrac{1}{n}, \ n \in \mathbb{N}^* \\ 0, & x \neq \dfrac{1}{n}, \ n \in \mathbb{N}^* \end{cases}$$

is integrable and $\displaystyle\int_0^1 f(x)\, dx = 0$.

12. Let $f : [0, 1] \rightarrow \mathbb{R}$ be increasing. Prove that
$$\left| \int_0^1 f(x)\, dx - \frac{1}{n} \sum_{k=1}^n f\left(\frac{k}{n}\right) \right| \leq \frac{f(1) - f(0)}{n},$$

for all positive integers n.

13. Let $a_1 = 0.5$, and $a_{n+1} = \sqrt{1 + na_n}$ for $n \geq 1$. Compute
$$\lim_{n\to\infty} \left(\frac{1}{n^2 + a_n} + \frac{2}{n^2 + 2a_n} + \cdots + \frac{n}{n^2 + na_n} \right).$$

14. Let $f, g : [a, b] \rightarrow \mathbb{R}$ be such that
$$|f(x) - f(y)| \leq |g(x) - g(y)|$$

for all x, y in $[a, b]$. Prove that f is integrable if g is integrable.

15. Let $f : [0, 1] \to \mathbb{R}$ be integrable. Prove that

$$\lim_{n \to \infty} \frac{\left(f\left(\frac{1}{n}\right) - f(0)\right)^2 + \left(f\left(\frac{2}{n}\right) - f\left(\frac{1}{n}\right)\right)^2 + \cdots + \left(f(1) - f\left(\frac{n-1}{n}\right)\right)^2}{n} = 0.$$

16. Let $f : [0, 1] \to \mathbb{R}$ be integrable and let $(a_n)_{n \geq 1}$ be a sequence such that

$$\left| \sum_{k=1}^{n} a_k \right| \leq 1,$$

for all positive integers n. Prove that

$$\lim_{n \to \infty} \frac{1}{n} \sum_{k=1}^{n} f\left(\frac{k}{n}\right) a_k = 0.$$

17. Let $f : [0, 1] \to \mathbb{R}$ be a continuously differentiable function. Prove that

$$\lim_{n \to \infty} n \left(\int_0^1 f(x)\, dx - \frac{1}{n} \sum_{k=0}^{n-1} f\left(\frac{k}{n}\right) \right) = \frac{f(1) - f(0)}{2}.$$

18. Compute the limit $\displaystyle \lim_{n \to \infty} n \left(\frac{\pi}{4} - \sum_{k=0}^{n-1} \frac{n}{n^2 + k^2} \right).$

19. Are there polynomials P, Q with real coefficients satisfying the equalities

$$\int_0^{\ln n} \frac{P(x)}{Q(x)}\, dx = 1 + \frac{1}{2} + \frac{1}{3} + \cdots + \frac{1}{n},$$

for each integer $n \geq 2$?

20. Let $f : (0, 1] \to \mathbb{R}$ be a continuous function such that

$$\lim_{x \to 0} \int_x^1 f(t)\, dt = \int_0^1 f(x)\, dx$$

exists and is finite. Does it follow that the sequence $(S_n)_{n \geq 1}$, where

$$S_n = \frac{1}{n} \sum_{k=1}^{n} f\left(\frac{k}{n}\right),$$

converges to $\int_0^1 f(x)\, dx$? What if f is decreasing and $\lim_{x \to 0} xf(x) = 0$?

21. Let f and g be two functions defined on a compact interval $[a, b]$, such that f is Riemann integrable on $[a, b]$, g has an antiderivative G on $[a, b]$, and $f(x) \leq g(x)$ for all $x \in [a, b]$.

   a) Prove that $\displaystyle\int_a^b f(x)\,dx \leq G(b) - G(a)$.

   b) Prove that if $\displaystyle\int_a^b f(x)\,dx = G(b) - G(a)$, then g is Riemann integrable on $[a, b]$.

22. Let $g : [\alpha, \beta] \to \mathbb{R}$ be a Riemann integrable function, and let $f : [a, b] \to [\alpha, \beta]$ be a continuous function having the property that for every $a' \in (a, b]$, there exists an $L > 0$ (depending on a') such that $|f(x) - f(y)| \geq L|x - y|$ for all $x, y \in [a', b]$. Prove that $g \circ f$ is Riemann integrable on $[a, b]$.

23. Let $A_n = \displaystyle\int_0^{\pi/2} \cos^{2n} x\,dx$ and $B_n = \displaystyle\int_0^{\pi/2} x^2 \cos^{2n} x\,dx$.

   a) Prove that $A_{n-1} - A_n = \dfrac{1}{2n-1}A_n$ and that $A_n = n(2n-1)B_{n-1} - 2n^2 B_n$ for all $n \geq 1$.

   b) Conclude that

$$\frac{1}{n^2} = 2\left(\frac{B_{n-1}}{A_{n-1}} - \frac{B_n}{A_n}\right)$$

   for all $n \geq 1$, and then deduce that

$$\zeta(2) = \sum_{n=1}^{\infty} \frac{1}{n^2} = \frac{\pi^2}{6}.$$

Solutions

1. a) If $f(x) \leq g(x)$ for all $x \in [a, b]$, it follows easily that $\sigma_{\Delta_n}\left(f, \xi_k^{(n)}\right) \leq \sigma_{\Delta_n}\left(g, \xi_k^{(n)}\right)$ for every $n \geq 1$, where Δ_n is a sequence of partitions of $[a, b]$ with $\lim_{n \to \infty} \|\Delta_n\| = 0$, and $\xi_k^{(n)}$ are some corresponding intermediate points. Thus the result follows by passing to the limit for $n \to \infty$:

$$\int_a^b f(x)\,dx = \lim_{n \to \infty} \sigma_{\Delta_n}\left(f, \xi_k^{(n)}\right) \leq \lim_{n \to \infty} \sigma_{\Delta_n}\left(g, \xi_k^{(n)}\right) = \int_a^b g(x)\,dx.$$

Note that, in particular, the integral of a nonnegative function is also nonnegative. Also, one can show that if continuity on $[a, b]$ is assumed for f

and g, the equality in the inequality that we proved is achieved if and only if $f = g$ on $[a, b]$.

b) Let $m = \inf\limits_{x \in [a,b]} f(x)$, and $M = \sup\limits_{x \in [a,b]} f(x)$ be the extrema of f on $[a, b]$. The extreme value theorem tells us that m and M are actually values assumed by f. The monotonicity of the Riemann integral applied to the functions $f - m$ and 0 to $f - M$ and 0, respectively, yields

$$m(b - a) \leq \int_a^b f(x)\, dx \leq M(b - a).$$

Thus

$$\frac{1}{b - a} \int_a^b f(x)\, dx$$

is between m and M—two values of f; now the intermediate value property of f ensures the existence of $c \in [a, b]$ such that

$$\frac{1}{b - a} \int_a^b f(x)\, dx = f(c).$$

We used this result in the proof of the fundamental theorem of calculus. So, now we know that if f is continuous on $[a, b]$, then the function

$$x \mapsto F(x) = \int_a^x f(t)\, dt$$

is an antiderivative of f. Thus the equality from the mean value theorem reads $F(b) - F(a) = (b - a)F'(c)$, that is, it can be inferred from Lagrange's mean value theorem applied to F. This shows that c can be actually chosen in the open interval (a, b).

2. For every integer $1 \leq k \leq n$, we have the estimations

$$\sqrt{n^2 + 1} \leq \sqrt{n^2 + k} \leq \sqrt{n^2 + n},$$

so

$$\frac{1}{\sqrt{n^2 + n}} \sum_{k=1}^{n} \sin \frac{k\pi}{n} \leq a_n \leq \frac{1}{\sqrt{n^2 + 1}} \sum_{k=1}^{n} \sin \frac{k\pi}{n}.$$

Now let us consider the function $f : [0, 1] \to \mathbb{R}, f(x) = \sin \pi x$. The function f is continuous, so it is integrable. It follows that

$$\lim_{n\to\infty} \frac{1}{n}\sum_{k=1}^{n}\sin\frac{k\pi}{n} = \lim_{n\to\infty}\frac{1}{n}\sum_{k=1}^{n}f\left(\frac{k}{n}\right) = \int_0^1 f(x)\,dx$$

$$= \int_0^1 \sin\pi x\,dx = -\frac{1}{\pi}\cdot\cos\pi x\bigg|_0^1 = \frac{2}{\pi}.$$

Because

$$\frac{n}{\sqrt{n^2+n}}\cdot\frac{1}{n}\sum_{k=1}^{n}\sin\frac{k\pi}{n} \le a_n \le \frac{n}{\sqrt{n^2+1}}\cdot\frac{1}{n}\sum_{k=1}^{n}\sin\frac{k\pi}{n}$$

it follows that $\lim_{n\to\infty} a_n = \dfrac{2}{\pi}$.

3. We have

$$a_n = \frac{1}{n}\sum_{k=1}^{n}\frac{1}{\frac{k}{n}+\sqrt{1+\frac{k}{n}+\frac{k}{n^2}}},$$

so

$$\frac{1}{n}\sum_{k=1}^{n}\frac{1}{\frac{k}{n}+\frac{k}{n^2}+\sqrt{1+\frac{k}{n}+\frac{k}{n^2}}} < a_n < \frac{1}{n}\sum_{k=1}^{n}\frac{1}{\frac{k}{n}+\sqrt{1+\frac{k}{n}}}.$$

The numbers

$$\frac{k}{n}+\frac{k}{n^2} \in \left[\frac{k}{n},\frac{k+1}{n}\right],\quad \frac{k}{n}\in\left[\frac{k}{n},\frac{k+1}{n}\right]$$

can be viewed as systems of intermediate points associated to the division

$$\Delta = \left(0 < \frac{1}{n} < \frac{2}{n} < \cdots < \frac{n-1}{n} < 1\right)$$

and the function $f : [0, 1] \to \mathbb{R}, f(x) = \dfrac{1}{x+\sqrt{1+x}}$. Thus

$$\lim_{n\to\infty} a_n = \int_0^1 \frac{dx}{x+\sqrt{1+x}}.$$

4. The answer is no. Supposing the contrary, consider the division

$$\Delta_n = \left(0 < \frac{1}{n} < \frac{2}{n} < \cdots < \frac{n}{n} = 1\right),\quad n \ge 1$$

of the interval $[0, 1]$ with the norm $||\Delta_n|| = \frac{1}{n} \to 0$, as $n \to \infty$. According to the hypothesis, we can find

$$\xi_k, \eta_k \in \left[\frac{k-1}{n}, \frac{k}{n}\right], \quad 1 \leq k \leq n$$

such that

$$f(\xi_k) = \xi_k^2, \quad f(\eta_k) = \eta_k^3.$$

Hence

$$\lim_{n\to\infty} \frac{1}{n} \sum_{k=1}^n f(\xi_k) = \lim_{n\to\infty} \frac{1}{n} \sum_{k=1}^n f(\eta_k) = \int_0^1 f(x)\, dx.$$

And yet, these limits are different, so the problem is solved. Indeed, consider the sum

$$\frac{1}{n} \sum_{k=1}^n f(\xi_k) = \frac{1}{n} \sum_{k=1}^n \xi_k^2, \quad \xi_k \in \left[\frac{k-1}{n}, \frac{k}{n}\right]$$

as a Riemann sum associated to the continuous function $\phi : [0, 1] \to \mathbb{R}$, $\phi(x) = x^2$. Therefore

$$\lim_{n\to\infty} \frac{1}{n} \sum_{k=1}^n \xi_k^2 = \int_0^1 \phi(x)\, dx = \int_0^1 x^2\, dx = \frac{1}{3}.$$

Similarly, if we consider the continuous function $\psi : [0, 1] \to \mathbb{R}$, $\psi(x) = x^3$, then

$$\lim_{n\to\infty} \frac{1}{n} \sum_{k=1}^n \eta_k^3 = \int_0^1 \psi(x)\, dx = \int_0^1 x^3\, dx = \frac{1}{4}.$$

5. First note that the sequence

$$a_n = \frac{1}{n} \sum_{k=1}^{n-1} f\left(\frac{k}{n}\right)$$

is convergent and

$$\lim_{n\to\infty} a_n = \int_0^1 f(x)\, dx.$$

If p is prime, then

$$0 \leq a_p = \frac{1}{p} \sum_{k=1}^{p-1} f\left(\frac{k}{p}\right) \leq \frac{1}{p} \sum_{k=1}^{p-1} \frac{1}{p-k} = \frac{1 + \frac{1}{2} + \cdots + \frac{1}{p-1}}{p}.$$

Using the Cesàro-Stolz theorem, we have

$$\lim_{p \to \infty} \frac{1 + \frac{1}{2} + \cdots + \frac{1}{p-1}}{p} = \lim_{p \to \infty} \frac{\frac{1}{p-1}}{p - (p-1)} = 0.$$

In conclusion, the convergent sequence $(a_n)_{n \geq 1}$ has a subsequence which converges to zero. Hence $a_n \to 0$ and $\int_0^1 f(x)\, dx = 0$.

6. We can prove by induction (using the same trick as the one Cauchy used to prove the AM-GM inequality) that

$$\frac{f(x_1) + f(x_2) + \cdots + f(x_n)}{n} \leq f\left(\sqrt[n]{x_1 x_2 \cdots x_n}\right),$$

for all positive integers n and all $x_1, x_2, \ldots, x_n$. Thus

$$a_n = \frac{1}{n} \sum_{k=1}^{n} f\left(\frac{k}{n}\right) \leq f\left(\sqrt[n]{\frac{1}{n} \cdot \frac{2}{n} \cdots \frac{n}{n}}\right) = f\left(\frac{\sqrt[n]{n!}}{n}\right).$$

Now, using the well known limit $\lim_{n \to \infty} \frac{\sqrt[n]{n!}}{n} = e^{-1}$, we obtain

$$\int_0^1 f(x)\, dx = \lim_{n \to \infty} a_n \leq \lim_{n \to \infty} f\left(\frac{\sqrt[n]{n!}}{n}\right) = f\left(e^{-1}\right).$$

7. For $\varepsilon > 0$, denote by

$$A_\varepsilon = \{x \in [a, b] \mid |f(x)| > \varepsilon\}$$

and assume that card $A_\varepsilon = n_\varepsilon$. Obviously, f is bounded so let

$$M = \sup_{x \in [a,b]} |f(x)|.$$

Let

$$\Delta = (a = x_0 < x_1 < \cdots < x_n = b)$$

be a division of the interval $[a, b]$ and let $\xi_k \in [x_{k-1}, x_k]$, $1 \le k \le n$ be a system of intermediate points. We have

$$\left| \sum_{k=1}^{n} f(\xi_k)(x_k - x_{k-1}) \right| \le \sum_{k=1}^{n} |f(\xi_k)| (x_k - x_{k-1})$$

$$= \sum_{\xi_k \in A_\varepsilon} |f(\xi_k)| (x_k - x_{k-1}) + \sum_{\xi_k \notin A_\varepsilon} |f(\xi_k)| (x_k - x_{k-1})$$

$$\le ||\Delta|| \cdot \sum_{\xi_k \in A_\varepsilon} |f(\xi_k)| + \varepsilon \cdot \sum_{\xi_k \notin A_\varepsilon} (x_k - x_{k-1}) \le M n_\varepsilon \cdot ||\Delta|| + \varepsilon(b - a).$$

Now, if $||\Delta|| < \dfrac{\varepsilon}{M n_\varepsilon}$, then

$$0 \le \left| \sum_{k=1}^{n} f(\xi_k)(x_k - x_{k-1}) \right| \le \varepsilon(b - a + 1)$$

and consequently, $\displaystyle\int_{0}^{1} f(x)\, dx = 0$.

8. For each integer $n \ge 2$, let us consider the equidistant division

$$\Delta_n = \left(0 < \frac{1}{n} < \frac{2}{n} < \cdots < \frac{n-1}{n} < 1 \right)$$

of the interval $[0, 1]$. Define the system of intermediate points

$$\xi_k^{(n)} \in \left[\frac{k-1}{n}, \frac{k}{n} \right], \quad \text{with} \quad f(\xi_k^{(n)}) = 0,$$

for all integers $1 \le k \le n$. The norm $||\Delta_n|| = \frac{1}{n}$ tends to zero, as $n \to \infty$, so the corresponding Riemann sum tends to $\displaystyle\int_{0}^{1} f(x)\, dx$ as $n \to \infty$:

$$\lim_{n \to \infty} \frac{1}{n} \sum_{k=1}^{n} f(\xi_k^{(n)}) = \int_{0}^{1} f(x)\, dx.$$

But the sums are identically zero, because all their terms are zero. Thus,

$$\int_{0}^{1} f(x)\, dx = 0.$$

9. Consider the division

$$\Delta_n = \left(0 < \frac{1}{n} < \frac{2}{n} < \cdots < \frac{n-1}{n} < 1\right)$$

of the interval $[0, 1]$. According to the hypothesis, we can find

$$\xi_k^n, \eta_k^n \in \left[\frac{k-1}{n}, \frac{k}{n}\right], \quad 1 \le k \le n,$$

so that $f(\xi_k^n) + f(\eta_k^n) = 2$. From the fact that f is integrable, it follows that

$$\lim_{n\to\infty} \frac{1}{n} \sum_{k=1}^{n} f(\xi_k^n) = \lim_{n\to\infty} \frac{1}{n} \sum_{k=1}^{n} f(\eta_k^n) = \int_0^1 f(x)\,dx.$$

Therefore

$$2\int_0^1 f(x)\,dx = \lim_{n\to\infty} \frac{1}{n} \sum_{k=1}^{n} f(\xi_k^n) + \lim_{n\to\infty} \frac{1}{n} \sum_{k=1}^{n} f(\eta_k^n)$$

$$= \lim_{n\to\infty} \frac{1}{n} \sum_{k=1}^{n} \left(f(\xi_k^n) + f(\eta_k^n)\right) = 2,$$

so $\int_0^1 f(x)\,dx = 1$.

10. Let us consider the divisions

$$\Delta_n = (0 < a_n < 2a_n < \cdots < na_n < 1), \quad n \ge 1$$

of the interval $[0, 1]$. Easily,

$$||\Delta_n|| = \max\{a_n, 1 - na_n\} \to 0, \quad \text{as} \quad n \to \infty.$$

The functions f, g are increasing, so they are integrable on $[0, 1]$. If we choose the points

$$\xi_k = ka_n \in [(k-1)a_n, ka_n], \quad 1 \le k \le n,$$

then

$$\lim_{n\to\infty} \sum_{k=1}^{n} f(ka_n)(ka_n - (k-1)a_n) = \lim_{n\to\infty} a_n \sum_{k=1}^{n} f(ka_n) = \int_0^1 f(x)\,dx$$

and

$$\lim_{n\to\infty}\sum_{k=1}^{n}g(ka_n)(ka_n-(k-1)a_n)=\lim_{n\to\infty}a_n\sum_{k=1}^{n}g(ka_n)=\int_0^1 g(x)\,dx.$$

Finally,

$$\lim_{n\to\infty}\frac{\displaystyle\sum_{k=1}^{n}f(ka_n)}{\displaystyle\sum_{k=1}^{n}g(ka_n)}=\lim_{n\to\infty}\frac{a_n\displaystyle\sum_{k=1}^{n}f(ka_n)}{a_n\displaystyle\sum_{k=1}^{n}g(ka_n)}=\frac{\displaystyle\int_0^1 f(x)\,dx}{\displaystyle\int_0^1 g(x)\,dx}.$$

11. The function f is bounded and continuous almost everywhere (continuous on $[0,1]\setminus\{1/n\mid n\in\mathbb{N}^*\}$). In order to compute the integral, we will choose the particular division

$$\Delta=\left(0<\frac{1}{n}<\frac{2}{n}<\cdots<\frac{n-1}{n}<1\right)$$

and the system of intermediate points

$$\xi_k\in\left[\frac{k-1}{n},\frac{k}{n}\right]\cap(\mathbb{R}\setminus\mathbb{Q}),\ 1\le k\le n.$$

We have $f(\xi_k)=0$ and

$$\int_0^1 f(x)\,dx=\lim_{n\to\infty}\frac{1}{n}\sum_{k=1}^{n}f(\xi_k)=0.$$

More generally, any integrable function that vanishes on a dense subset of $[a,b]$ has integral equal to 0 (as we have seen in a previous exercise).

12. We have

$$\left|\int_0^1 f(x)\,dx-\frac{1}{n}\sum_{k=1}^{n}f\left(\frac{k}{n}\right)\right|=\left|\sum_{k=1}^{n}\int_{\frac{k-1}{n}}^{\frac{k}{n}}f(x)\,dx-\frac{1}{n}\sum_{k=1}^{n}f\left(\frac{k}{n}\right)\right|$$

$$=\left|\sum_{k=1}^{n}\int_{\frac{k-1}{n}}^{\frac{k}{n}}\left(f(x)-f\left(\frac{k}{n}\right)\right)dx\right|$$

$$=\sum_{k=1}^{n}\int_{\frac{k-1}{n}}^{\frac{k}{n}}\left(f\left(\frac{k}{n}\right)-f(x)\right)dx$$

$$\leq \sum_{k=1}^{n} \int_{\frac{k-1}{n}}^{\frac{k}{n}} \left(f\left(\frac{k}{n}\right) - f\left(\frac{k-1}{n}\right) \right) dx$$

$$= \frac{f(1) - f(0)}{n}.$$

13. First we can prove by induction that $n - 2 < a_n < n$. Under this assumption, we have

$$a_{n+1} = \sqrt{1 + na_n} > \sqrt{1 + n(n-2)} = n - 1$$

and

$$a_{n+1} = \sqrt{1 + na_n} < \sqrt{1 + n^2} < n + 1,$$

so $n - 1 < a_{n+1} < n + 1$. These inequalities show that $\dfrac{a_n}{n}$ converges to 1. Let us consider the function $f : [0, 1] \to \mathbb{R}$, given by the formula $f(x) = \dfrac{x}{x+1}$, $x \in [0, 1]$ and the division

$$\Delta = \left(0 < \frac{a_n}{n^2} < \frac{2a_n}{n^2} < \cdots < \frac{na_n}{n^2} < 1 \right).$$

For the system of intermediate points

$$\xi_k = \frac{ka_n}{n^2} \in \left[\frac{(k-1)a_n}{n^2}, \frac{ka_n}{n^2} \right], \quad 1 \leq k \leq n,$$

we have

$$\int_0^1 f(x)\, dx = \lim_{n \to \infty} \sum_{k=1}^{n} f\left(\frac{ka_n}{n^2}\right) \left(\frac{ka_n}{n^2} - \frac{(k-1)a_n}{n^2} \right)$$

$$= \lim_{n \to \infty} \frac{a_n}{n^2} \sum_{k=1}^{n} \frac{ka_n}{n^2 + ka_n} = \lim_{n \to \infty} \frac{a_n^2}{n^2} \sum_{k=1}^{n} \frac{k}{n^2 + ka_n}.$$

In conclusion,

$$\lim_{n \to \infty} \sum_{k=1}^{n} \frac{k}{n^2 + ka_n} = \int_0^1 f(x)\, dx = 1 - \ln 2.$$

14. Let $\varepsilon > 0$. For every division

$$\Delta = (a = x_0 < x_1 < \cdots < x_n = b)$$

with norm $||\Delta|| < \delta(\varepsilon)$, we have $S_\Delta(g) - s_\Delta(g) < \varepsilon$. Let us denote

$$m_k^g = \inf_{x \in [x_{k-1}, x_k]} g(x), \quad M_k^g = \sup_{x \in [x_{k-1}, x_k]} g(x)$$

and

$$m_k^f = \inf_{x \in [x_{k-1}, x_k]} f(x), \quad M_k^f = \sup_{x \in [x_{k-1}, x_k]} f(x).$$

From the given inequality, we deduce

$$M_k^f - m_k^f \le M_k^g - m_k^g, \ 1 \le k \le n.$$

Therefore

$$S_\Delta(f) - s_\Delta(f) = \sum_{k=1}^n (M_k^f - m_k^f)(x_k - x_{k-1}) \le \sum_{k=1}^n (M_k^g - m_k^g)(x_k - x_{k-1})$$

$$= S_\Delta(g) - s_\Delta(g) < \varepsilon.$$

In conclusion, f is integrable. Alternatively, observe that the hypothesis and the fact that g is bounded easily imply that f is bounded: all we need is to note that $|f(x)| \le |f(a)| + |g(x)| + |g(a)|$. Also, for an $x_0 \in (a, b)$, the inequality $|f(x) - f(x_0)| \le |g(x) - g(x_0)|$ implies that f is continuous at x_0 if g is. Thus, the discontinuity points of f are included in the set of discontinuity points of g and we can apply Lebesgue's criterion in order to complete the solution.

15. The function f is integrable, so it is bounded. Assume that $|f(x)| \le M$ for all $x \in [0, 1]$ and for some $M > 0$. For every integer $1 \le k \le n$, we have

$$\left| f\left(\frac{k}{n}\right) - f\left(\frac{k-1}{n}\right) \right| \le \left| f\left(\frac{k}{n}\right) \right| + \left| f\left(\frac{k-1}{n}\right) \right| \le 2M$$

and thus

$$\left(f\left(\frac{k}{n}\right) - f\left(\frac{k-1}{n}\right) \right)^2 \le 2M \cdot \left| f\left(\frac{k}{n}\right) - f\left(\frac{k-1}{n}\right) \right|.$$

By adding these inequalities, we obtain

$$\frac{1}{n} \sum_{k=1}^n \left(f\left(\frac{k}{n}\right) - f\left(\frac{k-1}{n}\right) \right)^2 \le \frac{2M}{n} \sum_{k=1}^n \left| f\left(\frac{k}{n}\right) - f\left(\frac{k-1}{n}\right) \right|.$$

Next,

$$\frac{1}{n}\sum_{k=1}^{n}\left|f\left(\frac{k}{n}\right)-f\left(\frac{k-1}{n}\right)\right| = \sum_{k=1}^{n}\left|f\left(\frac{k}{n}\right)-f\left(\frac{k-1}{n}\right)\right|\cdot\left(\frac{k}{n}-\frac{k-1}{n}\right)$$

$$\leq S_\Delta(f) - s_\Delta(f) \to 0,$$

where $S_\Delta(f)$ and $s_\Delta(f)$ are the Darboux sums corresponding to the equidistant division

$$\Delta = \left(0 < \frac{1}{n} < \frac{2}{n} < \cdots < \frac{n-1}{n} < 1\right).$$

Thus

$$\frac{1}{n}\sum_{k=1}^{n}\left|f\left(\frac{k}{n}\right)-f\left(\frac{k-1}{n}\right)\right| \to 0$$

and the problem is solved.

16. Let us define the sequence $(b_n)_{n\geq 0}$ by the formula $b_n = \sum_{k=1}^{n} a_k$, with $b_0 = 0$.
By using the Abel-Dirichlet summation method, we have

$$\frac{1}{n}\left|\sum_{k=1}^{n}f\left(\frac{k}{n}\right)a_k\right| = \frac{1}{n}\left|\sum_{k=1}^{n}f\left(\frac{k}{n}\right)(b_k - b_{k-1})\right|$$

$$= \frac{1}{n}\left|b_1\left[f\left(\frac{1}{n}\right)-f\left(\frac{2}{n}\right)\right] + b_2\left[f\left(\frac{2}{n}\right)-f\left(\frac{3}{n}\right)\right] + \cdots\right.$$

$$\left. + b_{n-1}\left[f\left(\frac{n-1}{n}\right)-f(1)\right] + b_n f(1)\right|$$

$$\leq \frac{1}{n}\left[|b_1|\cdot\left|f\left(\frac{1}{n}\right)-f\left(\frac{2}{n}\right)\right| + |b_2|\cdot\left|f\left(\frac{2}{n}\right)-f\left(\frac{3}{n}\right)\right| + \cdots\right.$$

$$\left. + |b_{n-1}|\cdot\left|f\left(\frac{n-1}{n}\right)-f(1)\right| + |b_n|\cdot|f(1)|\right]$$

$$\leq \frac{1}{n}\left[\left|f\left(\frac{1}{n}\right)-f\left(\frac{2}{n}\right)\right| + \left|f\left(\frac{2}{n}\right)-f\left(\frac{3}{n}\right)\right| + \cdots\right.$$

$$\left. + \left|f\left(\frac{n-1}{n}\right)-f(1)\right| + |f(1)|\right],$$

which tends to zero, as $n \to \infty$, because, as we have seen in the previous problem,

$$\lim_{n \to \infty} \frac{\left|f\left(\frac{2}{n}\right) - f\left(\frac{1}{n}\right)\right| + \left|f\left(\frac{3}{n}\right) - f\left(\frac{2}{n}\right)\right| + \cdots + \left|f(1) - f\left(\frac{n-1}{n}\right)\right|}{n} = 0.$$

17. We have

$$x_n = n\left(\int_0^1 f(x)\,dx - \frac{1}{n}\sum_{k=1}^{n} f\left(\frac{k-1}{n}\right)\right)$$

$$= n\left(\sum_{k=1}^{n}\int_{\frac{k-1}{n}}^{\frac{k}{n}} f(x)\,dx - \frac{1}{n}\sum_{k=1}^{n} f\left(\frac{k-1}{n}\right)\right)$$

$$= n\sum_{k=1}^{n}\int_0^{\frac{1}{n}}\left(f\left(x + \frac{k-1}{n}\right) - f\left(\frac{k-1}{n}\right) - xf'\left(\frac{k-1}{n}\right)\right)dx$$

$$+ \frac{1}{2n}\sum_{k=0}^{n-1} f'\left(\frac{k}{n}\right).$$

Pick $\epsilon > 0$. Because f' is continuous on $[0, 1]$, it is uniformly continuous, and so there exists n_0 such that for all $n \geq n_0$ and all x, y such that $|x - y| \leq \frac{1}{n}$, we have $|f'(x) - f'(y)| \leq \epsilon$. From now on, we consider $n \geq n_0$. For a fixed $x \in \left[0, \frac{1}{n}\right]$, Lagrange's theorem asserts the existence of a point $c_{k,n,x} \in \left[\frac{k-1}{n}, x + \frac{k-1}{n}\right]$ such that

$$f\left(x + \frac{k-1}{n}\right) - f\left(\frac{k-1}{n}\right) = xf'(c_{k,n,x}).$$

Because $\left|c_{n,k,x} - \frac{k-1}{n}\right| \leq \frac{1}{n}$, we deduce that

$$\left|f\left(x + \frac{k-1}{n}\right) - f\left(\frac{k-1}{n}\right) - xf'\left(\frac{k-1}{n}\right)\right| < \epsilon \cdot x$$

for all k, x. By integrating this between 0 and $\frac{1}{n}$, we deduce that

$$\left|\int_0^{\frac{1}{n}}\left(f\left(x + \frac{k-1}{n}\right) - f\left(\frac{k-1}{n}\right) - xf'\left(\frac{k-1}{n}\right)\right)dx\right| < \frac{\epsilon}{2n^2},$$

which shows (by summation) that the term

$$n \sum_{k=1}^{n} \int_0^{\frac{1}{n}} \left(f\left(x + \frac{k-1}{n}\right) - f\left(\frac{k-1}{n}\right) - xf'\left(\frac{k-1}{n}\right) \right) dx$$

converges to 0. Because the term $\dfrac{1}{2n} \displaystyle\sum_{k=0}^{n-1} f'\left(\dfrac{k}{n}\right)$ converges to

$$\int_0^1 \frac{f'(x)dx}{2} = \frac{f(1) - f(0)}{2},$$

the conclusion follows.

18. Let us consider the function $f : [0, 1] \to \mathbb{R}$, given by $f(x) = \frac{1}{1+x^2}$, $x \in [0, 1]$. According to the previous problem,

$$\lim_{n \to \infty} n \left(\int_0^1 f(x)\, dx - \frac{1}{n} \sum_{k=0}^{n-1} f\left(\frac{k}{n}\right) \right) = \frac{f(1) - f(0)}{2},$$

which can be written as

$$\lim_{n \to \infty} n \left(\frac{\pi}{4} - \sum_{k=0}^{n-1} \frac{n}{n^2 + k^2} \right) = -\frac{1}{4}.$$

19. The answer is no. We have

$$\int_0^{\ln n} \left(\frac{P(x)}{Q(x)} - 1 \right) dx = 1 + \frac{1}{2} + \frac{1}{3} + \cdots + \frac{1}{n} - \ln n,$$

or

$$\int_0^{\ln n} \frac{R(x)}{Q(x)}\, dx = c_n,$$

where $R(x) = P(x) - Q(x)$ and

$$c_n = 1 + \frac{1}{2} + \frac{1}{3} + \cdots + \frac{1}{n} - \ln n.$$

It can be easily proved that

$$\lim_{n \to \infty} n^2(c_{n+1} - c_n) = -\frac{1}{2},$$

using

$$\ln\left(1+\frac{1}{n}\right) = \frac{1}{n} - \frac{1}{2n^2} + o\left(\frac{1}{n^2}\right).$$

Further,

$$\int_{\ln n}^{\ln(n+1)} \frac{R(x)}{Q(x)}dx = c_{n+1} - c_n$$

and, according to Leibniz-Newton's theorem, there exists $\xi_n \in (\ln n, \ln(n+1))$ so that

$$(\ln(n+1) - \ln n) \cdot \frac{R(\xi_n)}{Q(\xi_n)} = c_{n+1} - c_n.$$

Equivalently,

$$\frac{R(\xi_n)}{Q(\xi_n)} \cdot \ln\left(1+\frac{1}{n}\right) = c_{n+1} - c_n,$$

or

$$\frac{nR(\xi_n)}{Q(\xi_n)} = \frac{n^2(c_{n+1} - c_n)}{\ln\left(1+\frac{1}{n}\right)^n},$$

so

$$\lim_{n\to\infty} \frac{nR(\xi_n)}{Q(\xi_n)} = -\frac{1}{2}.$$

With $k = \deg Q - \deg R$, we have $k \geq 1$, because

$$\lim_{n\to\infty} \frac{R(\xi_n)}{Q(\xi_n)} = 0.$$

Furthermore, the limit

$$\lim_{n\to\infty} \frac{\xi_n^k R(\xi_n)}{Q(\xi_n)}$$

is finite and nonzero. It follows that

$$-\frac{1}{2} = \lim_{n\to\infty} \frac{nR(\xi_n)}{Q(\xi_n)} = \lim_{n\to\infty} \frac{\xi_n^k R(\xi_n)}{Q(\xi_n)} \cdot \lim_{n\to\infty} \frac{n}{\xi_n^k},$$

hence the limit $\lim_{n \to \infty} \frac{n}{\xi_n^k}$ is finite and nonzero. But this is a contradiction, because

$$\frac{n}{\xi_n^k} \geq \frac{n}{\ln^k(n+1)}.$$

20. The answer to the first question is negative. Indeed, for positive integers n, the intervals

$$I_n = \left[\frac{1}{2^n} - \frac{1}{2 \cdot 4^n}, \frac{1}{2^n} + \frac{1}{2 \cdot 4^n} \right]$$

are pairwise disjoint; thus one can define a continuous function which is piecewise linear on these intervals, zero between them, and such that

$$f\left(\frac{1}{2^n}\right) = 2^n$$

and f vanishes at the endpoints of I_n. Then f is positive,

$$S_{2^n} \geq \frac{1}{2^n} f\left(\frac{1}{2^n}\right) \geq 1,$$

and one can easily check that

$$\int_{\frac{1}{2^n} - \frac{1}{2 \cdot 4^n}}^{1} f(x)dx = \sum_{i=1}^{n} \frac{1}{2^{i+1}},$$

which converges to $\frac{1}{2}$. As expected, the second question has a positive answer. The fact that f is decreasing implies the inequality

$$\int_{\frac{k-1}{n}}^{\frac{k}{n}} f(x)dx \geq \frac{1}{n} f\left(\frac{k}{n}\right) \geq \int_{\frac{k}{n}}^{\frac{k+1}{n}} f(x)dx.$$

This implies the inequality

$$\frac{1}{n} f\left(\frac{1}{n}\right) + \int_{\frac{1}{n}}^{1} f(x)dx \geq S_n \geq \frac{1}{n} f(1) + \int_{\frac{1}{n}}^{1} f(x)dx,$$

which combined with the hypothesis $\lim_{n \to \infty} \frac{1}{n} f\left(\frac{1}{n}\right) = 0$ gives the desired result.

21. (Sorin Rădulescu, District Mathematical Olympiad, 1984)

a) Let

$$\Delta_n = (a = x_0^n < x_1^n < \cdots < x_{k_n-1}^n < x_{k_n}^n = b)$$

be a sequence of divisions of the interval $[a, b]$, with $\lim_{n\to\infty} ||\Delta_n|| = 0$. For every positive integer n and every $1 \le i \le k_n$, choose in the interval $[x_{i-1}^n, x_i^n]$ the point ξ_i^n (whose existence is ensured by Lagrange's mean value theorem) having the property that $G(x_i^n) - G(x_{i-1}^n) = (x_i^n - x_{i-1}^n)g(\xi_i^n)$. We then have, for each $n \ge 1$,

$$\sigma_{\Delta_n}(f, \xi^n) = \sum_{i=1}^{k_n}(x_i^n - x_{i-1}^n)f(\xi_i^n) \le \sum_{i=1}^{k_n}(x_i^n - x_{i-1}^n)g(\xi_i^n) =$$

$$= \sum_{i=1}^{n}(G(x_i^n) - G(x_{i-1}^n)) = G(b) - G(a).$$

Therefore, by passing to the limit and using the definition of the Riemann integral,

$$\int_a^b f(x)\,dx = \lim_{n\to\infty} \sigma_{\Delta_n}(f, \xi^n) \le G(b) - G(a),$$

as we intended to prove.

b) Clearly, one can prove, in the exact same way as above, that

$$\int_{a'}^{b'} f(x)\,dx \le G(b') - G(a')$$

for all $a \le a' \le b' \le b$. Thus, for $a \le \alpha \le \beta \le b$,

$$\int_a^\alpha f(x)\,dx \le G(\alpha) - G(a),$$

$$\int_\alpha^\beta f(x)\,dx \le G(\beta) - G(\alpha),$$

$$\int_\beta^b f(x)\,dx \le G(b) - G(\beta).$$

But we also have, by hypothesis,

$$\int_a^\alpha f(x)\,dx + \int_\alpha^\beta f(x)\,dx + \int_\beta^b f(x)\,dx = \int_a^b f(x)\,dx = G(b) - G(a)$$

$$= G(\alpha) - G(a) + G(\beta) - G(\alpha) + G(b) - G(\beta);$$

therefore all the above inequalities must be, in fact, equalities. This shows that, actually, $\int_{a'}^{b'} f(x)dx = G(b') - G(a')$ for all $a \le a' \le b' \le b$.

Now, for such a' and b', and any (arbitrarily chosen, but fixed for the moment) $x_0 \in [a', b']$ and any $x \in [a', b']$, we have that

$$\frac{G(x) - G(x_0)}{x - x_0} = \frac{\int_{x_0}^{x} f(t)dt}{x - x_0}$$

is between the bounds of f in the interval $[a', b']$. Thus, the limit of this ratio for $x \to x_0$ (which is $g(x_0)$; if x_0 is one end of the interval $[a', b']$, we only take a one-sided limit) is also between these bounds; therefore

$$\inf_{x \in [a',b']} f(x) \le g(x_0) \le \sup_{x \in [a',b']} f(x)$$

for every $x_0 \in [a', b']$. We deduce that

$$\inf_{x \in [a',b']} f(x) \le \inf_{x \in [a',b']} g(x) \le \sup_{x \in [a',b']} g(x) \le \sup_{x \in [a',b']} f(x)$$

for any subinterval $[a', b']$ of $[a, b]$. (Of course, the first of these inequalities also follows from the hypothesis $f(x) \le g(x)$ for all $x \in [a, b]$, but that is not the case with the last inequality.) Now it's an immediate consequence of the above inequalities that

$$s_\Delta(f) \le s_\Delta(g) \le S_\Delta(g) \le S_\Delta(f) \Rightarrow S_\Delta(g) - s_\Delta(g) \le S_\Delta(f) - s_\Delta(f),$$

for any partition Δ of $[a, b]$, and the Darboux criterion for Riemann integrability shows that the Riemann integrability of f implies the Riemann integrability of g, finishing the proof.

22. **Solution I.** First we see that f is injective on $(a, b]$. Indeed, for some $x, y \in (a, b]$ choose an $a' > a$ such that $x, y \in [a', b]$, and let L be the corresponding constant for this interval. If we assume $f(x) = f(y)$, the inequality $|f(x) - f(y)| \ge L|x - y|$ shows that we necessarily have $x = y$. Being continuous and injective, f is strictly monotone on $(a, b]$, and, by continuity, it is actually strictly monotone on $[a, b]$. We can assume, without loss of generality, that f is strictly increasing and that $[\alpha, \beta]$ is actually the image of $[a, b]$ under the continuously increasing map f (so that $f(a) = \alpha$ and $f(b) = \beta$).

Consider an arbitrary $\epsilon > 0$, and let $M > 0$ be such that $|g(x)| \le M$ for all $x, y \in [\alpha, \beta]$. Let $a' = \min\{a + \epsilon/(4M), b\}$ and let $\alpha' = f(a')$. Consider $L > 0$ with property that $|f(x) - f(y)| \ge L|x - y|$ for all $x, y \in [a', b]$.

Because g is integrable on $[\alpha, \beta]$, it is also integrable on $[\alpha', \beta]$, hence we can find a division $\Delta' = (\alpha' = y_1 < \cdots < y_n = \beta)$ of the interval $[\alpha', \beta]$ such that $S_{\Delta'}(g) - s_{\Delta'}(g) < L\epsilon/2$. Of course, for each y_i, there exists a unique $x_i \in [\alpha, \beta]$ such that $f(x_i) = y_i$ and, in fact, $a' = x_1 < x_2 < \cdots < x_n = b$.

Now we consider the partition $\Delta = (a = x_0 < x_1 < \cdots < x_n = b)$ of $[a, b]$ and compute

$$S_\Delta(g \circ f) - s_\Delta(g \circ f)$$

$$= (a' - a)(M_1(g \circ f) - m_1(g \circ f)) + \sum_{i=2}^{n}(x_i - x_{i-1})(M_i(g \circ f) - m_i(g \circ f)),$$

where $M_i(g \circ f)$ and $m_i(g \circ f)$ are the respective upper and lower bounds of $g \circ f$ in the interval $[x_{i-1}, x_i]$. One sees immediately that $M_i(g \circ f) = M_i(f)$ and $m_i(g \circ f) = m_i(f)$, where $M_i(f)$ and $m_i(f)$ are the corresponding bounds of f in the interval $[y_{i-1}, y_i]$, for $i \geq 2$. For such i we have $x_i - x_{i-1} \leq (1/L)(f(x_i) - f(x_{i-1}) = (1/L)(y_i - y_{i-1})$; therefore the above expression of the difference between the upper and lower Darboux sums of $g \circ f$ on the interval $[a, b]$ can be evaluated as follows:

$$S_\Delta(g \circ f) - s_\Delta(g \circ f) \leq (a' - a)2M + \frac{1}{L}\sum_{i=2}^{n}(y_i - y_{i-1})(M_i(g) - m_i(g))$$

$$= (a' - a)2M + \frac{1}{L}(S_{\Delta'}(g) - s_{\Delta'}(g))$$

$$< \frac{\epsilon}{4M} \cdot 2M + \frac{1}{L} \cdot \frac{L\epsilon}{2} = \epsilon.$$

Thus, for every $\epsilon > 0$, there exists a partition Δ of $[a, b]$ such that $S_\Delta(g \circ f) - s_\Delta(g \circ f) < \epsilon$, which means that $g \circ f$ is Riemann integrable on $[a, b]$, finishing our (first) proof.

Solution II. We use the Lebesgue criterion for Riemann integrability. As in the previous solution, we may assume that f is one to one from $[a, b]$ onto $[\alpha, \beta]$ and strictly increasing. Of course, $g \circ f$ is bounded, so all we need to prove is that $g \circ f$ is continuous almost everywhere. We observe that $D_{g \circ f} \subseteq f^{-1}(D_g)$, where by D_h we denote the set of discontinuities of the function h. (This is just another way to put the fact that if g is continuous at some point of the form $f(x_0)$, $x_0 \in [a, b]$, then $g \circ f$ is continuous at x_0.) Thus it suffices to show that $f^{-1}(D_g)$ is a null set.

Let $\epsilon > 0$ be given, and let $c = \min\{a + \epsilon/2, b\}$. Consider $\gamma = f(c) \Leftrightarrow c = f^{-1}(\gamma)$, and $L > 0$ such that $|f(x) - f(y)| \geq L|x - y|$ for all $x, y \in [c, b]$. It follows that $|f^{-1}(x) - f^{-1}(y)| \leq (1/L)|x - y|$ for all $x, y \in [\gamma, \beta]$. The set D_g can be covered by the union of intervals $[a_i, b_i]$ whose sum of lengths is less than $L\epsilon/2$, and we can assume that each and every such interval is included in $[\alpha, \beta]$ (because D_g is included in this interval; therefore it can also be covered by the union of $[a_i, b_i] \cap [\alpha, \beta]$—whose sum of lengths is at most equal to the sum of lengths of $[a_i, b_i]$). We have

$$D_g \subseteq \bigcup_i [a_i, b_i] \Rightarrow f^{-1}(D_g) \subseteq \bigcup_i [f^{-1}(a_i), f^{-1}(b_i)]$$

$$\Rightarrow f^{-1}(D_g) \subseteq [a, c] \cup \left(\bigcup_i ([f^{-1}(a_i), f^{-1}(b_i)] \cap [c, b]) \right).$$

Each intersection $[f^{-1}(a_i), f^{-1}(b_i)] \cap [c, b]$ is either empty, or reduced to one single point, or $[c, f^{-1}(b_i)] = [f^{-1}(\gamma), f^{-1}(b_i)]$, or the whole interval $[f^{-1}(a_i), f^{-1}(b_i)]$. In the first two cases, the length of such an interval is zero, in the third case, its length is

$$f^{-1}(b_i) - f^{-1}(\gamma) \leq \frac{1}{L}(b_i - \gamma) \leq \frac{1}{L}(b_i - a_i),$$

and in the fourth case, its length is again at most $(1/L)(b_i - a_i)$. Thus we managed to cover $f^{-1}(D_g)$ with a set of intervals whose sum of lengths does not exceed

$$c - a + \frac{1}{L} \sum_i (b_i - a_i) \leq \frac{\epsilon}{2} + \frac{1}{L} \cdot \frac{L\epsilon}{2} = \epsilon,$$

which is what we intended to do.

Remarks. 1) The result remains true if f satisfies the weaker condition: there exists $L > 0$ such that $|f(x) - f(y)| \geq L|x - y|$ for all $x, y \in [a, b]$. Any of the above proofs can be adapted to prove it in this form, with an obviously simpler wording of the demonstration.

2) We have already seen that the composition $g \circ f$ of a continuous g with an integrable f is integrable. This result shows that we can ensure the integrability of $g \circ f$ when g is integrable, and f is continuous, if we add some extra condition for f (some kind of a reverse Lipschitzian property).

3) When $g : [0, 1] \rightarrow \mathbb{R}$ is an integrable function, the function $h : [0, 1] \rightarrow \mathbb{R}$ given by $h(x) = g(x^2)$, for all $x \in [0, 1]$ is also integrable. We had this problem, proposed by Sorin Rădulescu, in the National Mathematics Olympiad, as 12th grade students in 1986.

23. (See, for example, Daniel Daners: *A Short Elementary Proof of* $\sum 1/k^2 = \pi^2/6$, in *Mathematics Magazine*, 5/2012, pp 361–364, where many other good references are given.) This is actually *not* about Riemann sums (or Darboux sums)—but the Riemann integral appears so beautifully in the evaluation of the celebrated sum of the inverses of the squares of positive integers that we can't miss it.

a) The first part is well known. Integration by parts solves it immediately:

$$A_{n-1} - A_n = \int_0^{\pi/2} \cos^{2n-2} x \sin^2 x \, dx$$

$$= -\frac{1}{2n-1} \int_0^{\pi/2} (\cos^{2n-1} x)' \sin x \, dx$$

$$= -\frac{1}{2n-1} \left(\cos^{2n-1} x \sin x \Big|_0^{\pi/2} - \int_0^{\pi/2} \cos^{2n-1} x \cdot \cos x \, dx \right)$$

$$= \frac{1}{2n-1} A_n.$$

Note that $A_{n-1} = \dfrac{2n}{2n-1} A_n$ follows. Integrating by parts (twice), we can get the second formula. Namely, we have,

$$A_n = x \cos^{2n} x \Big|_0^{\pi/2} - \int_0^{\pi/2} x \cdot 2n \cos^{2n-1} x \cdot (-\sin x) \, dx$$

$$= n \int_0^{\pi/2} (x^2)' \cos^{2n-1} x \sin x \, dx$$

$$= nx^2 \cos^{2n-1} x \sin x \, dx \Big|_0^{\pi/2} - n \int_0^{\pi/2} x^2 (\cos^{2n} x - (2n-1) \cos^{2n-2} x \sin^2 x) \, dx$$

$$= -nB_n + n(2n-1)(B_{n-1} - B_n) = n(2n-1)B_{n-1} - 2n^2 B_n.$$

b) By dividing the second equation from the first part by $n^2 A_n$, and by using

$$\frac{2n}{2n-1} A_n = A_{n-1},$$

we get exactly

$$\frac{1}{n^2} = 2 \left(\frac{B_{n-1}}{A_{n-1}} - \frac{B_n}{A_n} \right).$$

Summing up from $n = 1$ to $n = N$ we get (by telescoping)

$$\sum_{n=1}^N \frac{1}{n^2} = 2 \left(\frac{B_0}{A_0} - \frac{B_N}{A_N} \right) = 2 \left(\frac{\pi^2}{12} - \frac{B_N}{A_N} \right).$$

Now we use the well-known inequality $x < \frac{\pi}{2}\sin x$, valid for $x \in (0, \pi/2)$ (it becomes an equality at the endpoints of the interval; the inequality actually expresses the geometric fact that the graph of the concave function $x \mapsto \sin x$ is situated above the chord that connects the points $(0,0)$ and $(\pi/2, 1)$). Accordingly, we have

$$B_n = \int_0^{\pi/2} x^2 \cos^{2n} x\, dx < \frac{\pi^2}{4} \int_0^{\pi/2} \sin^2 x \cos^{2n} x\, dx$$

$$= \frac{\pi^2}{4} \int_0^{\pi/2} (1 - \cos^2 x) \cos^{2n} x\, dx$$

$$= \frac{\pi^2}{4}(A_n - A_{n+1}) = \frac{\pi^2}{4} \cdot \frac{1}{2n+2} A_n.$$

Putting together the last two results, we get

$$0 < \frac{\pi^2}{6} - \sum_{n=1}^{N} \frac{1}{n^2} < \frac{\pi^2}{4(N+1)},$$

and the desired conclusion follows by making N tend to infinity. This is one classical approach to evaluate the value $\zeta(2)$ of the Riemann zeta function (the celebrated Basel problem, solved by Euler in 1735, at the age of twenty-eight).

Note that if, instead of the integrals A_n and B_n, one would rather work with

$$C_n = \int_0^{\pi/2} \cos^{2n+1} x\, dx \quad \text{and} \quad D_n = \int_0^{\pi/2} x^2 \cos^{2n+1} x\, dx,$$

then one would obtain

$$\sum_{n=0}^{N} \frac{1}{(2n+1)^2} = \frac{\pi^2}{8} - \frac{1}{2} \cdot \frac{D_N}{C_N}.$$

The ratio D_N/C_N can be shown to approach zero, as N goes to infinity (in the same manner, we proceeded with B_N/A_N), and hence

$$\sum_{n=0}^{\infty} \frac{1}{(2n+1)^2} = \frac{\pi^2}{8}$$

can be proved in this way. This is no surprise: the latter equation is well known to be equivalent to $\zeta(2) = \pi^2/6$.

Chapter 16
Antiderivatives

An *antiderivative* of the function $f : I \subseteq \mathbb{R} \to \mathbb{R}$ is a differentiable function $F : I \to \mathbb{R}$, such that $F' = f$. If F is an antiderivative of f, then $F + c$ is also an antiderivative of the function f, for every real constant c. In fact, when I is an interval (and in most cases it is) these functions $F + c, c \in \mathbb{R}$, are all the antiderivatives of f. (This is because if F and G are two antiderivatives for the same function f on the interval I, then the derivative of $G - F$ vanishes on I; therefore the difference $G - F$ must be a constant. Note the importance of the fact that I is an interval.) We denote by

$$\int f(x)dx = \{F + c \mid c \in \mathbb{R}\}$$

the set of all antiderivatives of the function f. Starting with the classical formula

$$(fg)' = f'g + fg'$$

one obtains the formula of integration by parts,

$$\int f'(x)g(x)dx = f(x)g(x) - \int f(x)g'(x)dx.$$

We recall the formula of change of the variable, which can be deduced from the formula of derivation of composition of two functions,

$$[f(g(x))]' = f'(g(x))g'(x).$$

More precisely,

$$\int f'(g(x))g'(x)dx = (f \circ g)(x) + c.$$

© Springer Science+Business Media LLC 2017
T. Andreescu et al., *Mathematical Bridges*, DOI 10.1007/978-0-8176-4629-5_16

The existence of antiderivatives of a given function cannot be taken for granted—in general—and we will see below examples of functions that fail to have antiderivatives. However, as an immediate corollary of the fundamental theorem of calculus (see the previous chapter), it follows that continuous functions have antiderivatives. Namely, if $f : I \to \mathbb{R}$ (where I is an interval) is continuous, then any F defined on I by

$$F(x) = \int_a^x f(t) \, dt + c,$$

(with $a \in I$, and $c \in \mathbb{R}$) is an antiderivative of f. Note that, although the above expression of F may give the impression that we can evaluate the antiderivatives of any continuous function, this is not at all the case. For example, the function

$$F(x) = \int_0^x e^{-t^2} \, dt$$

is definitely an antiderivative of $x \mapsto e^{-x^2}$. However, there is no way to express this (or any other antiderivative of $x \mapsto e^{-x^2}$) by means of elementary functions. Nevertheless, such expressions of the antiderivatives of continuous functions can be useful in various problems.

Problem. Let $f, g : I \subseteq \mathbb{R} \to \mathbb{R}$, f with antiderivatives and $g \in C^1$, that is, g is differentiable with continuous derivative. Prove that the function fg has antiderivatives.

Solution. Let F be an antiderivative of the function f. Then by the formula

$$(Fg)' = fg + Fg'$$

we deduce that

$$fg = (Fg)' - Fg'.$$

The function Fg' is continuous, so it has antiderivatives. Hence fg has antiderivatives as a difference of two functions with antiderivatives. $\square$

Problem. Let $f, g : \mathbb{R} \to \mathbb{R}$ be functions with the following properties:

i) f is continuous and bijective.
ii) g is nonnegative and has antiderivatives.

Prove that fg also has antiderivatives.

Solution. Let G be an antiderivative for g and let H be an antiderivative for $G(f^{-1}(x))$. We claim that the function

$$F(x) = f(x)G(x) - H(f(x))$$

is an antiderivative for fg. Consider a real number x_0. Let

$$S(x) = \frac{F(x) - F(x_0)}{x - x_0}.$$

Observe that we can also write

$$S(x) = f(x) \cdot \frac{G(x) - G(x_0)}{x - x_0} + G(x_0) \cdot \frac{f(x) - f(x_0)}{x - x_0} - \frac{H(f(x)) - H(f(x_0))}{x - x_0}.$$

Using the mean value theorem, we can write

$$\frac{H(f(x)) - H(f(x_0))}{x - x_0} = H'(f(c(x))) \cdot \frac{f(x) - f(x_0)}{x - x_0}$$

for some $c(x)$ between x_0 and x. The choice of H gives $H'(f(c(x))) = G(c(x))$. In order to prove that $S(x)$ tends to $f(x_0)g(x_0)$ when x tends to x_0, it is enough to show that

$$\frac{G(x) - G(c(x))}{x - x_0}$$

is bounded in a neighborhood of x_0 (we use here the continuity of f). However, this is not difficult, since G is increasing (because $G' = g \geq 0$) and

$$\left| \frac{G(x_0) - G(c(x))}{x - x_0} \right| \leq \left| \frac{G(x_0) - G(x)}{x - x_0} \right|.$$

The last quantity is bounded as x tends to x_0, by the differentiability of G at x_0. This finishes the proof. $\square$

Problem. Let $f : \mathbb{R} \to \mathbb{R}$ be a continuous function such that the following limit exists:

$$\lim_{|x| \to \infty} \frac{1}{x} \int_0^x f(t)dt = I(f).$$

Prove that the function g defined by $g(0) = I(f)$ and $g(x) = f(\frac{1}{x})$ if $x \neq 0$ has antiderivatives.

Solution. Let F be an antiderivative for f. Then we clearly have for all $x \neq 0$

$$\left(x^2 F\left(\frac{1}{x}\right) \right)' = 2xF\left(\frac{1}{x}\right) - g(x).$$

Now, consider the function h defined by $h(0) = 2I(f)$ and $h(x) = 2xF(\frac{1}{x})$ for nonzero x. The hypothesis of the problem implies that h is continuous, so it has an antiderivative H. We can write g as the difference of two functions having antiderivatives $g = H' - U'$, where U is defined by $U(x) = x^2 F(\frac{1}{x})$ if $x \neq 0$ and $U(0) = 0$. $\square$

We have seen several conditions under which a function has antiderivatives. But how can we prove that a function does not have antiderivatives? The easiest and most practical way is the observation that any function which has antiderivatives also has the intermediate value property. This follows from Darboux's theorem, stating that for any differentiable function F, its derivative has the intermediate value property. The proof of this result has been presented in Chapter 14. There are countless examples of situations in which this criterion works very well, but we will limit ourselves to one problem. Before that, note however that there are functions having the intermediate value property which do not have antiderivatives. We let the reader check that such an example is given by the function $f(x) = \frac{1}{x}\cos(\frac{1}{x})$ for $x \neq 0$ and $f(0) = 0$.

Problem. Prove that any function $f : I \to \mathbb{R}$ defined on a nontrivial interval I and satisfying $f(f(x)) = -x$ does not have antiderivatives.

Solution. Suppose that f has antiderivatives. Therefore, f has the intermediate value property. Note however that f is injective, because if $f(x) = f(y)$, then $-x = f(f(x)) = f(f(y)) = -y$ and so $x = y$. Therefore f is monotone. This implies that $f(f(x))$ is increasing, which is impossible, because $g(x) = -x$ is decreasing. This contradiction shows that f cannot have antiderivatives. $\square$

Here is a more subtle problem, which needs several arguments.

Problem. Prove that there exists no function $f : [0, 1] \to [0, 1]$ having an antiderivative $F : [0, 1] \to [0, 1]$ such that $F(f(x)) = x$ for all x.

Solution. Suppose, by way of contradiction, that f is such a function. Clearly, f is injective because $f(x) = f(y) \Rightarrow x = F(f(x)) = F(f(y)) = y$. Because f is injective and has the intermediate value property, f is strictly monotone and continuous. Because $F' = f \geq 0$, F is increasing. We also deduce that f is increasing.

Suppose now that $f(0) > 0$, thus $F(f(0)) > F(0)$, thus $F(0) < 0$, a contradiction. Thus $f(0) = 0$, so $0 = F(f(0)) = F(0)$. Similarly, $f(1) = F(1) = 1$. Now take $x \in (0, 1)$; then Lagrange's theorem gives $c_1, c_2 \in (0, 1)$ such that

$$\begin{cases} F(x) - F(0) = xf(c_1) \\ F(1) - F(x) = (1 - x)f(c_2) \end{cases}$$

Thus $F(x) < x$ and $1 - F(x) < 1 - x$, which is impossible. Therefore such a function cannot exist. $\square$

We continue with a functional equation involving antiderivatives.

Problem. Let a be a nonzero real number. Find all functions $f : (0, \infty) \to \mathbb{R}$ having an antiderivative F such that

$$f(x)F\left(\frac{1}{x}\right) = \frac{a}{x}.$$

Solution. Replacing x with $\frac{1}{x}$, we also obtain

$$f\left(\frac{1}{x}\right)F(x) = ax.$$

Thus

$$f\left(\frac{1}{x}\right)F(x) = x^2 f(x)F\left(\frac{1}{x}\right) \Rightarrow \frac{f(x)}{F(x)} = \frac{1}{x^2}\frac{f\left(\frac{1}{x}\right)}{F\left(\frac{1}{x}\right)}.$$

This implies

$$(\ln|F(x)|)' = -\left[\ln F\left(\frac{1}{x}\right)\right]',$$

so there exists a constant c such that $F(x)F\left(\frac{1}{x}\right) = c$. Replace this in the statement of the problem to obtain

$$\frac{F'(x)}{F(x)} = \frac{a}{cx} \Rightarrow F(x) = bx^{\frac{a}{c}}$$

for some b. Because $F\left(\frac{1}{x}\right) = bx^{-\frac{a}{c}}$, we deduce that $b = \sqrt{c}$. Thus the solution of the problem is $f(x) = \frac{1}{\sqrt{c}}x^{\frac{a}{c}-1}$ for some constant $c > 0$. $\square$

Problem. Let $f : \mathbb{R} \to \mathbb{R}$ be a continuous differentiable function, strictly monotone, such that $f(0) = 0$. Prove that the function $g : \mathbb{R} \to \mathbb{R}$

$$g(x) = \begin{cases} f'(x)\sin\dfrac{1}{f(x)}, & x \neq 0 \\ 0, & x = 0 \end{cases}$$

has antiderivatives.

Solution. Let us observe that

$$\left(f^2(x)\cos\frac{1}{f(x)}\right)' = 2f(x)f'(x)\cos\frac{1}{f(x)} + f'(x)\sin\frac{1}{f(x)}$$

which can be also written as

$$f'(x)\sin\frac{1}{f(x)} = \left(f^2(x)\cos\frac{1}{f(x)}\right)' - 2f(x)f'(x)\cos\frac{1}{f(x)}.$$

Consider

$$h(x) = \begin{cases} 2f(x)f'(x)\cos\dfrac{1}{f(x)}, & x \neq 0 \\ 0, & x = 0 \end{cases},$$

which is clearly continuous. Take H to be an antiderivative of h and consider

$$G(x) = \begin{cases} f^2(x)\cos\dfrac{1}{f(x)} - H(x), & x \in \mathbb{R}^* \\ -H(0), & x = 0 \end{cases}.$$

Note that G is continuous because f is and $f(0) = 0$. Also,

$$\lim_{x\to 0}\frac{G(x)-G(0)}{x} = \lim_{x\to 0}\frac{f^2(x)\cos\dfrac{1}{f(x)} - H(x) + H(0)}{x}$$

$$= \lim_{x\to 0}\frac{f(x)}{x}f(x)\cos\frac{1}{f(x)} - \frac{H(x)-H(0)}{x}$$

$$= f'(0)\lim_{x\to 0}f(x)\cos\frac{1}{f(x)} - h(0) = 0.$$

Thus G is differentiable and $G' = g$. $\square$

Proposed Problems

1. Find the integrals

$$I = \int\frac{\sqrt{\sqrt{x}+\sqrt{x-1}}}{1+\sqrt{x}}dx, \quad J = \int\frac{\sqrt{\sqrt{x}-\sqrt{x-1}}}{1+\sqrt{x}}dx, \quad x \in (1,\infty).$$

2. Let $F : \mathbb{R} \to \mathbb{R}$ be an antiderivative of the function $f : \left[\dfrac{\pi}{3}, \dfrac{2\pi}{3}\right] \to \mathbb{R}$ with

$$f(x) = \frac{x}{\sin x}.$$

Prove that

$$F\left(\frac{2\pi}{3}\right) - F\left(\frac{\pi}{3}\right) = \frac{\pi}{2} \ln 3.$$

3. Let $f : [0, 2\pi] \to \mathbb{R}$ be continuous and decreasing such that $f(\pi) = 0$. Prove that for every antiderivative F of the function f, we have

$$\int_0^{2\pi} F(x) \cos x \, dx \le 0.$$

4. Show that if $f : \mathbb{R} \to \mathbb{R}$ has antiderivatives, then the same is true for $h : \mathbb{R} \to \mathbb{R}$ defined by $h(x) = |x| f(x)$ for all x.

5. Prove that the function $f : \mathbb{R} \to \mathbb{R}$ given by the formula

$$f(x) = \begin{cases} \sin \frac{1}{x}, & x \neq 0 \\ 0, & x = 0 \end{cases}$$

has antiderivatives.

6. Let $f : \mathbb{R} \to \mathbb{R}$ be such that one of its antiderivatives F has the property that $\lim\limits_{x \to -\infty} F(x) = 0$. Prove that the function $g : [0, \infty) \to \mathbb{R}$ given by the formula

$$g(x) = \begin{cases} f(\ln x), & x > 0 \\ 0, & x = 0 \end{cases}$$

has antiderivatives.

7. Let $f : \mathbb{R} \to \mathbb{R}$ be such that the functions $g, h : \mathbb{R} \to \mathbb{R}$,

$$g(x) = f(x) \sin x, \quad h(x) = f(x) \cos x$$

have antiderivatives. Prove that f has antiderivatives.

8. For a function $f : \mathbb{R} \to \mathbb{R}$ and for every positive integer n, denote by f_n the restriction of the function f to the interval $(-n, \infty)$. Decide if the following assertions are true:

a) If the functions f_n have the intermediate value property, for all positive integers n, then the function f has the intermediate value property.

b) If the functions f_n have antiderivatives, for all positive integers n, then the function f has antiderivatives.

c) Answer similar questions for a function $f : [0, \infty) \to \mathbb{R}$ with its restrictions f_n on the interval $[1/n, \infty), n \in \mathbb{N}^*$.

9. Find an antiderivative of the function $g : (0, \infty) \to \mathbb{R}$, $g(x) = e^{x^2 + \frac{1}{x^2}}$, if we know an antiderivative F of the function $f : \mathbb{R} \to \mathbb{R}, f(x) = e^{x^2}$.

10. Find an antiderivative of the function $f : \mathbb{R} \to \mathbb{R}, f(x) = e^{x^2}$, if we know an antiderivative G of the function $g : (0, \infty) \to \mathbb{R}$, $g(x) = e^{x^2 + \frac{1}{x^2}}$.

11. Let $f : (-\pi/2, \pi/2) \to [0, 1]$ and let us define the function $g : (-\pi/2, \pi/2) \to \mathbb{R}$, by the formula

$$g(x) = f(x) \tan x.$$

Prove that:

a) If f has antiderivatives, then g has antiderivatives.

b) If g has antiderivatives and f is continuous at zero, then f has antiderivatives.

12. Are there differentiable functions $f : \mathbb{R} \to (0, \infty)$ such that $\ln f$ is an antiderivative of $\ln F$, for some antiderivative F of f?

13. Let $f : \mathbb{R} \to \mathbb{R}$ be a function with antiderivatives such that

$$f(x) = f(f(x) - x),$$

for all real numbers x. Prove that the function f is continuous.

14. Let $f : [0, \infty) \to \mathbb{R}$, with $f(0) = 0$, continuous on $(0, \infty)$, satisfying the following properties:

a) there exist positive real numbers r and M such that $|f(x)| \le M$, for all real numbers $x \in [0, r]$;

b) there exists a decreasing sequence $(x_n)_{n \ge 0}$, converging to zero, with $\lim\limits_{n \to \infty} \dfrac{x_{n+1}}{x_n} = 1$, and such that

$$\int_{x_n}^{x_{n+1}} f(x)dx = 0,$$

for all positive integers n. Prove that the function f has antiderivatives.

15. Consider the function $f : [0, \infty) \to \mathbb{R}$, with $f(0) = 0$, continuous on $(0, \infty)$ and linear on each interval of the form $\left[\dfrac{1}{n+1}, \dfrac{1}{n} \right], n \ge 1$ such that

$$f\left(\frac{1}{n}\right) = (-1)^n,$$

for all positive integers n. Prove that the function f has antiderivatives.

16. Let f be a real function with an antiderivative F on $\mathbb{R}$. Prove that if

$$f(x) \le \frac{|x|}{1 + |x|}$$

for all $x \in \mathbb{R}$, then F has a unique fixed point (i.e., there exists precisely one $x_0 \in \mathbb{R}$ such that $F(x_0) = x_0$).

17. Let $f, g : \mathbb{R} \to \mathbb{R}$ be two continuously differentiable functions such that f' is bounded, $\lim_{x \to \pm\infty} f(x)/x = 0$, $g(x) = 0$ if and only if $x = 0$, and $g'(0) \ne 0$. Prove that $h : \mathbb{R} \to \mathbb{R}$ defined by

$$h(x) = \begin{cases} f'\left(\frac{1}{g(x)}\right), & \text{for } x \ne 0 \\ 0, & \text{for } x = 0 \end{cases}$$

has antiderivatives on $\mathbb{R}$.

18. (Jarnik's theorem) Let $f, g : \mathbb{R} \to \mathbb{R}$ be real functions that have antiderivatives and such that g is nonzero on $\mathbb{R}$. Show that f/g has the intermediate value property.

Solutions

1. First let us compute

$$I + J = \int \frac{\sqrt{\sqrt{x} + \sqrt{x-1}} + \sqrt{\sqrt{x} - \sqrt{x-1}}}{1 + \sqrt{x}} dx$$

With the notation

$$f(x) = \sqrt{\sqrt{x} + \sqrt{x-1}} + \sqrt{\sqrt{x} - \sqrt{x-1}},$$

we have

$$f^2(x) = (\sqrt{x} + \sqrt{x-1}) + (\sqrt{x} - \sqrt{x-1}) + 2\sqrt{(\sqrt{x} + \sqrt{x-1})(\sqrt{x} - \sqrt{x-1})}$$

$$= 2\sqrt{x} + 2,$$

so $f(x) = \sqrt{2(1 + \sqrt{x})}$. It follows that

$$I + J = \int \frac{\sqrt{2(1 + \sqrt{x})}}{1 + \sqrt{x}} dx = \sqrt{2} \int \frac{dx}{\sqrt{1 + \sqrt{x}}}.$$

By changing the variable $\dfrac{1}{\sqrt{1 + \sqrt{x}}} = t$, we obtain

$$I + J = \frac{4\sqrt{2}}{3}(\sqrt{x} - 2)\sqrt{1 + \sqrt{x}} + C.$$

Next,

$$I - J = \int \frac{\sqrt{\sqrt{x} + \sqrt{x - 1}} - \sqrt{\sqrt{x} - \sqrt{x - 1}}}{1 + \sqrt{x}}\,dx.$$

With the notation

$$g(x) = \sqrt{\sqrt{x} + \sqrt{x - 1}} - \sqrt{\sqrt{x} - \sqrt{x - 1}},$$

we have

$$g^2(x) = 2\sqrt{x} - 2 \Rightarrow g(x) = \sqrt{2(\sqrt{x} - 1)}.$$

Therefore

$$I - J = \int \frac{\sqrt{2(\sqrt{x} - 1)}}{1 + \sqrt{x}}\,dx = \sqrt{2}\int \frac{\sqrt{\sqrt{x} - 1}}{\sqrt{x} + 1}\,dx$$

$$= \frac{4(\sqrt{x} - 4)\sqrt{\sqrt{x} - 1}}{3} - 4\sqrt{2}\arctan \frac{\sqrt{\sqrt{x} - 1}}{\sqrt{2}} + c.$$

2. By the Leibniz-Newton formula,

$$I = F\left(\frac{2\pi}{3}\right) - F\left(\frac{\pi}{3}\right) = \int_{\pi/3}^{2\pi/3} \frac{x}{\sin x}\,dx$$

then with $x = \pi - t$, it follows that

$$I = \int_{\pi/3}^{2\pi/3} \frac{\pi - t}{\sin t}\,dt = \int_{\pi/3}^{2\pi/3} \frac{\pi - x}{\sin x}\,dx.$$

(Note the general useful result that hides behind, namely, the formula

$$\int_a^b f(x)\,dx = \int_a^b f(a + b - x)\,dx.$$

We mentioned this in Chapter 1, too.) Hence

$$2I = \int_{\pi/3}^{2\pi/3} \frac{\pi}{\sin x} dx = \pi \ln\left(\tan\frac{x}{2}\right)\Big|_{\pi/3}^{2\pi/3} = \pi \ln 3.$$

3. For $x \in [0, \pi]$, $f(x) \geq 0$ and for $x \in [\pi, 2\pi]$, $f(x) \leq 0$. Thus $f(x) \sin x \geq 0$, for all $x \in [0, 2\pi]$. By integrating by parts, we have

$$\int_0^{2\pi} F(x) \cos x \, dx = \int_0^{2\pi} F(x)(\sin x)' dx$$

$$= F(x)\sin x\Big|_0^{2\pi} - \int_0^{2\pi} f(x) \sin x \, dx$$

$$= -\int_0^{2\pi} f(x) \sin x \, dx \leq 0.$$

4. Let F be an antiderivative of f; thus $F' = f$ on $\mathbb{R}$. Because

$$\int xf(x)dx = \int xF'(x)dx = xF(x) - \int F(x)dx$$

the idea is to consider an antiderivative G of F (F is differentiable, thus continuous, thus it has antiderivatives), then note that $x \mapsto xF(x) - G(x)$ is an antiderivative of $x \mapsto xf(x)$. Finally glue together two such antiderivatives of $x \mapsto -xf(x)$ (for $x < 0$) and of $x \mapsto xf(x)$ (for $x > 0$) in order to obtain a continuous function. Namely, consider $H : \mathbb{R} \to \mathbb{R}$ defined by

$$H(x) = \begin{cases} -xF(x) + G(x), & \text{for } x < 0 \\ xF(x) - G(x) + 2G(0), & \text{for } x \geq 0. \end{cases}$$

We have $H'(x) = |x|f(x)$ for all $x \neq 0$, and, by the corollary of Lagrange's theorem, the differentiability of H at the origin follows; moreover,

$$H'(0) = \lim_{x \to 0} H'(x) = \lim_{x \to 0} h(x) = h(0)$$

follows; therefore H is an antiderivative of h on $\mathbb{R}$.

5. Let us define the function

$$h(x) = \begin{cases} x^2 \cos\frac{1}{x}, & x \neq 0 \\ 0, & x = 0 \end{cases}.$$

For every nonzero real x, we have

$$h'(x) = 2x \cos\frac{1}{x} + \sin\frac{1}{x}$$

and

$$\lim_{x \to 0} \frac{h(x) - h(0)}{x - 0} = \lim_{x \to 0} \left(x \cos \frac{1}{x} \right) = 0.$$

Hence the function h is differentiable with

$$h'(x) = \begin{cases} 2x \cos \frac{1}{x} + \sin \frac{1}{x}, & x \neq 0 \\ 0, & x = 0 \end{cases}.$$

Moreover, if we denote

$$u(x) = \begin{cases} 2x \cos \frac{1}{x}, & x \neq 0 \\ 0, & x = 0 \end{cases},$$

then u is continuous, and

$$h'(x) = u(x) + f(x).$$

Hence $f(x) = h'(x) - u(x)$ has antiderivatives, as the difference of two functions with antiderivatives. Observe that this is a special case of a problem discussed in the theoretical part.

6. Let us define the function $h : [0, \infty) \to \mathbb{R}$ by the formula

$$h(x) = \begin{cases} xF(\ln x), & x > 0 \\ 0, & x = 0 \end{cases}.$$

For $x > 0$, we have

$$h'(x) = F(\ln x) + f(\ln x).$$

Also

$$\lim_{x \to 0} \frac{h(x) - h(0)}{x - 0} = \lim_{x \to 0} F(\ln x) = \lim_{y \to -\infty} F(y) = 0,$$

so h is differentiable with

$$h'(x) = \begin{cases} F(\ln x) + f(\ln x), & x > 0 \\ 0, & x = 0 \end{cases}.$$

If $u : [0, \infty) \to \mathbb{R}$,

$$u(x) = \begin{cases} F(\ln x), & x > 0 \\ 0, & x = 0 \end{cases}$$

then u is continuous, and $h'(x) = u(x) + g(x)$. From here,

$$g(x) = h'(x) - u(x),$$

so the function g has antiderivatives, as a difference of two functions that have antiderivatives.

7. We will use the fact that the product of a C^1-function and a function with antiderivatives is also a function with antiderivatives. Here, $g(x)$ has antiderivatives, and the functions sin and cos are C^1, so the functions $g(x) \sin x$ and $h(x) \cos x$ have antiderivatives. We have

$$g(x) \sin x = f(x) \sin^2 x, \quad h(x) \cos x = f(x) \cos^2 x.$$

Their sum also has antiderivatives,

$$f(x) \sin^2 x + f(x) \cos^2 x = f(x).$$

8. a) True. For $a < b$, we prove that $f(I)$ is an interval, where $I = (a, b)$. For $c < a, c < 0, f$ has the intermediate value property on (c, ∞), so $f(I)$ is an interval.

b) True. If F_n is the antiderivative of f_n with $F_n(1) = 0$, then the function $F : \mathbb{R} \to \mathbb{R}$ given by the formula

$$F(x) = F_n(x), \quad x \in (-n, \infty)$$

is well defined, and it is an antiderivative of the function f.

c) False. One example is the function $f : [0, \infty) \to \mathbb{R}$ given by the formula

$$f(x) = \begin{cases} \sin \frac{1}{x}, & x \in (0, \infty) \\ 2007, & x = 0 \end{cases}.$$

9. We have

$$\left[F\left(x + \frac{1}{x}\right) \right]' = f\left(x + \frac{1}{x}\right)\left(1 - \frac{1}{x^2}\right) = \left(1 - \frac{1}{x^2}\right) e^{\left(x + \frac{1}{x}\right)^2},$$

so

$$\left[F\left(x + \frac{1}{x}\right) \right]' = e^2 \left(1 - \frac{1}{x^2}\right) e^{x^2 + \frac{1}{x^2}}. \tag{16.1}$$

Then

$$\left[F\left(x - \frac{1}{x}\right) \right]' = f\left(x - \frac{1}{x}\right)\left(1 + \frac{1}{x^2}\right) = \left(1 + \frac{1}{x^2}\right) e^{\left(x - \frac{1}{x}\right)^2}$$

and

$$\left[F\left(x-\frac{1}{x}\right)\right]' = e^{-2}\left(1+\frac{1}{x^2}\right)e^{x^2+\frac{1}{x^2}}. \qquad (16.2)$$

From relations (16.1) and (16.2), we obtain

$$\begin{cases} e^{-2}\left[F\left(x+\frac{1}{x}\right)\right]' = \left(1-\frac{1}{x^2}\right)e^{x^2+\frac{1}{x^2}} \\ e^{2}\left[F\left(x-\frac{1}{x}\right)\right]' = \left(1+\frac{1}{x^2}\right)e^{x^2+\frac{1}{x^2}} \end{cases}$$

and by adding, we deduce that

$$2e^{x^2+\frac{1}{x^2}} = e^{-2}\left[F\left(x+\frac{1}{x}\right)\right]' + e^{2}\left[F\left(x-\frac{1}{x}\right)\right]'.$$

Hence an antiderivative of the function g is

$$G(x) = \frac{1}{2}\left[e^{-2}F\left(x+\frac{1}{x}\right) + e^{2}F\left(x-\frac{1}{x}\right)\right].$$

10. We have

$$\left[G(x) - G\left(\frac{1}{x}\right)\right]' = g(x) - g\left(\frac{1}{x}\right)\left(-\frac{1}{x^2}\right) = g(x) + \frac{1}{x^2}\cdot g\left(\frac{1}{x}\right)$$

$$= e^{x^2+\frac{1}{x^2}} + \frac{1}{x^2}\cdot e^{x^2+\frac{1}{x^2}} = \left(1+\frac{1}{x^2}\right)e^{x^2+\frac{1}{x^2}}$$

$$= e^2\left(x-\frac{1}{x}\right)' e^{(x-\frac{1}{x})^2} = e^2\left[F\left(x-\frac{1}{x}\right)\right]',$$

so

$$\left[G(x) - G\left(\frac{1}{x}\right)\right]' = e^2\left[F\left(x-\frac{1}{x}\right)\right]'.$$

By integration,

$$F\left(x-\frac{1}{x}\right) = e^{-2}\left[G(x) - G\left(\frac{1}{x}\right)\right].$$

With the notation $x - \dfrac{1}{x} = y$, we have

$$x = \frac{\sqrt{y^2+4}+y}{2}$$

and so

$$F(y) = e^{-2} \left[G\left(\frac{\sqrt{y^2 + 4} + y}{2} \right) - G\left(\frac{\sqrt{y^2 + 4} - y}{2} \right) \right].$$

11. a) The function f has antiderivatives, and the tangent function (defined on $(-\pi/2, \pi/2)$) is differentiable, with continuous derivative, so g has antiderivatives.

b) We have

$$f(x) = g(x) \cot x,$$

for all $x \in (-\pi/2, 0) \cup (0, \pi/2)$, so f has antiderivatives on the intervals $(-\pi/2, 0)$ and $(0, \pi/2)$, as the product between a function with antiderivatives and a C^1-function. Let F_1, F_2 be antiderivatives of the function f on the intervals $(-\pi/2, 0)$ and $(0, \pi/2)$, respectively. The functions F_1 and F_2 are monotone, so there exist the limits

$$\lim_{x \to 0} F_1(x) = l_1, \quad \lim_{x \to 0} F_2(x) = l_2.$$

These limits are finite, because according to Lagrange's theorem,

$$\left| F_1(x) - F_1\left(-\frac{\pi}{4} \right) \right| = \left| f(c_x) \left(x + \frac{\pi}{4} \right) \right| \le \frac{\pi}{4}$$

and similarly for F_2. Now let us define the function $F : (-\pi/2, \pi/2) \to \mathbb{R}$, by

$$F(x) = \begin{cases} F_1(x) - l_1, & x \in (-\pi/2, 0) \\ 0, & x = 0 \\ F_2(x) - l_2, & x \in (0, \pi/2) \end{cases}.$$

Then F is continuous on $(-\pi/2, \pi/2)$ and differentiable on $(-\pi/2, \pi/2) \setminus \{0\}$, with $F'(x) = f(x)$. Thus, F is an antiderivative of the function f.

12. The answer is yes. One example is

$$f(x) = e^{1 + x + e^x}.$$

We have

$$(\ln f)' = \ln F \Rightarrow \frac{f'}{f} = \ln F \Rightarrow f' = f \ln F$$

or

$$f' = (F \ln F - F)'.$$

Therefore the functions f and $F \ln F - F$ differ by a constant. We try to find a function for which $f = F \ln F - F$ or

$$\frac{f}{F \ln F - F} = 1,$$

if we are lucky to find a function for which the denominator is not zero. Then $[G(F(x))]' = 1$, where G is an antiderivative of

$$\frac{1}{x \ln x - x}.$$

Let us put $y = \ln x$, then

$$\int \frac{dx}{x \ln x - x} = \int \frac{dx}{x(\ln x - 1)} = \int \frac{dy}{y - 1} = \ln(\ln x - 1) + C.$$

Thus

$$\ln(\ln F(x) - 1) = x \Rightarrow \ln F(x) - 1 = e^x \Rightarrow F(x) = e^{1 + e^x},$$

so $f(x) = e^{1 + x + e^x}$.

13. Let us define the function $g : \mathbb{R} \to \mathbb{R}$ by the formula $g(x) = f(x) - x$. The function g has antiderivatives, as the difference of two functions which have antiderivatives. We have

$$g(g(x)) = g(f(x) - x) = f(f(x) - x) - (f(x) - x) = x,$$

so $g \circ g = 1_{\mathbb{R}}$. From this it follows that g is injective. But any injective function with antiderivatives is continuous.

Finally, the function $f(x) = x + g(x)$ is continuous, as the sum of two continuous functions.

14. We will prove that the function $F : [0, \infty) \to \mathbb{R}$ given by the formula

$$F(x) = \begin{cases} \int_{x_0}^x f(t)dt, & x > 0 \\ 0, & x = 0 \end{cases},$$

for a fixed $x_0 > 0$ is an antiderivative of the function f, i.e., $F' = f$. For $x > 0$,

$$F'(x) = \left(\int_{x_0}^x f(t)dt \right)' = f(x).$$

Let now $(y_n)_{n\geq 0}$ be a sequence convergent to 0 and let k_n be such that $y_n \in$ $(x_{k_n+1}, x_{k_n}]$ for all n. Then the hypothesis implies

$$\int_{x_0}^{y_n} f(t)dt = \int_{x_0}^{x_1} f(t)dt + \cdots + \int_{x_{k_n-1}}^{x_{k_n}} f(t)dt + \int_{x_{k_n}}^{y_n} f(t)dt = \int_{x_{k_n}}^{y_n} f(t)dt,$$

so

$$\left| \frac{F(y_n) - F(0)}{y_n} \right| \leq M \left(\frac{x_{k_n}}{x_{k_n+1}} - 1 \right).$$

This last quantity being convergent to 0, it follows that F is differentiable at 0 and $F'(0) = f(0) = 0$. This finishes the proof.

15. The problem follows from the previous one, by taking $x_n = \dfrac{1}{n+1}$, $n \in \mathbb{N}$.

Being linear on each interval $\left[\dfrac{1}{n+1}, \dfrac{1}{n} \right]$ with

$$f\left(\frac{1}{n+1} \right) = (-1)^{n+1}, \quad f\left(\frac{1}{n} \right) = (-1)^n,$$

it follows that

$$\int_{1/(n+1)}^{1/n} f(t)dt = 0,$$

for all nonnegative integers n. Now we are under the hypothesis of the previous problem.

16. The relation

$$f(x) \leq \frac{x}{1+x} = 1 - \frac{1}{1+x}$$

for all $x \geq 0$ can also be expressed by saying that the derivative of the function $G(x) = F(x) - x + \ln(1+x)$ is nonpositive on $[0, \infty)$. Thus G decreases on $[0, \infty)$, consequently, $G(x) \leq G(0)$ for all $x \geq 0$, that is,

$$F(x) - x \leq F(0) - \ln(1+x)$$

for $x \geq 0$. Letting x go to infinity, we get $\lim\limits_{x \to \infty} (F(x) - x) = -\infty$.
Similarly, we have

$$f(x) \leq \frac{-x}{1-x} = 1 - \frac{1}{1-x}$$

for all $x \leq 0$, which yields the nonpositivity of the derivative of

$$H(x) = F(x) - x - \ln(1-x),$$

then

$$F(x) - x \geq F(0) + \ln(1 - x)$$

for $x \leq 0$, and, further, $\lim_{x \to -\infty} (F(x) - x) = \infty$.

Now, $x \mapsto F(x) - x$ is a continuous function whose limits at $\pm\infty$ have opposite signs. By the intermediate value theorem, there must be a point at which this function has zero value, that is, there exists a fixed point for F.

However, two such points cannot exist. For, if we assumed that $a \neq b$, $F(a) = a$, and $F(b) = b$, then we would get the contradiction

$$1 = \frac{F(b) - F(a)}{b - a} = f(c) \leq \frac{|c|}{1 + |c|} \Leftrightarrow 1 + |c| \leq |c|$$

by using Lagrange's mean value theorem (c is the intermediate point between a and b whose existence is assured by the theorem). (Actually, the condition for f shows that F is a contraction; hence it cannot have more than one fixed point.)

This is a problem proposed by Sorin Rădulescu for the Romanian Mathematical Olympiad in the year 1986. (Yes, we were there.)

17. Let A, b, c be defined by

$$A(x) = \begin{cases} g^2(x)f\left(\frac{1}{g(x)}\right), & \text{for } x \neq 0 \\ 0, & \text{for } x = 0 \end{cases}$$

$$b(x) = \begin{cases} 2g(x)g'(x)f\left(\frac{1}{g(x)}\right), & \text{for } x \neq 0 \\ 0, & \text{for } x = 0 \end{cases}$$

and

$$c(x) = \begin{cases} (g'(x) - g'(0))f'\left(\frac{1}{g(x)}\right), & \text{for } x \neq 0 \\ 0, & \text{for } x = 0. \end{cases}$$

One immediately sees that $A'(x) = b(x) - c(x) - g'(0)h(x)$ for all $x \neq 0$.
Because

$$\lim_{x \to 0} g(x)f\left(\frac{1}{g(x)}\right) = \lim_{x \to 0} \frac{f(1/g(x))}{1/g(x)} = \lim_{t \to \pm\infty} \frac{f(t)}{t} = 0$$

and $g(x)/x$ has a finite limit ($g'(0)$) at the origin, we also have

$$A'(0) = \lim_{x \to 0} \frac{A(x)}{x} = \lim_{x \to 0} \frac{g(x)}{x} \cdot g(x)f\left(\frac{1}{g(x)}\right) = 0;$$

hence the equality $A'(x) = b(x) - c(x) - g'(0)h(x)$ actually holds for all $x \in \mathbb{R}$.

Because (as we have just seen) $\lim_{x \to 0} g(x)f(1/g(x)) = 0$ and because $\lim_{x \to 0} g'(x) = g'(0)$ is finite, the continuity of b at the origin follows. Similarly, since $\lim_{x \to 0} (g'(x) - g'(0)) = 0$ and f' is bounded, c is continuous at the origin. Also, the fact that b and c are continuous at nonzero points is a consequence of the conditions from the hypothesis—thus b and c are continuous functions on $\mathbb{R}$. Being continuous, they surely have antiderivatives, while A' obviously has antiderivative A. All the above facts yield the conclusion that

$$h = \frac{1}{g'(0)}(b - c - A')$$

has antiderivatives, too, which finishes the proof.

This is a problem proposed by Ion Chiţescu for the Romanian Mathematical Olympiad in the year 1990. (We were not there anymore.)

18. Let $a < b$ be reals such that $f(a)/g(a)$ and $f(b)/g(b)$ are distinct and let y be between $f(a)/g(a)$ and $f(b)/g(b)$. Note that g has the intermediate value property, because it has antiderivatives. Since g is nonzero, it must have a constant sign on $\mathbb{R}$. It follows that $g(a)$ and $g(b)$ have the same sign; hence $f(a) - yg(a)$ and $f(b) - yg(b)$ have opposite signs. Thus for the function h defined by $h(t) = f(t) - yg(t)$, for $t \in [a, b]$, we have $h(a)h(b) < 0$. On the other hand, because f and g both have antiderivatives, h has also antiderivatives; therefore h has the intermediate value property. Consequently, there exists $x \in (a, b)$ such that $h(x) = 0$, meaning that $(f/g)(x) = y$.

Bibliography

1. M. Aigner, G.M. Ziegler, *Proofs from the Book* (Springer, Berlin, New York, 2003)
2. T. Andreescu, G. Dospinescu, *Problems from the Book*, 2nd edn. (XYZ Press, Dallas, 2010)
3. T. Andreescu, G. Dospinescu, *Straight from the Book* (XYZ Press, Dallas, 2012)
4. B. Bollobás, *The Art of Mathematics - Coffee Time in Memphis* (Cambridge University Press, New York, 2006)
5. W. Burnside, *Theory of Groups of Finite Order* (Cambridge University Press, Cambridge, 1897)
6. I.N. Herstein, *Topics in Algebra*, 2nd edn. (Xerox College Publishing, Lexington, Massachusetts, Toronto, 1975)
7. A.I. Kostrikin, *Introduction à l'algèbre* (Mir, Moscow, 1981)
8. A.G. Kurosh, *Higher Algebra* (Mir, Moscow, 1984)
9. S. Lang, *A First Course in Calculus*, 5th edn. Undergraduate Texts in Mathematics (Springer, New York, Berlin, Heidelberg, 1986)
10. V.V. Prasolov, *Essays on Numbers and Figures*. Mathematical World (American Mathematical Society, Providence, RI, 2000)
11. W. Rudin, *Principles of Mathematical Analysis* (McGraw-Hill, Inc., New York, St. Louis, San Francisco, 1976)
12. M. Spivak, *Calculus*, 3rd edn. (Publish or Perish, Inc., Houston, 1994)

© Springer Science+Business Media LLC 2017
T. Andreescu et al., *Mathematical Bridges*, DOI 10.1007/978-0-8176-4629-5

Printed by Printforce, the Netherlands